Lecture Notes in Mathematics

Edited by A. Dold, F. Takens and B. Teissier

Editorial Policy
for the publication of monographs

1. Lecture Notes aim to report new developments in all areas of mathematics – quickly, informally and at a high level. Monograph manuscripts should be reasonably self-contained and rounded off. Thus they may, and often will, present not only results of the author but also related work by other people. They may be based on specialized lecture courses. Furthermore, the manuscripts should provide sufficient motivation, examples and applications. This clearly distinguishes Lecture Notes from journal articles or technical reports which normally are very concise. Articles intended for a journal but too long to be accepted by most journals, usually do not have this "lecture notes" character. For similar reasons it is unusual for doctoral theses to be accepted for the Lecture Notes series.

2. Manuscripts should be submitted (preferably in duplicate) either to one of the series editors or to Springer-Verlag, Heidelberg. In general, manuscripts will be sent out to 2 external referees for evaluation. If a decision cannot yet be reached on the basis of the first 2 reports, further referees may be contacted: the author will be informed of this. A final decision to publish can be made only on the basis of the complete manuscript, however a refereeing process leading to a preliminary decision can be based on a pre-final or incomplete manuscript. The strict minimum amount of material that will be considered should include a detailed outline describing the planned contents of each chapter, a bibliography and several sample chapters.
Authors should be aware that incomplete or insufficiently close to final manuscripts almost always result in longer refereeing times and nevertheless unclear referees' recommendations, making further refereeing of a final draft necessary.
Authors should also be aware that parallel submission of their manuscript to another publisher while under consideration for LNM will in general lead to immediate rejection.

3. Manuscripts should in general be submitted in English.
Final manuscripts should contain at least 100 pages of mathematical text and should include

- a table of contents;
- an informative introduction, with adequate motivation and perhaps some historical remarks: it should be accessible to a reader not intimately familiar with the topic treated;
- a subject index: as a rule this is genuinely helpful for the reader.

Lecture Notes in Mathematics 1714

Springer
Berlin
Heidelberg
New York
Barcelona
Hong Kong
London
Milan
Paris
Singapore
Tokyo

O. Diekmann R. Durrett
K.P. Hadeler P. Maini H.L. Smith

Mathematics Inspired by Biology

Lectures given at the 1st Session of the Centro Internazionale Matematico Estivo (C.I.M.E.) held in Martina Franca, Italy, June 13–20, 1997

Editors: V. Capasso, O. Diekmann

Fondazione
C.I.M.E.

Authors and Editors

Vincenzo Capasso
Department of Mathematics
University of Milan
Via C. Saldini, 50
20133 Milano, Italy
e-mail: capasso@miriam.mat.unimi.it

Odo Diekmann
Mathematisch Instituut
Universiteit Utrecht
P.O.Box 80.010
3508 TA Utrecht, The Netherlands
e-mail: o.diekmann@math.uu.nl

Richard Durrett
Department of Mathematics
523 Malott Hall
Cornell University
Ithaca, NY 14853, USA
e-mail: rtdl@cornell.edu

Karl Peter Hadeler
Department of Matheamtics
University of Tübingen
Auf der Morgenstelle 10
72076 Tübingen, Germany
e-mail: k.p.hadeler@uni-tuebingen.de

Philip K. Maini
Mathematical Institute
Oxford University
24–29 St. Giles
Oxford, OX1 3LB, UK
e-mail: maini@maths.ox.ac.uk

Hal Smith
Department of Mathematics
Arizona State University
Tempe, AZ 85287-1804, USA
e-mail: halsmith@asu.edu

Cataloging-in-Publication Data applied for

Die Deutsche Bibliothek - CIP-Einheitsaufnahme

Mathematics inspired by biology : held in Martina Franca, Italy, June 13 - 20, 1997 / Fondazione CIME. O. Diekmann ... Ed.: V. Capasso ; O. Diekmann. - Berlin ; Heidelberg ; New York ; Barcelona ; Hong Kong ; London ; Milan ; Paris ; Singapore ; Tokyo : Springer, 1999
(Lectures given at the ... session of the Centro Internazionale Matematico Estivo (CIME) ... ; 1997,1) (Lecture notes in mathematics ; Vol. 1714 : Subseries: Fondazione CIME)
ISBN 3-540-66522-6

Mathematics Subject Classification (1991): 34C, 34D, 35B, 35K, 60H, 60K, 92-06, 92B, 92D, 92-02, 92C15, 92D25, 60K35, 34C35, 35Q80

ISSN 0075-8434
ISBN 3-540-66522-6 Springer-Verlag Berlin Heidelberg New York

Typesetting: Camera-ready TEX output by the authors/editors
SPIN: 10700254 41/3143-543210 - Printed on acid-free paper

Preface

The Centro Internazionale Matematico Estivo (CIME), organised on June 13 - 20 1997 a summer course on MATHEMATICS INSPIRED BY BIOLOGY in the Ducal Palace of Martina Franca (a nice baroque village in Apulia, Italy).

Progress in applied mathematics often derives from co-ordinated efforts to bring the challenges of specific applications into agreement with the quest of intrinsic and general mathematical structure. In this spirit, the aim of the course was to demonstrate how mathematical problems, ideas, methods and results arise from attempts to describe, analyse and understand the world of living systems. A wide spectrum of mathematical and biological subjects was covered. The intention, to give to beginning researchers a stimulating impression of the current state-of-the-art in a selected number of fields, is echoed in the present lecture notes.

In this book you find the written account of five series of six lectures each. The common theme is the role of structure in shaping transient and ultimate dynamics. But the type of structure ranges from spatial (Hadeler and Maini in the deterministic setting, Durrett in the stochastic setting) to physiological (Diekmann) and order (Smith). Each contribution sketches the present state of affairs while, by including some wishful thinking, pointing at open problems that deserve attention. No doubt this book will be superseded before many years have passed, but most likely it will itself catalyse the process by which our knowledge and understanding is enhanced and extended.

The co-ordinators are most grateful to CIME for giving us the opportunity to gather excellent lecturers and a surprisingly large crowd of bright and enthusiastic students, coming from about 15 countries and ranging from Mathematicians, Physicists, Biologists, Physicians, Engineers, in a truely ideal Apulian setting. No doubt the organisation could have been painful; but the assistance of Daniela Morale and Alessandra Micheletti was so much beyond the call of duty that it was actually a pleasure. To them, to the wonderful students and (last but certainly not least) to the authors that showed so much sense of responsibility for producing a readable text, we extend our warmest thanks.

Financial support from CIME and European Union is gratefully acknowledged. Finally we like to thank the Mayor, the Secretary General, and all the staff of the Town Council of Martina Franca, that offered the XVII's century Ducal Palace as the site of the course, for the warm hospitality and continuous assistance.

Milano and Utrecht, July 8 1998,

Vincenzo Capasso and Odo Diekmann

Table of Contents

Modeling and analysing physiologically structured populations

Odo Diekmann
Mathematisch Instituut
Universiteit Utrech
P.O. Box 80.010 3508 TA Utrecht
The Netherlands

Contents

1 Introduction

Physiologically structured population models are individual based, in the sense that processes are modelled at the i-level (i for individual). They are deterministic at the p-level (p for population), in the sense that stochastic fluctuations are neglected by invoking informally a law of large numbers. The aim of their analysis is to unravel, by systematic bookkeeping and analytical investigation, how phenomena at the p-level are related to mechanisms at the i-level. Usually, and certainly in this exposition, the complications of sex are avoided by pretending that mothers produce daughters without any contribution by males (while, of course, having in mind that males are sufficiently abundant to fertilise all eligible females).

Mathematics is often helpful in deriving biological understanding and insight. Conversely, the biological interpretation can be quite helpful in guessing identities or estimates and even suggesting quick and elegant proofs. In stochastic models, where individuals are in the foreground, this is usually more prominent than in deterministic models. Applied analysts have a tendency to try their tools on the equation at hand, while postponing the interpretation till after the result is obtained. The present paper will emphasise throughout that the biological interpretation is very helpful, if not essential, when analysing structured population models. This is possible since our construction will for a long time concern individuals. Only at a late stage do we advance to the p-level, simply by adding i-contributions.

The linear theory is in good shape, such in sharp contrast to the nonlinear theory. This paper intends to make propaganda for the idea to model dependence among individuals in a two-step procedure: first introduce environmental interaction variables which, when given, guarantee independence of individuals; next model the feedback process that determines the interaction variables in terms of the extant population. The result is a fixed point problem that has to be analysed.

In many ways this paper is an update of the more mathematical aspects of structured population models as presented in [Metz & Diekmann, 1986]. Much has changed in twelve years !

As a warming-up to the cumulative formulation of general models, we discuss the relatively simple case of an age-structured population in section 2. We first introduce the basic reproduction number R_0, the intrinsic rate of natural increase r and the stable age distribution and show how these can be determined from so-called model ingredients that specify the reproduction and survival. We then use cannibalism in an age-structured population to explain what environmental interaction variables are and how they can be used when looking for steady states.

Section 3.1 is devoted to an exposition of the linear theory for general structured models. The added complication, relative to the case of age-structure , is mostly of a notational/bookkeeping type and I hope that the starter Section 2.1 makes this main course easy to digest. The proof of the main result in this section is new (it is based on an idea of Haiyang Huang). The two examples presented in Section 3.2 are, respectively, a size structured cell population and juvenile dispersal in space. How to determine steady states directly from the ingredients is described in Section 3.3 and this is then elaborated for a restricted class of problems in Section 3.4 in such a way that numerical implementation is relatively straightforward. The final section sketches how a nonlinear semigroup can be constructed by combining the linear construction with the feedback condition. It ends with many open problems.

Acknowledgements Without the interaction over a long period of time with Mats Gyllenberg, Hans Metz and Horst Thieme, this paper would never have been written. More recently it was a pleasure to brainstorm with Haiyang Huang and Markus Kirkilionis. At the CIME meeting Enzo Capasso, Daniela Morale and Alessandra Micheletti created an atmosphere which was at the same time serious, relaxed and friendly. To them, as well as all participants, goes my gratitude for the opportunity to lecture under such ideal conditions. Finally, I am much indebted to Sybren Botma for typing the manuscript and, meanwhile, correcting my English.

2 Age-structured populations revisited: a tribute to W.Feller

2.1 Independent individuals in a constant environment (the linear autonomous case)

Our aim is to derive relations between a statistical description of survival and offspring production of individuals on the one hand, and the growth and age composition on the other hand. To do so, we first introduce two so-called modeling ingredients that should capture all relevant aspects of individual behaviour:

L(a):= expected number of daughters produced in the time interval $(0, a]$ after birth (1)

$\mathcal{F}(a) :=$ the probability to be alive a time units after birth (i.e., $\mathcal{F}$ is the survival function). (2)

We assume that L is monotone non-decreasing, bounded, right-continuous with $L(0) = 0$ and that $\mathcal{F}$ is monotone non-increasing with $\mathcal{F}(0) = 1$ and $\mathcal{F}(\infty) = 0$.

To decide about growth or decline, and to determine the **rate** of growth or decline, we only need L. The **basic reproduction number** R_0 is defined by

$$R_0 = L(\infty) \tag{3}$$

and is, therefore, the expected life time production of daughters. So, by first thinking about growth or decline on the basis of **generations**, we see that R_0 has threshold value one:

-if $R_0 > 1$ the population will grow,

if $R_0 < 1$ it will go extinct

(This strict dichotomy is a deterministic idealisation: one needs a stochastic description in terms of branching processes to deal with the possibility of extinction in the supercritical case $R_0 > 1$ [Jagers,1991 and preprint].)

In quest of the growth rate, we introduce the Laplace-Stieltjes transform

[Widder,1946] of L:

$$\hat{L}(z) = \int_{(0,\infty)} e^{-za} L(da) \tag{4}$$

Note that this integral is the limit of the sums

$$\sum_{i=1}^{\infty} e^{-za_{i+1/2}} (L(a_{i+1}) - L(a_i))$$

for partition width going to zero and with $a_{i+1/2}$ symbolic notation for a point in between a_i and a_{i+1}. $\hat{L}(z)$ counts daughters, while discounting them with a factor e^{-za} if the mother has age a at the moment of giving birth. An intuitive consistency argument, due to Fisher [Fisher,1958], claims that the **relative** contribution to a population of an individual and of its total offspring should be the same. So if the population grows like e^{rt}, we have to discount daughters that are born a time units after the birth of their mother by e^{-ra}. In other "words", the rate r should satisfy the so-called **Euler-Lotka equation**

$$\hat{L}(r) = 1 \tag{5}$$

(A financial analogy: if the interest is fixed and wealth should stay constant, one can determine the prevailing rate of inflation.)

As a function of a real variable z, $\hat{L}$ is a smooth and monotone decreasing function with limit zero at infinity. So there will be at most one solution to (5) and, to be assured that one exists, we only have to find a z for which $\hat{L}(z) \geq 1$. This is automatic if $R_0 \geq 1$, while if $R_0 < 1$ and L has behaviour for $a \to \infty$ which guarantees that (4) makes sense for all real z, it follows from $\hat{L}(z) \uparrow \infty$ for $z \downarrow -\infty$. In the following we shall, without further mentioning it, assume that r exists. So we shall disregard the possibility that the integral in (4) diverges for $z < \alpha$ while $\hat{L}(\alpha) < 1$ (see [Jagers, 1983, Shurenkov, preprint 1] for this issue in a more general context; if such a case arises, one has to be contented with exponential estimates). Note that the monotonicity of $\hat{L}$ implies at once that

$$\text{sign } (R_0 - 1) = \text{sign } r \tag{6}$$

which is reassuring: populations grow in real time if and only if they grow from generation to generation).

Exercise 1 Show that, when comparing different L's, the ordering of R_0 does not induce any ordering of r. Hint: consider organisms that produce c daughters exactly when reaching age T and play with the parameters c and T. The point is that **late** reproduction makes for **slow** population growth in real time.

What can we say about the age composition ?

Individuals aged a at time t were born at time $t-a$ and so they are represented by

$$b(t-a)\mathcal{F}(a)$$

where $b(t)$ is the population birth rate at time t (we use the somewhat vague "represented" to express that this is a density: only when we integrate over an age interval do we get a number of individuals).
Inserting

$$b(t) \sim e^{rt}$$

we find that the **stable age distribution** is described by the density

$$e^{-ra}\mathcal{F}(a) \tag{7}$$

(It is the factor e^{-ra} that makes for steep age pyramids in fast growing populations.)

Thus we have, in a rather informal (but consequently very quick) way, obtained answers to our questions: to decide about growth or decline, compute R_0 as in (3); to find the population growth rate r (often called the Malthusian parameter), solve (5); to determine the stable age distribution, compute (7). Before going into a more mathematical formulation and justification of our assertions, which will expose a further and rather subtle case distinction, we briefly comment on the nature of our modeling ingredients. The functions L and $\mathcal{F}$ correspond closely to the kind of information one could obtain from the civil registration. Yet often an age-structured model is specified in terms of an age-specific **birth rate** $\beta(a)$ and an age-specific **death rate** $\mu(a)$. In that case one can derive L and $\mathcal{F}$ by solving the ordinary differential equations.

$$\begin{aligned} \frac{d\mathcal{F}}{da}(a) &= -\mu(a)\mathcal{F}(a) \quad , \ \mathcal{F}(0)=1 \\ \frac{dL}{da}(a) &= \beta(a)\mathcal{F}(a) \quad\ \ , \ L(0)=0 \end{aligned} \tag{8}$$

and proceed with these as described above. This is the recommended procedure whenever one has mechanistic arguments for the choice of μ and/or β. If, on the other hand, one starts from data it seems more appropriate to define the rates in terms of $\mathcal{F}$ and L, i.e. to put

$$\begin{aligned} \mu(a) &= -\tfrac{d}{da}\ln\mathcal{F}(a) \\ \beta(a) &= \tfrac{1}{\mathcal{F}(a)}\tfrac{d}{da}L(a) \end{aligned} \tag{9}$$

while realising that the inaccuracy of numerical differentiation makes this into a precarious venture.

We now establish a constructive procedure, called **generation expansion**, to calculate the expected clan size of an individual on the basis of the given function L. The expected number of **granddaughters** that an individual has by the time it reaches age a, is given by

$$\int_{(0,a]} L(a-\alpha)L(d\alpha) \quad ,$$

since granddaughters are daughters of daughters, and the latter can be "parametrised" by the age of the mother at the moment of giving birth. Introducing the notation

$$(f\otimes g)(a) = \int_{(0,a]} f(a-\alpha)g(d\alpha) \tag{10}$$

$$f^{k\otimes} = f^{(k-1)\otimes}\otimes f \qquad \text{for } k\geq 2, \tag{11}$$

with the convention $f^{1\otimes}=f$, we can say that the expected number of granddaughters equals $L^{2\otimes}(a)$ and, inductively, that the expected number of k-th generation offspring is $L^{k\otimes}(a)$. We define the **clan kernel** R by

$$R = \sum_{k=1}^{\infty} L^{k\otimes} \tag{12}$$

(the nonnegativity of L guarantees that we do not have to worry about convergence to give the sum meaning; yet one should realise that the interpretation more or less dictates that L is zero in a right neighbourhood of $a=0$, in which case the sum has, for any given finite a, only finitely many terms; a more technical approach can be based on exponential estimates for L; we refer to [Diekmann e.a., 1998] for further details in a much more general setting).

Theorem 1 *R defined by (12) is the unique solution of the* **renewal equations**

$$R = L + L \otimes R \tag{13}$$

$$R = L + R \otimes L \tag{14}$$

Proof To show that R is a solution, we just substitute (12) at both the right hand side and the left hand side, and ascertain that this yields an identity. Now suppose that $Q = L + L \otimes Q$. Then, using the associativity and distributivity of the $\otimes$ product and (14), we deduce that $R \otimes Q = R \otimes L + R \otimes L \otimes Q = R \otimes L + (R - L) \otimes Q$ and hence $R \otimes L = L \otimes Q$ which, when substituted in the equation for Q, yields $Q = L + R \otimes L = R$. If, on the other hand, $Q = L + Q \otimes L$ then a similar calculation, now using left $\otimes$ multiplication by R and using (13), leads again to $Q = R$. □

Equation (13) tells us that a clan member is either a daughter or a daughter of a clan member, while (14) asserts that a clan member is either a daughter or a clan member of one of the daughters. The mathematical jargon [see Gripenberg e.a., 1990] is that R is the **resolvent** of L.

Definition 1 L is a **lattice kernel** if L is constant, except for jump discontinuities which all occur at integer multiples of some number $d > 0$. The largest d with this property is then called the **span** of L.
(In [Feller, 1966] a lattice kernel is called arithmetic.)

The celebrated **Renewal Theorem** (see [Feller, 1966, Ch.XI]) asserts that Eve's expected clan grows with rate r:

Theorem 2 *Assume that L is non-lattice. For every $h > 0$*

$$\lim_{a \to \infty} \int_{(a-h,a]} e^{-r\tau} R(d\tau) = \frac{h}{\sigma} \tag{15}$$

where

$$\sigma = \int_{(0,\infty)} \tau e^{-r\tau} L(d\tau) \tag{16}$$

(with the convention that $\frac{1}{\sigma} = 0$ whenever $\sigma = \infty$).
When L is lattice, the assertion is still true if we restrict h to be a multiple of the span d.

At this point it becomes helpful to consider $\hat{L}$ as an analytic function of the complex variable z, defined on a right half plane. Since $\hat{L}(z) < 1$ for real $z > r$ and for general z we have $|\hat{L}(z)| \leq \hat{L}(\text{Re } z)$, we see that roots of the equation

$$\hat{L}(z) = 1 \tag{17}$$

are confined to the region Re $z \leq r$. When L is lattice with span d, the function $\hat{L}$ is periodic with period $\frac{2\pi}{d}$ along lines Re z = constant. So in particular there is a group of roots of the equation (17) on the line Re $z = r$.

Definition 2 The Malthusian parameter r is called **strictly dominant** if all other roots of (17) satisfy Re $z < r$ and **strongly dominant** if they satisfy Re $z < r - \varepsilon$ for some $\varepsilon > 0$.

Exercise 2 Consider the organisms described in the hint of Exercise 1. Show that L is lattice with span T. Next consider organisms that produce one daughter upon reaching age $T < 1$ and again one daughter upon reaching age 1, after which no further reproduction occurs. Show that L is lattice when T is rational and non-lattice when T is irrational. Verify that in the latter case L is quasi-periodic along lines Re z = constant. Make plausible (or better still: prove) that in that case r is strictly, but not strongly, dominant. For inspiration you may wish to consult [Gyllenberg, 1986].

Exercise 3 Use the lemma of Riemann-Lebesgue to show that r is strongly dominant if L is absolutely continuous (i.e., the integral of an L_1-function).

Theorem 3 *If L is absolutely continuous, so is R and*

$$R'(t) = \sigma e^{rt} + K(t) \tag{18}$$

with K such that $t \mapsto K(t)e^{-(r+\varepsilon)t}$ belongs to $L_1([0,\infty))$ for all $\varepsilon > 0$ and with σ defined by (16).

Sketch of the proof: From (14) and $R(0) = L(0) = 0$ it follows that

$$R' = L' + R' * L'.$$

A famous theorem of Paley and Wiener states that R' belongs to $L_1([0,\infty))$ whenever all roots of the characteristic equation $\hat{L}(z) = 1$ lie in the left half plane (Gripenberg e.a., 1990, Thm. 2.4.1).
To achieve a more general result we can either work with weighted L_1 spaces or, which is basically equivalent, shift the contour in the Laplace inversion formula to the left of the pole at $z = r$ while splitting off the residue. See [Gripenberg e.a., 1990, Thm. 7.2.4] for details. □

Thus we demonstrated that for absolutely continuous kernels one can use Laplace transformation to deduce that the expected clan size behaves like e^{rt} for large t.

Let us now turn attention to the age composition. An individual which is alive at age a will be alive t units of time later with probability

$$\frac{\mathcal{F}(a+t)}{\mathcal{F}(a)} \tag{19}$$

and its expected production of offspring during this time interval equals

$$L_a(t) := \frac{L(a+t) - L(a)}{\mathcal{F}(a)} \tag{20}$$

To find the expected clan production, we only need R as an additional ingredient (since birth is a renewal event in which all information about the mother is lost):

$$Q_a(t) = L_a(t) + \int_{(0,t]} R(t-\tau) L_a(d\tau) \tag{21}$$

Exercise 4 Why is the analogue of (20), that is

$$\frac{R(a+t) - R(a)}{\mathcal{F}(a)}$$

not the right quantity ?

Let the population composition at time zero be given by the measure m_0. Adding contributions of all individuals we find the expressions

$$F(t) = \int_{(0,\infty)} L_a(t) m_0(da) \tag{22}$$

$$B(t) = \int_{(0,\infty)} Q_a(t) m_0(da) \tag{23}$$

for, respectively, all children and all clan members.

Lemma 1 *B and F are related to each other by the renewal equation*

$$B = F + L \otimes B \tag{24}$$

and therefore also by the resolvent representation

$$B = F + R \otimes F \tag{25}$$

Lemma 2 lemma:1

Proof. The equivalence of (24) and (25) is shown by formula manipulation as in the proof of Theorem 1. Substitution of (21) into (23) yields, upon an interchange of the order of integration, (25). □

The above lemma is somewhat redundant, as all we shall need below is B, and the combination of (23), (21) and (20) yields already an explicit recipe of how to compute B, given L, $\mathcal{F}$, R and m_0. But the lemma illustrates how our construction of R as an object related to an individual fits in with the more usual approach of formulating the renewal equation (24) at the population level and then defining B as its solution, given quasi-explicitly by (25), with R defined as the resolvent of L.

The population composition at time t is given in terms of $\mathcal{F}$, m_0 and B by the formula:

$$m(t)(\omega) = \int_{(0,\infty)} \chi_\omega(a+t) \frac{\mathcal{F}(a+t)}{\mathcal{F}(a)} m_0(da) + \int_{(0,t]} \chi_\omega(t-\sigma)\mathcal{F}(t-\sigma)B(d\sigma) \tag{26}$$

where χ_ω is the indicator function of the set ω.
The two terms correspond, respectively, to those already present at time zero and surviving to time t and those born in this time window and surviving until time t.
When L is absolutely continuous, so is R and so is B. Moreover, the analogue of Theorem 3 holds, with σ replaced by a constant that depends on m_0 (see

[Gripenberg e.a., 1990, Thm.7.2.5]). The second term of (26) can be rewritten as

$$\int_{(0,t]} \chi_\omega(\sigma)\mathcal{F}(\sigma)B'(t-\sigma)d\sigma \tag{27}$$

and for large t this equals

$$\int_\omega \mathcal{F}(\sigma)B'(t-\sigma)\, d\sigma \tag{28}$$

which shows that this term is absolutely continuous as well, with density $\mathcal{F}(a)B'(t-a)$ (note that B' was called b in the first part of this section). Substitution of the analogue of (18) then leads to the conclusion that (7) does indeed describe the stable age distribution.
When L is not absolutely continuous, one can use Theorem 2 as a basis for a further analysis of the second term of (26) and clearly it will then make a big difference whether or not L is a lattice kernel.
We refer to Section 3.2 and [Diekmann, Gyllenberg, Thieme & Verduyn Lunel, preprint] for some more details in this spirit.

A more traditional approach for linear age-dependent population problems is to start from the first order partial differential equation

$$\frac{\partial n}{\partial t} + \frac{\partial n}{\partial a} = -\mu n \tag{29}$$

with boundary condition

$$n(t,0) = \int_0^\infty \beta(\alpha)n(t,\alpha)d\alpha \tag{30}$$

and interpret these as defining a generator of a strongly continuous semigroup of bounded linear operators on $L_1([0,\infty))$. Next one determines the spectrum of the generator and uses this information to deduce the asymptotic large time behaviour of the semigroup [Webb, 1985, Nagel, 1986].

In the setting presented above, the definition

$$m(t) = T(t)m_0$$

yields a semigroup of bounded linear operators on the space of measures, but $t \mapsto T(t)m_0$ is not continuous with respect to the norm topology but only with respect to the weak * topology. We have completely avoided to consider the infinitesimal generator, but have defined $T(t)$ constructively (from L we constructed R and everything else was explicit in terms of L, $\mathcal{F}$, R and m_0). We have discussed the asymptotic behaviour in terms of Laplace transforms. Recalling the identity

$$\int_0^\infty e^{-zt}T(t)dt = (zI - A)^{-1}$$

(where A is the generator) we observe that this amounts to determining the singularities of the resolvent, hence the spectrum of A. So after all there is very little difference between the presented approach and the traditional one. Yet we emphasise that we have only dealt with bounded operators and that, in particular, we never needed to determine the domain of definition of the generator. It is this aspect which provides enormous technical advantages when we consider general structure and, in particular, time-dependent or nonlinear problems. Before turning to general structure, we shall introduce several aspects of nonlinear problems in the simplest setting, which is an age-structured population.

2.2 An example of density dependence described by feedback via environmental interaction variables: cannibalism

Several lakes exist in which one finds a predatory fish species, such as pike or perch, and yet no prey fish species. In such a situation the small individuals, which eat zooplankton, are in a sense part of the foraging apparatus of the big individuals.

In an interesting paper with the title "Cannibalism as a life boat mechanism" [Van den Bosch, De Roos and Gabriel, 1988] a less extreme situation is considered in which prey are available at a constant density which may, however, be insufficient to sustain the predator. Assuming that juveniles feed on a resource inaccessible to the adults, they investigated under which conditions cannibalism provides enough energy gain to allow the predator to reach a positive steady state density. Even though "size" is the more natural individual-state variable, we shall use "age" instead, since this simplifies the modeling. So assume that

$$\frac{d\mathcal{F}}{da}(a) \quad = \quad -(\mu + h(a)I_1)\ \mathcal{F}(a) \tag{31}$$

$$\frac{dL}{da}(a) = \zeta(Z + I_2)\,\mathcal{F}(a)\,H(a - \bar{a}) \tag{32}$$

Here μ (assumed to be independent of both age and time) is the probability per unit of time to die of any cause different from cannibalism, while $h(a)I_1$ corresponds to death as a result of cannibalism. The function h is given and reflects the vulnerability. The factor I_1 is still to be determined and reflects the cannibalistic predation pressure. In general, I_1 will be a function of time, but since we are going to restrict attention to steady states, we take I_1 to be a constant. We call I_1 an **environmental interaction variable**. Such variables are chosen such that, pretending that they are known, we arrive at a situation in which individuals are independent and hence the population model is linear. In a second step we shall then realise that in fact their determination is part of our job and so we shall pay attention to the feedback relations that tell us how they relate to the population composition. Thus we arrive at a fixed point problem, which we have to analyse.

In equation (32), ζ is a conversion factor that transforms the energy intake rate into reproductive output (in other words, an amount ζ^{-1} of energy is needed to produce one young). The parameter $\bar{a}$ marks the transition from infertile juvenile to reproducing adult. Z denotes the energy intake rate due to prey consumption and I_2 is the energy intake rate due to consumption of conspecifics. So I_2 is another environmental interaction variable.

To specify how individuals contribute to I_1, we need to say how the cannibalistic attack tendency depends on age. To avoid the introduction of yet other parameters, we simply assume that this attack tendency is zero for $a < \bar{a}$ and one for $a \geq \bar{a}$.

For the determination of I_2 we need, in addition to the vulnerability h, the function E that specifies how the energy content depends on age. Concerning h we assume that its support, which we call the vulnerability window, is contained in the interval $[0, \bar{a})$.

This ends the model specification and we can begin the analysis.

The steady state condition reads $R_0 = 1$. Since $R_0 = L(\infty)$ we have to integrate (32). To do so, we first observe that, since the support of h does not extend to $a > \bar{a}$, (31) implies that for $a > \bar{a}$

$$\mathcal{F}(a) = \mathcal{F}(\bar{a})e^{-\mu(a-\bar{a})} \tag{33}$$

Hence $L(\infty) = \mu^{-1}\zeta(Z + I_2)\,\mathcal{F}(\bar{a})$ where

$$\mathcal{F}(\bar{a}) = e^{-\mu\bar{a} - I_1 \int_0^{\bar{a}} h(\alpha)d\alpha} \tag{34}$$

Let b denote the (unknown) steady state population birth rate. Then the steady age density function equals $b\mathcal{F}(a)$ and so

$$I_1 = \int_{\bar{a}}^{\infty} b\mathcal{F}(\alpha)d\alpha = \frac{b\mathcal{F}(\bar{a})}{\mu} \quad (35)$$

$$I_2 = \int_0^{\bar{a}} h(\alpha)E(\alpha)b\mathcal{F}(\alpha)d\alpha = b\, int_0^{\bar{a}} h(\alpha)E(\alpha)e^{-\mu\alpha - I_1 \int_0^{\alpha} h(\tau)d\tau}d\alpha \quad (36)$$

Together with the steady state condition

$$1 = \frac{\zeta(Z+I_2)}{\mu}e^{-\mu\bar{a} - I_1 \int_0^{\bar{a}} h(\alpha)d\alpha} \quad (37)$$

these are three equations in the three unknowns b, I_1 and I_2.

In general, it is hard to study such nonlinear problems analytically and it is best to resort to numerical continuation methods. In [Kirkilionis e.a., preprint] and Section 3.4 a method is described to assemble the analogue of (35)-(37) numerically for rather general structured population models and then to study these equations by continuation techniques. In the present and rather simple situation we can, fortunately, proceed analytically.
Eliminating b and I_2 we arrive at one equation

$$Z = \mu e^{\mu\bar{a} + CI_1}\left\{\frac{1}{\zeta} - I_1 \int_0^{\bar{a}} h(\alpha)E(\alpha)e^{-\mu\alpha - I_1 \int_0^{\alpha} h(\tau)d\tau}d\alpha\right\} \quad (38)$$

in one unknown, viz. I_1. Here

$$C := \int_0^{\bar{a}} h(\alpha)d\alpha \quad (39)$$

We will consider μ, ζ, $\bar{a}$, h and E as given fixed parameters and Z as a variable (or, bifurcation) parameter. The question is: for which Z does (38) admit a nontrivial (i.e., positive) solution I_1. Note that (38) provides an explicit expression for Z as a function of I_1. Our task is to invert this function.
We define the critical prey density Z_c by

$$Z_c = \mu e^{\mu\bar{a}}\frac{1}{\zeta} \quad (40)$$

that is, by inserting $I_1 = 0$ at the right hand side of (38). Note that Z_c is the prey density that makes $R_0 = 1$ in the absence of any cannibalistic activity. By obvious monotonicity, the zero population steady state is stable for $Z < Z_c$ and unstable for $Z > Z_c$. We expect that a nontrivial steady state exists for $Z > Z_c$, but does not exist for $Z < Z_c$. Do we indeed expect this ?

Let us first consider the limiting case of a very narrow vulnerability window by letting h turn into a δ "function" concentrated at $a = a_v$. If we first rewrite (38) in the form

$$Z = \mu e^{\mu\bar{a}+CI_1}\left\{\frac{1}{\zeta} + \int_0^{\bar{a}} E(\alpha)e^{-\mu\alpha}d_\alpha\left(e^{-I_1\int_0^\alpha h(\tau)d\tau}\right)\right\}$$

we see that in this limit the equation becomes

$$Z = \mu e^{\mu\bar{a}+CI_1}\left\{\frac{1}{\zeta} + E(a_v)e^{-\mu a_v}\left(e^{-CI_1} - 1\right)\right\}$$

from which we deduce immediately that

$$\frac{dZ}{dI_1} = C\mu e^{\mu\bar{a}+CI_1}\left\{\frac{1}{\zeta} - E(a_v)e^{-\mu a_v}\right\}. \tag{41}$$

Clearly then there is a dichotomy:

- when $\zeta\ E(a_v)\ e^{-\mu a_v} > 1$, cannibalism acts as a positive feedback mechanism, since the "harvesting" of a juvenile provides enough energy to produce more than one offspring, even if we correct for the fact that this offspring should at least reach age a_v to replace the cannibalised juvenile. As a consequence a steady state exists for all $Z < Z_c$ and presumably its stable manifold separates the domains of attraction of the stable zero population steady state and unbounded growth to infinite population size (as we did not model any limitations in the supply of the alternative food for juveniles, unbounded growth is indeed a possibility).

- when $\zeta\ E(a_v)\ e^{-\mu a_v} < 1$, cannibalism acts as a negative feedback mechanism, since the loss of an individual due to cannibalism is not completely compensated by increased offspring production. As a consequence, cannibalism keeps population growth in check for $Z > Z_c$, where the population would grow exponentially in the absence of cannibalism. The steady state that exists for $Z > Z_c$ may be stable (it certainly is so for Z slightly larger than Z_c, according to the principle of the exchange of stability) or unstable, in which case oscillations occur (a characteristic feature of delayed negative feedback, see [Diekmann e.a., 1995, Ch.XV]).

Returning to general h, we observe that one can deduce from (38) a local version of the above dichotomy by computing that

$$\frac{dZ}{dI_1}|_{I_1=0} = \mu e^{\mu\bar{a}} \left\{ \frac{C}{\zeta} - \int_0^{\bar{a}} h(\alpha)E(\alpha)e^{-\mu\alpha}d\alpha \right\}$$

and noting that $\frac{1}{C}\int_0^{\bar{a}} E(\alpha)h(\alpha)e^{-\mu\alpha}d\alpha$ is the mean energy yield of a cannibalised juvenile when we disregard the effect of cannibalism on the stable age distribution. As cannibalism does affect the age distribution, the mean yield is not constant along the branch of nontrivial solutions. When $\zeta E(a)e^{-\mu a} - 1$ changes sign on the support of h, we are in a mixed feedback situation and a quantitative balance determines the direction of the branch. In particular, the branch may show one or more turns. We refer to [Van den Bosch e.a., 1988] for examples/figures.

We conclude that the effect of cannibalism can, within the context of the present model, best be understood in terms of energy gain or loss when we discount energy put into reproduction according to the probability that the newborn will survive to replace the cannibalised juvenile.

We hope that this example serves to explain the idea of environmental interaction variables. In this section we freely used some terminology of nonlinear dynamical systems, without providing any formal justification. For age-structured models such a justification can be found in [Webb, 1985, Clement e.a., 1989, part III].

In Section 3.5 we will sketch an approach for general structured population models which is built on the notion of interaction variables. The precise elaboration of this approach is in slow progress.

3 General structure: the cumulative formulation

3.1 Independent individuals in a variable environment (the linear non-autonomous case)

Age is a very predictable feature: it is, by definition, zero at birth and it keeps pace with time. Size, in contrast, varies at birth and its rate of increase depends on food supply. In this subsection we think of environmental interaction variables, such as food supply, as given functions of time. Our aim is to extend the bookkeeping considerations of section 2.1 in two directions: general i-states (and i-state development) and time dependence for the basic ingredients.

We denote i-state by x etc. and assume $x \in \Omega$ where Ω, the i-state space, is a measurable space. The basic ingredients now pertain to an individual which at time t has i-state x. They provide information about offspring and about survival and i-state development. Concerning offspring, we have to specify the production as well as their i-state at birth as well as their birth time. The latter we will do in a relative way, by specifying the time elapsed since t (this is a matter of choice and the alternative, to specify the absolute time, is completely equivalent; we have chosen the "relative" option since this allows us to simply drop arguments when we consider the case of a constant environment, for which absolute time labels are irrelevant). We will describe the combination of state-at-birth of a newborn and the time elapsed since the reference time as "birth coordinates". So birth coordinates correspond to points in $\Omega \times \mathbb{R}_+$. We will use the symbol A to denote measurable subsets of $\Omega \times \mathbb{R}_+$.

Our first model ingredient describes reproduction:

$\Lambda(t,x)(A) :=$ the expected number of children, with relative birth coordinates in A produced by an individual which at time t has i-state x (42)

Once we explain the notation

$$A_{-\tau} := \{(\sigma,\xi) \in \mathbb{R}_+ \times \Omega : (\sigma+\tau,\xi) \in A\} \quad (43)$$

the formula

$$\Lambda^{2\otimes}(t,x)(A) = \int_{\mathbb{R}_+\times\Omega} \Lambda(t+\tau,\xi)(A_{-\tau})\Lambda(t,x)(d\tau\times d\xi) \tag{44}$$

should need no further explanation after our presentation of the age-structured case in (10), (11) etc. . Here we have assumed that $\Lambda : \mathbb{R}\times\Omega \to M_+(\mathbb{R}_+\times\Omega)$, with $M_+(\mathbb{R}_+\times\Omega)$ the space of positive measures on $\mathbb{R}_+\times$ *Omega*, is such that for any measurable $A \subset \mathbb{R}_+\times\Omega$ the function $(t,x) \mapsto \Lambda(t,x)(A)$ is measurable. The analogue of R we now denote by Λ^c and define it by

$$\Lambda^c = \sum_{k=1}^{\infty} \Lambda^{k\otimes} \tag{45}$$

(the c stands for clan). All our remarks about positivity, finitely many terms, exponential weights and renewal equations carry over immediately. But a new element is that the set Ω may be much larger than the set of possible birth states (for instance, in the age-structured case only $a = 0$ is a possible birth state).

Definition 3 Let Ω_b be such that $\Lambda(t,x)(A) = 0$ whenever $A \cap (\mathbb{R}_+\times\Omega_b) = \emptyset$. Then we say that Ω_b represents the birth states and define $\Lambda_b(t,x)$ for $x \in \Omega_b$ by $\Lambda_b(t,x)(A) = \Lambda(t,x)(A)$ for $A \subset \mathbb{R}_+\times\Omega_b$.

In formulas like (44) and (45) one can now add an index b and restrict to $A \subset \mathbb{R}_+\times\Omega_b$ to achieve a much bigger computational efficiency whenever Ω_b has lower dimension compared to Ω (like in the age-structured case, where $\Omega = \mathbb{R}_+$ and $\Omega_b = \{0\}$; note that our choice to work with measures rather than densities allows us to describe this kind of reduction very easily). A similar reduction is possible whenever a (preferably lower-dimensional) subset can be identified such that any individual that ever reproduces must necessarily have an i-state belonging to this subset at some moment prior to reproduction. We then call this subset a renewal set. We refer to [Diekmann e.a., 1998] for details and examples.

Whenever $\Lambda(t,x)(A)$ is, for all x and A, independent of t, we say that the problem is **autonomous** and omit t everywhere in the notation. How do we define R_0 and r in such a situation ? As individuals differ in their state-at-birth, we can no longer define R_0 as the expected number of offspring without deliberating upon the question "for which individual ?". Since the reproduction process itself determines the distribution of state-at-birth, this should somehow be taken into account. Now note that a generation is related to the preceding generation by a recursion:

$$w_{n+1}(\omega) = \int_{\Omega_b} \Lambda_b(x)(\mathbb{R}_+ \times \omega) w_n(dx) \tag{46}$$

(Here we distinguish individuals of one generation according to their state-at-birth.) Accordingly we expect that in the long run the size of subsequent generations will differ by a factor R_0, where R_0 is the positive/dominant eigenvalue of the integral operator at the right hand side of (46). When

$$\sup_{x \in \Omega_b} \Lambda_b(x)(\mathbb{R}_+ \times \Omega) < \infty \tag{47}$$

the right hand side of (46) defines a **bounded** linear operator on the space of measures on Ω_b and R_0 is its spectral radius.

Under fairly weak conditions R_0 is a simple eigenvalue with positive eigenvector. The normalised eigenvector describes the distribution of state-at-birth that sets in in the long run. So it is the stable distribution for state-at-birth of individuals sampled from one and the same generation (which is **not** the same as the distribution of state-at-birth in a sample of individuals born at approximately the same time).

The discounting argument explained in Section 2.1 motivates the introduction of the family of integral operators assigning to $w \in M(\Omega_b)$ the measure

$$\int_{\mathbb{R}_+ \times \Omega_b} e^{-zt} \Lambda(x)(dt \times \omega) w(dx). \tag{48}$$

This family is parametrised by z, where z ranges in a right half plane of the complex plane. For real z the operator is positive and considerations presented above for $z = 0$ carry over. We claim that r is the value of z on the real line which makes the spectral radius (= dominant eigenvalue) equal to one. The justification of this claim is based on the Laplace transform and on the construction of the population process that we present below.

The second model ingredient describes (possibly stochastic) i-state development and survival:

$u(t, x; s)(\omega) =$ probability that an individual which has state x at time t is alive s time units later and then has a state in $\omega \subset \Omega$ (49)

We assume that

(i) $(t,x;s) \longmapsto u(t,x;s)(\omega)$ is measurable for all ω

(ii) $u(t,x;s)(\Omega) \leq 1$

(iii) life expectancy is finite: $\sup_{(t,x)\in\mathbb{R}_+\times\Omega} \int_{\mathbb{R}_+} s\, u(t,x;ds)(\Omega) < \infty$

The big advantage of our non-infinitesimal description (49) is that stochastic i-state development is included at no extra technical cost. The price we pay for this is that a certain consistency condition has to be imposed on u. Indeed, in addition to i)-iii) we assume that u satisfies the **Chapman-Kolmogorov** condition that for $0 < \sigma < s$

(iv) $$u(t,x;s)(\omega) = \int_{\Omega} u(t+\sigma,\xi;s-\sigma)(\omega)\; u(t,x;\sigma)(d\xi)$$

which says no more and no less than: on the way from t to $t+s$ an imaginary stop at $t+\sigma$ should not matter. That same idea underlies the fifth and final assumption, which we call the RSG (for Reproduction-Survival-Growth) consistency condition, that for all A with $A \cap ([0,s]\times\Omega) = \emptyset$

(v) $$\Lambda(t,x)\;(A) = \int_{\Omega} \Lambda(t+s,\xi)\;(A_{-s})\; u(t,x;s)(d\xi).$$

The contribution of an individual, which at time t has i-state x, to the expected population size and composition at time $t+s$ has two components, viz. the individual itself and its offspring. For the offspring, we use time of birth and state-at-birth to characterise them, but we have to determine from this information the probability that they are alive at $t+s$ and, if so, the distribution of their i-state at that time. These considerations take the form

$$u^c(t,x;s)(\omega) = u(t,x;s)(\omega) + \int_{(0,s]\times\Omega} u(t+\tau,\xi;s-\tau)(\omega)\Lambda^c(t,x)(d\tau\times d\xi) \quad (50)$$

when expressed as a formula (here the index c refers again to "clan"). On the basis of the interpretation we expect that u^c thus defined will satisfy the Chapman-Kolmogorov consistency condition. Our aim is now to show that this is indeed the case. So we have to show that the conditions iv) and v) on u and Λ together with the constructive definitions of Λ^c and u^c in, respectively, (45) and (50) imply that u^c satisfies the Chapman-Kolmogorov condition.

The proof that we present below differs from the one given in [Diekmann e.a., 1998] and is based on an idea of Haiyang Huang (personal communication). It has the advantage that the biological interpretation underlies and motivates the various auxiliary results that we need as intermediate steps.

Definition 4 The actual distribution at time $t+s$ of k-th generation offspring of an individual which at time t has i-state x is given by

$$w^k(t,x;s)(\omega) = \int_{(0,s]\times\Omega} u(t+\tau,\xi;s-\tau)(\omega)\ \Lambda^{k\otimes}(t,x)\ (d\tau \times d\xi) \tag{51}$$

for $k \geq 1$; for notational convenience we add to this the natural convention

$$w^0(t,x;s) = u(t,x;s) \tag{52}$$

such that, by (50) and (45), we can write

$$u^c(t,x;s) = \sum_{k=0}^{\infty} w^k(t,x;s) \tag{53}$$

The fact that a k-th generation offspring born after time $t+s$ is, for some j less than k, the $(k-j)$-th generation offspring of a j-th generation offspring alive at time $t+s$, is reflected in the following identity.

Lemma 3 *For all A such that $A \cap ((0,s]\times\Omega) = \emptyset$ we have*

$$\Lambda^{k\otimes}(t,x)(A) = \sum_{j=0}^{k-1} \int_{\Omega} \Lambda^{(k-j)\otimes}(t+s,\xi)(A_{-s})\ w^j(t,x;s)(d\xi) \tag{54}$$

Proof. For $k=1$ this is v), which is true by assumption. We now assume that the identity has been verified for some k and then check that, as a consequence, it also holds for $k+1$.
To do so we use

$$\Lambda^{(k+1)\otimes}(t,x)(A) = \int_{\mathbf{R}_+\times\Omega} \Lambda(t+\tau,\xi)\ (A_{-\tau})\ \Lambda^{k\otimes}(t,x)\ (d\tau \times d\xi)$$

and rewrite this, on the basis of the decomposition $\mathbb{R}_+ = (0,s] \cup (s,\infty)$ as the sum of two terms, the first of which is, by v)

$$\int\limits_{(0,s]\times\Omega}\int\limits_{\Omega} \Lambda(t+s,\eta)\,(A_{-s})\,u(t+\tau,\xi;s-\tau)(d\eta)\,\Lambda^{k\otimes}(t,x)(d\tau\times d\xi) =$$

$$\int\limits_{\Omega} \Lambda(t+s,\eta)\,(A_{-s})\,w^k(t,x;s)\,(d\eta)$$

while the second equals

$$\int\limits_{(s,\infty)\times\Omega} \Lambda(t+\tau,\xi)(A_{-\tau})\sum_{j=0}^{k-1}\int\limits_{\Omega} \Lambda^{(k-j)\otimes}(t+s,\eta)\,(d_\tau(\tau-s)\times d\xi)\,w^j(t,x;s)(d\eta) =$$

$$\sum_{j=0}^{k-1}\int\limits_{\Omega}\int\limits_{\mathbb{R}_+\times Omega} \Lambda(t+\sigma+s,\xi)(A_{-\sigma-s})\,\Lambda^{(k-j)\otimes}(t+s,\eta)\,(d\sigma\times d\xi)\,w^j(t,x;s)(d\eta) =$$

$$\sum_{j=0}^{k-1}\int\limits_{\Omega} \Lambda^{(k-j+1)\otimes}(t+s,\eta)(A_{-s})\,w^j(t,x;s)(d\eta)$$

So by adding the two we see that the identity holds for $k+1$. □

Corollary 1 *The combination Λ^c, u^c satisfies the RSG consistency condition, i.e.*

$$\Lambda^c(t,x)(A) = \int\limits_{\Omega} \Lambda^c(t+s,\xi)(A_{-s})\,u^c(t,x;s)\,(d\xi) \tag{55}$$

Proof. Simply add the identity (54) with respect to k from 1 to ∞. □

For any $\sigma \in (0,s)$, a k-th generation offspring at time $t+s$ is a $(k-j)$-th generation offspring of a j-th generation offspring alive at time $t+\sigma$, for some j less than or equal to k. This observation is expressed in the next identity.

Lemma 4

$$w^k(t,x;s)(\omega) = \sum_{j=0}^{k}\int\limits_{\Omega} w^{k-j}(t+\sigma,\xi;s-\sigma)(\omega)\,w^j(t,x;\sigma)(d\xi) \tag{56}$$

Proof. We now use the decomposition $(0,s] = (0,\sigma] \cup (\sigma, s]$ in the defining formula (51) to arrive at two terms.
By the Chapman-Kolmogorov identity for u, the first of these takes the form

$$\int_{\Omega} u(t+\sigma,\eta;s-\sigma) \int_{(0,\sigma]\times\Omega} u(t+\tau,\xi;\sigma-\tau)(d\eta)\ \Lambda^{k\otimes}(t,x)\ (d\tau\times d\xi) =$$

$$\int_{\Omega} u(t+\sigma,\eta;s-\sigma)\ w^k(t,x;\sigma)(d\eta).$$

For the second we use (54) to write it as

$$\int_{(\sigma,s]\times\Omega} u(t+\tau,\xi;s-\tau)\sum_{j=0}^{k-1}\int_{\Omega} \Lambda^{(k-j)\otimes}(t+\sigma,\eta)(d_\tau(\tau-\sigma)\times d\xi)\ w^j(t,x;\sigma)(d\eta) =$$

$$\sum_{j=0}^{k-1}\int_{\Omega}\int_{(0,s-\sigma]\times\Omega} u(t+a+\sigma,\xi;s-a-\sigma)\ \Lambda^{(k-j)\otimes}(t+\sigma,\eta)(da\times d\xi)\ w^j(t,x;\sigma)(d\eta) =$$

$$\sum_{j=0}^{k-1}\int_{\Omega} w^{k-j}(t+\sigma,\eta;s-\sigma)\ w^j(t,x;\sigma)(d\eta)$$

So by adding the two we arrive at the stated identity. □

Adding the identities (56) with respect to k from 0 to ∞ we find as a direct corollary the Chapman-Kolmogorov identity for u^c; we state this as a theorem.

Theorem 4 *The following identity holds:*

$$u^c(t,x;s)(\omega) = \int_{\Omega} u^c(t+\sigma,\xi;s-\sigma)(\omega)\ u^c(t,x;\sigma)(d\xi) \tag{57}$$

Exercise 5 Prove that for any $j \in \{1,\ldots,k\}$ the identity

$$w^k(t,x;s)(\omega) = \int_{(0,s]\times\Omega} w^{k-j}(t+\tau,\xi;s-\tau)(\omega)\ \Lambda^{j\otimes}(t,x)(d\tau\times d\xi) \tag{58}$$

holds and give its interpretation.

To go from the individual to the population level is now a matter of adding contributions.

Definition 5 The two-parameter family U of linear operators on the space of measures on Ω is defined by

$$U(t+s,t)m_0 = \int_\Omega u^c(t,x;s)\, m_0(dx) \tag{59}$$

As a direct consequence of Theorem 14 we have that this family forms an evolutionary system, i.e. the identity

$$U(t+s_1+s_2, t+s_1)U(t+s_1,t) = U(t+s_1+s_2,t) \tag{60}$$

holds.

So starting from two modeling ingredients, Λ and u, we have, through a constructive procedure, defined a kernel operator on the space of measures which maps the expected population size and composition at some time onto the expected population size and composition at any later time. That is, we have shown how Λ and u govern the population dynamics.

In the next section we present two examples to illustrate the approach and, in particular, to show how to arrive at Λ and u for models which at first are formulated completely differently. In Section 3.3 we give a characterisation of equilibria in terms of the given ingredients Λ and u (thus avoiding the construction of Λ^c and u^c). In Section 3.4 it is outlined how this characterisation may be exploited to build a numerical algorithm for the constructive determination of an equilibrium. And in the final section, 3.5, we sketch how the idea of interaction via environmental variables can be mathematically formulated. We are then pretty much at the research frontier (a very slow moving boundary between established results and wishful thinking) and consequently end with an invitation to open problems.

3.2 Two examples

The first example concerns a size-structured cell population reproducing by fission into two equal parts. A shorthand description of the model is provided by the first order partial-functional differential equation

$$\frac{\partial n}{\partial t}(t,x)+\frac{\partial}{\partial x}(g(x)n(t,x)) = -\mu(x)n(t,x)-b(x)n(t,x)+4b(2x)n(t,2x) \quad (61)$$

Here x denotes the size of cells and $n(t,\cdot)$ the cell population size density at time t. However, we do not take (61) as our starting point but instead build on the i-level description of growth, death and division.

Cells grow according to

$$\frac{dx}{da} = g(x) \quad (62)$$

where a denotes age. Cells of size x have a probability per unit of time $\mu(x)$ of dying and a probability per unit of time $b(x)$ of dividing. We assume that $b(x) = 0$ for $x < \alpha$ which implies that the minimal cell size is $\frac{\alpha}{2}$ (note that we implicitly assumed that g is positive). We also introduce a maximal size, say $x = 1$ (accordingly the last term of (61) has to be interpreted as zero whenever $2x > 1$). Survival (in the extended sense of neither dying nor dividing) is described by

$$\frac{d\mathcal{F}}{da} = -(\mu(x(a)) + b(x(a))\mathcal{F} \quad (63)$$

but we can rewrite this as

$$\frac{d\mathcal{F}}{dx} = -\frac{\mu(x)+b(x)}{g(x)}\mathcal{F} \quad (64)$$

and then solve to obtain the expression

$$\mathcal{F}(x,x_0) = e^{-\int_{x_0}^{x} \frac{\mu(\xi)+b(\xi)}{g(\xi)}d\xi} \quad (65)$$

for the probability to reach size x, given that the cell has reached size x_0. To achieve that $x = 1$ is the maximal size, the function $\frac{\mu+b}{g}$ should have a non-integrable singularity at $x = 1$.

For later use we also introduce, on the basis of (62),

$$G(x) = \int_{\frac{\alpha}{2}}^{x} \frac{d\xi}{g(\xi)} \quad (66)$$

which gives the time needed to grow from size $\frac{\alpha}{2}$ to size x.
As the problem is autonomous, we leave out the argument t in $\Lambda(t,x)$ etc. We take $\Omega = (\frac{\alpha}{2}, 1)$. We denote by $X(a;x)$ the solution of the differential equation (62) satisfying the initial condition $X(0;x) = x$. Then

$$u(x;s)(\omega) = \delta_{X(s;x)}(\omega)\ \mathcal{F}(X(s;x);x) \tag{67}$$

and

$$\Lambda(x)((0,t]\times\omega) = \int_{(0,t]} 2b(X(\tau;x))\ \mathcal{F}(X(\tau;x);x)\ \chi_\omega(\frac{X(\tau;x)}{2})\ d\tau \tag{68}$$

which we rewrite as

$$\Lambda(x)((0,t]\times\omega) = 2\int_x^{X(t;x)} \frac{b(\xi)}{g(\xi)}\ \mathcal{F}(\xi;x)\ \chi_\omega(\frac{\xi}{2})d\xi \tag{69}$$

Now assume that $\alpha > \frac{1}{2}$, which means that the greatest daughter is smaller than the smallest mother. Then we can, in the spirit of Definition 3, restrict x as argument of Λ to values less than α and take α as the lower boundary of the integration interval in (69). The function $X(t;x)$ either reaches 1 in finite time or approaches 1 for $t \to \infty$. If we now also observe that for $x < \alpha < \xi$

$$\mathcal{F}(\xi;x) = \mathcal{F}(\xi;\alpha)\ \mathcal{F}(\alpha;x) \tag{70}$$

we can let $t \to \infty$ in (69) and rewrite it as

$$\Lambda(x)(\mathbb{R}_+\times\omega) = 2\ \mathcal{F}(\alpha;x)\int_\alpha^1 \frac{b(\xi)\ \mathcal{F}(\xi;\alpha)}{g(\xi)}\ \chi_\omega(\frac{\xi}{2})\ d\xi \tag{71}$$

The key feature of this expression is that, considered as a function of x and ω, it factors into the product of a function of x and a function of ω. In more biological terms this reflects that there is but one state-at-birth in a stochastic sense, viz. that the probability distribution of size-at-birth of a daughter is completely independent of the size-at-birth of the individual that we consider (but note, for completeness, that the probability that this individual does divide, rather than die, does depend on her size at birth).
As a consequence of this key feature the operator defined by the right hand side of (46) has one-dimensional range spanned by the absolutely continuous measure with density

$$\frac{b(2x)\mathcal{F}(2x;\alpha)}{g(2x)}$$

with support in the interval $[\frac{\alpha}{2},\frac{1}{2}]$. Applying the operator to this measure we find its one and only nonzero eigenvalue

$$R_0 = 4\int_{\frac{\alpha}{2}}^{\frac{1}{2}} \frac{b(2x)\mathcal{F}(2x;\alpha)}{g(2x)}\,\mathcal{F}(\alpha;x)\,dx$$

or

$$R_0 = 2\int_{\alpha}^{1} \frac{b(x)\,\mathcal{F}(x;\frac{x}{2})}{g(x)}\,dx \qquad (72)$$

(this expression allows a straightforward biological interpretation for which we refer to [Metz & Diekmann, 1986]).
To find the intrinsic rate of natural increase, we have to replace $\Lambda(x)(\mathbb{R}_+\times\omega)$ by

$$\int_{\mathbb{R}_+} e^{-z\tau}\,2b(X(\tau;x))\,\mathcal{F}(X(\tau;x);x)\,\chi_\omega(\frac{X(\tau;x)}{2})\,d\tau =$$

$$2\int_{\alpha}^{1} e^{-z[G(\xi)-G(x)]}\,\frac{b(\xi)\,\mathcal{F}(\xi;x)}{g(\xi)}\,\chi_\omega(\frac{\xi}{2})\,d\xi$$

which, by (70), again factors as the product of the function

$$x \longmapsto e^{zG(x)}\,\mathcal{F}(\alpha;x)$$

and a measure with density

$$\xi \longmapsto 2\,e^{-zG(\xi)}\frac{b(\xi)\mathcal{F}(\xi;\alpha)}{g(\xi)}$$

The condition that the corresponding operator has spectral radius one therefore leads to the characteristic equation

$$2\int_{\alpha}^{1} e^{-z(G(\frac{\xi}{2})-G(\xi))}\,\frac{b(\xi)\,\mathcal{F}(\xi;\frac{\xi}{2})}{g(\xi)} = 1 \qquad (73)$$

Unless $G(\frac{\xi}{2})-G(\xi)\equiv$ constant, there is a strongly dominant real root r. But when actually this function is identically constant, we have a lattice case and periodic behaviour occurs. This is worked out in detail in [Metz & Diekmann, 1986] and [Diekmann, Gyllenberg, Thieme & Verduyn Lunel, preprint] where it is also explained that this condition (which one can rewrite as $g(\xi) = 2g(\frac{\xi}{2})$) guarantees that "equal size" is an inheritable property, which then brings about that infinite-dimensional information about the initial condition remains manifest for all time.

I hope this example demonstrated two things. The first is that one can construct u and Λ from more "primitive" model ingredients (like μ, b and g) without much effort. The second is that one is greatly rewarded for such preliminary work, since the determination of R_0 and (an equation for) r is subsequently as easy as it can ever be.

In the second example, which concerns juvenile dispersal in space, we only deal with the first aspect. We choose this example to show that the determination of u and Λ from more primitive model ingredients may involve the solution of a partial differential equation.

Consider individuals which are characterised by their age and by their spatial position $y \in \mathbb{R}^2$. So $\Omega = \mathbb{R}_+ \times \mathbb{R}^2$. Assume that the probability per unit of time of dying, $\mu(a)$, and of giving birth, $\beta(a)$ do not depend on y. Assume that space is homogeneous and isotropic and that individual dispersal is described by a diffusion process with age-independent diffusion coefficient $D(a)$. We call individuals with $a < \bar{a}$ juveniles and those with $a \geq \bar{a}$ adults.
The difference is that juveniles don't reproduce ($\beta(a) = 0$ for $a < \bar{a}$) and that adults don't move ($D(a) = 0$ for $a \geq \bar{a}$).
The so-called fundamental solution w is defined by

$$\begin{cases} \frac{\partial w}{\partial a} = D(a)\,\Delta w - \mu(a) w \\ w(0,y) = \delta(y) \end{cases} \tag{74}$$

and it is the density for u with respect to the space variable, i.e.

$$u((0,y);s)(Q \times \Sigma) = \delta_s(Q) \int_{\Sigma} w(s, \eta - y)\, d\eta. \tag{75}$$

And $\Lambda_b(y)$ has density (for $a > \bar{a}$)

$$(a,\eta) \longmapsto w(\bar{a}, \eta - y)\, \beta(a)\, e^{-\int_{\bar{a}}^{a} mu(\alpha)\, d\alpha}$$

that is, for $t > \bar{a}$,

$$\Lambda_b(y)((0,t] \times \Sigma) = \int_{\Sigma} w(\bar{a}, \eta - y)\, d\eta \int_{\bar{a}}^{t} \beta(\alpha)\, e^{-\int_{\bar{a}}^{\alpha} \mu(\sigma)\, d\sigma}\, d\alpha. \tag{76}$$

For such kernel descriptions much is known about traveling waves and asymptotic speed of propagation, even for nonlinear problems (see [Mollison, 1991, Van den Bosch e.a., 1990, Metz & Van den Bosch, 1995]).
In infinite domains the asymptotic speed of propagation gives a more accurate description of initial population growth than the intrinsic rate of natural increase r.
A related point is that the Perron root, as described in [Jagers, 1991, preprint , Shurenkov, preprint 2], distinguishes between population growth in a given bounded area and population growth while being blown away by the wind to infinity; the dominant eigenvalue R_0 does not make such a distinction; for compact Ω the two are equal.
Admittedly the presentation of this second example is very superficial. All I did was show that such problems of age-dependent population growth and dispersal fit in the framework without much technical difficulty. In contrast, a semigroup approach that attempts to characterise the domain of the generator that is formally described by

$$\begin{cases} \frac{\partial n}{\partial t} + \frac{\partial n}{\partial a} = D(a)\ \Delta n - \mu(a)n \\ n(t,0,y) = \int_{\bar{a}}^{\infty}\ \beta(a)\ n(t,\alpha,y)\ d\alpha \end{cases} \tag{77}$$

is bound to encounter technical difficulties of various sorts, which do not add to our understanding of the dynamics.

3.3 Steady states

When looking for steady states of a dynamical system generated by the differential equation

$$\frac{dX}{dt} = F(X) \tag{78}$$

we don't have to analyse solutions of the initial value problem but instead can study the equation

$$F(X) = 0 \tag{79}$$

which is fully determined by the "ingredient" F.
What is the analogue in the context of our constructive definition of a dynamical system in terms of Λ and u ?

Theorem 5 *i) Assume that* $R_0 = 1$, *i.e.* $b \in M_+(\Omega_b)$ *can be found such that*

$$\int_{\Omega_b} \Lambda(x)\ (\mathbb{R}_+ \times \omega)\ b(dx) = b(\omega) \tag{80}$$

then

$$m(\omega) = \int_{\mathbf{R}_+ \times \Omega} u(\xi; s)(\omega)\ b(d\xi)\ ds \tag{81}$$

is a steady state for the linear semigroup, i.e.

$$\int_{\Omega} u^c(\xi; t)(\omega)\ m(d\xi) = m(\omega) \tag{82}$$

for all $t \geq 0$.
Conversely, let m satisfy (82), then b defined by

$$b(\omega) = \frac{1}{t} \int_{\Omega} \Lambda^c(\xi)((0, t] \times \omega)\ m(d\xi) \tag{83}$$

is independent of t and satisfies (80).

For the proof we refer to [Diekmann e.a., 1998]. The assumption iii) on u is crucial to guarantee that the integral in (81) converges. And indeed, when individuals have eternal lives, one may have a steady state on a generation basis while the extant population grows linearly.
Note that (80) and (81) are fully determined by the ingredients Λ and u. The converse part ii) is mainly added to demonstrate that we don't miss a steady state when restricting the analysis to (80) and then use (81).
In the next section we turn attention to a nonlinear version in a more restricted setting.

3.4 Numerical construction of steady states

In this section we briefly outline the approach of [Kirkilionis e.a., preprint]. When the state-at-birth is fixed the problem simplifies considerably, in particular when growth (i.e. movement in Ω) is deterministic. So call the state-at-birth x_0 and assume that for given and constant environmental conditions I the fate of an individual is described by

$$\frac{dx}{da}(a) = g(x(a), I), \qquad x(0) = x_0 \qquad \text{(growth)} \tag{84}$$

$$\frac{d\mathcal{F}}{da}(a) = -\mu(x(a), I)\ \mathcal{F}(a), \qquad \mathcal{F}(0) = 1 \qquad \text{(survival)} \tag{85}$$

$$\frac{dL}{da}(a) = \beta(x(a), I)\ \mathcal{F}(a), \qquad L(0) = 0 \qquad \text{(reproduction)} \tag{86}$$

then the steady state condition is given by

$$R_0 = L(\infty; I) = 1 \tag{87}$$

which we consider as a condition on I. The steady population composition is then described by

$$a \longmapsto b\, \mathcal{F}(a; I)$$

with b the (unknown) population birth rate. So if I is actually determined by the population size and composition through a weighing function $\gamma(x, I)$ we have to add to (87) the feedback condition

$$I = b \int_0^\infty \gamma(x(a; I), I)\, \mathcal{F}(a; I)\, da \tag{88}$$

which we can rewrite as

$$I = b\, \theta(\infty; I) \tag{89}$$

when we define, in the spirit of (84)-(86), θ by

$$\frac{d\theta}{da}(a) = \gamma(x(a), I)\, \mathcal{F}(a)\,, \qquad \theta(0) = 0 \qquad \text{(feedback)} \tag{90}$$

The equations (87), (88) are $\dim I + 1$ equations in as many unknowns, viz . I and b. So we may single out some parameter and study these equations by continuation methods, incorporating the numerical solution of (84)-(86), (90) to constructively define the equations. This idea is implemented in BASE, a programme package, by Markus Kirkilionis [Kirkilionis e.a., preprint]. With some more effort one can also determine stability boundaries and other bifurcation curves in a two-dimensional parameter plane.

Hopefully the presentation above is clear enough to also convey the message that one can **couple** different populations via I. In that case one has as many steady state conditions (87) as there are populations. So generically at most $\dim I$ populations can coexist in steady state. And when $\dim I$ equals the number of populations I is fully determined by the extended (87) and subsequently the b's can be determined from the analogue of the linear (in b) feedback condition (88). That is, there is an effective decoupling in the determination of I and b. Finally, I mention that one can extend this approach to the situation where state-at-birth is distributed, simply by making a **discretisation** of state-at-birth, letting x_0 in (84) take all possible values, and making a matrix version of (87).

3.5 Construction of the nonlinear semigroup

Our approach is to incorporate nonlinearity as interaction via "environmental" variables, by which we mean such quantities as predation pressure and food supply (recall the cannibalism model of Section 2.2). Let us denote these interaction variables by $I(t)$ and let us consider them, for the time being, as given.

The ingredients Λ and u will depend on I and consequently so will the evolutionary system of operators U introduced in Definition 5. We incorporate this dependence into the notation by writing

$$U_{I(\cdot)}(t, t_0)$$

with the implicit convention that we think of I as defined on, at least, the interval specified by the two arguments of U.

We assume that all initial time-dependence derives from the dependence on I, which we express mathematically in the form of the assumption that for all $s \in \mathbb{R}$

$$U_{I(\cdot-s)}(t+s, t_0+s) = U_{I(\cdot)}(t, t_0) \tag{91}$$

(that is, if we translate both the input I and the time interval, nothing changes).

The fact that I is not given, but rather determined by the population size and composition, is now expressed by the **feedback** consistency condition

$$I(t) = \int_{\Omega} \gamma(x, I(t)) \left(U_{I(\cdot)}(t, 0)\, m_0\right)(dx) \tag{92}$$

where γ is a **third modeling ingredient**. Under appropriate conditions on Λ, u and γ this equation (which is like a Volterra integral equation) will have for each positive measure m_0 a unique solution I, which we will denote by $I(t; m_0)$ to stress the dependence on m_0. We then define

$$S(t)\, m_0 = U_{I(\cdot; m_0)}(t, 0)\, m_0 \tag{93}$$

and first demonstrate the correctness of the

$$\text{"Lemma"} \quad I(t+s; m_0) = I(t; S(s)\, m_0) \tag{94}$$

and from this the more fundamental result:

$$\text{"Theorem"} \quad S(t+s) = S(t)\, S(s). \tag{95}$$

We invite the reader to check the formula manipulations leading to those assertions. A paper giving precise assumptions and statements is in statu nascendi [Diekmann e.a., in preparation].
So the feedback equation (92) is the key to extend the constructive definition of future-population-state-mappings to the nonlinear realm. Can we prove the principle of linearised stability in this setting ? Do steady states have stable , centre and unstable manifolds ? Does the Hopf bifurcation theorem hold ? Is there a global attractor ? Can periodic orbits be continued numerically ?
Questions of this type have been answered for less general classes of models in, for instance, [Webb, 1985, Tucker & Zimmermann, 1988, Thieme, 1988, Calsina & Saldaña, 1995, 1997, Clément e.a., 1989, Huyer, 1997]

Hopefully more and more results and tools from the general theory of dynamical systems will in the near future be extended to general physiologically structured population models as described in this paper.

References

[1] Calsina, À & J. Saldaña. 1995. A model of physiologically structured population dynamics with a nonlinear individual growth rate. *J. Math. Biol.* **33**: 335-364.

[2] Calsina, À & J. Saldaña. 1997. Asymptotic behaviour of a model of hierarchically structured population dynamics. *J. Math. Biol.* **35**: 967-987.

[3] Clément, Ph., O. Diekmann, M. Gyllenberg, H.J.A.M. Heijmans & H.R. Thieme. 1989. Perturbation theory for dual semigroups. III Nonlinear Lipschitz continuous perturbations in the sun-reflexive case. *Volterra Integro-Differ ential Equations in Banach Spaces and Applications*, G. Da Prato & M. Ianelli, Pitman Research Notes in Maths **190**: 67-89.

[4] Diekmann, O., M. Gyllenberg, J.A.J. Metz & H.R. Thieme. 1998. On the formulation and analysis of general deterministic structured population models. I. Linear Theory *J. Math. Biol.* **36**: 349-388.

[5] DIEKMANN, O., M. GYLLENBERG, H. HUANG, M. KIRKILIONIS, J.A.J. METZ & H.R. THIEME. II. Nonlinear Theory. In preparation.

[6] DIEKMANN, O. & J.A.J. METZ. 1994. On the reciprocal relationship between life histories and population dynamics. *Frontiers in Mathematical Biology*, S.A. LEVIN, Springer LNiB **100**: 263-279.

[7] DIEKMANN, O., M. GYLLENBERG, H.R. THIEME & S.M. VERDUYN LUNEL. (preprint). A cell-cycle model revisited.

[8] DIEKMANN, O., H.J.A.M. HEIJMANS & H.R. THIEME. 1984. On the stability of the cell size distribution. *J. Math. Biol.* **19**: 227–248.

[9] DIEKMANN, O., S.A. VAN GILS, S.M. VERDUYN LUNEL & H.-O. WALTHER. 1995. *Delay Equations: Functional- Complex and Nonlinear Analysis.* Springer Verlag.

[10] FELLER, W. 1966. *An Introduction to Probability Theory and Its Applications.* Vol. II. Wiley.

[11] FISHER, R.A. 1958. *The Genetical Theory of Natural Selection*, 2nd rev. ed., Dover.

[12] GRIPENBERG, G., S-O. LONDEN & O. STAFFANS. 1990. *Volterra Integral and Functional Equations.* Cambridge Univ. Press.

[13] GYLLENBERG, M. 1986. The size and scar distribution of the yeast *Saccharonymes cerevisiae. J. Math. Biol.* **24**: 81-101.

[14] H.HUANG (personal communication)

[15] HUYER, B. 1997. On periodic cohort solutions of a size-structured population model *J. Math. Biol.* **35**: 908–934.

[16] JAGERS, P. 1991. The growth and stabilization of populations. *Statistical Sience* **6**: 269–283.

[17] JAGERS, P. preprint. The deterministic evolution of general branching populations.

[18] JAGERS, P. 1983. On the Malthusianness of general branching processes in abstract type spaces. In: Probability and Mathematical Statistics. Essays in Honour of Carl-Gustav Esseen. Dep. Mathematics, Uppsala University.

[19] KIRKILIONIS, M.A., O. DIEKMANN, B. LISSER, M. NOOL, A.M. DE ROOS & B.P. SOMMEIJER (preprint) Numerical continuation of equilibria of physiologically structured population models. I. Theory *CWI-Report MAS 9714*. 1997.

[20] METZ, J.A.J. & O. DIEKMANN (eds.) 1986. *Dynamics of Physiologically Structured Populations*. Lecture Notes in Biomath. **68**. Springer Verlag.

[21] METZ, J.A.J. & F. VAN DEN BOSCH. 1995. Velocities of epidemic spread. *Epidemic Models: Their Structure and Relation to Data*. D. MOLLISON. Cambridge University Press.

[22] MOLLISON, D.. 1991. Dependence of epidemic and population velocities on basic parameters. *Math. Biosc.* **107**: 255-287.

[23] NAGEL, R. 1986. One Parameter Semigroups of Positive Operators. Springer LNiM 1184.

[24] DE ROOS, A.M., O. DIEKMANN & J.A.J. METZ. 1992. Studying the dynamics of structured population models: a versatile technique and its application to *Daphnia*. *Am. Nat.* **139**: 123–147.

[25] DE ROOS, A.M.. 1997. A gentle introduction to physiologically structured population models. In: TULJAPURKAR, S. & H. CASWELL 1997. Structured-population models in marine, terrestrial, and freshwater systems. Chapman & Hall: 119-204.

[26] SHURENKOV, V.M. (preprint 1). On the existence of a Malthusian parameter.

[27] SHURENKOV, V.M. (preprint 2) On the relationship between spectral radii and Perron roots.

[28] SHURENKOV, V.M. 1984. On the theory of Markov renewal. *Theory Prob. Appl.* **29**: 247–265.

[29] SHURENKOV, V.M. 1992. *Markov renewal theory and its applications to Markov ergodic processes*. Lect. Notes Dept. of Math., Chalmers Univ. of Technology.

[30] THIEME, H.R. 1988. Well-posedness of physiologically structured population models for *Daphnia magna*. *J. Math. Biol.* **26**: 299–317.

[31] TUCKER, S.L. & S.O. ZIMMERMANN. 1988. A nonlinear model of population dynamics containing an arbitrary number of continuous structure variables. SIAM. *J. Appl. Math.* **48**: 549-591.

[32] VAN DEN BOSCH, F., A.M. DE ROOS & W. GABRIEL. 1988. Cannibalism as a life boat meachanism. *J. Math. Biol.* **26**: 619-633.

[33] VAN DEN BOSCH, F., J.A.J. METZ & O. DIEKMANN. 1990. The velocity of spatial population expansion. *J. Math. Biol.* **28**: 529-565.

[34] WEBB, G.F.. 1985. Theory of Nonlinear Age-Dependent Population Dynamics. Marcel Dekker, New York.

[35] WIDDER, D.V. 1946. *The Laplace Transform*. Princeton Univ. Press.

Stochastic Spatial Models

Rick Durrett
Cornell University
Ithaca NY 14853 USA

Contents

Introduction and summary.

In a stochastic spatial model space is represented by a grid of sites, usually the d-dimensional integer lattice $\mathbf{Z}^d$. Each site can be in one of a set of states S and changes its state at a rate that depends on the states of neighboring sites. This framework is appropriate for a large number of situations in biology, so it has seen a diverse range of applications. See Durrett and Levin (1994a) and references therein.

In six 45 minute lectures it would be impossible to survey the field, so we will concentrate on two aspects. The first is the process that mathematicians call *the voter model with mutation*, but which biologists would call a spatial version of the Wright-Fisher or Moran models. Adopting the first viewpoint we can describe the dynamics by saying that (i) each voter at rate one decides to change its mind and adopts the opinion of a randomly chosen neighbor, (ii) each voter mutates at rate α creating a new type. To implement (ii) it is convenient to take S to be the unit interval and choose the new types at random from S.

The homogenizing force of (i) and the introduction of new types in (ii) create a stochastic equilibrium state for the process. Since in many situations the rate at which new types enter the system due to migration or genetic mutation is small, it is interesting to investigate the limiting behavior of this equilibrium state as $\alpha \to 0$. Here, we will concentrate on two aspects motivated by classical questions in biology: species-area curves and the abundance of species.

To define the species-area curves for the equilibrium state, let N_r be the number of different types in the box of side L^r centered at the origin. Since most research suggests a power law relationship between species and area, we plot log species versus log area, considering $\varphi_L(r) = \log N_r / \log(L^2)$. Results on the limiting behavior of this curve when $L = 1/\sqrt{\alpha}$ and $\alpha \to 0$ are given in Section 2 and the relationship with data on species area curves in nature is discussed.

In addition to the number of species seen in a viewing window, one can be concerned about their relative abundances. In this case experimental work suggests that we should look at abundances on a logarithmic scale, and many theoretical papers suggest that when we do this the result will be a log normal distribution. In Section 3 we show that the species abundance distribution for our model, when viewed on logarithmic scale, has a non-normal limit which is similar to Hubbell's data from Barro Colorado Island in having an over abundance of rare species when compared with the normal distribution.

In the last four lectures, we switch from the detailed study of one model to a much broader perspective. Durrett and Levin (1994b) proposed that the behavior of stochastic spatial models could be determined from the properties of the mean field ODE (ordinary differential equation), i.e., the equations for the densities that result from pretending that adjacent sites are independent. In their scheme there are three cases depending upon the properties of the ODE:

Case 1. *One attracting fixed point with all components positive.* There will be coexistence in the ODE and in the stochastic spatial model.

Case 2. *Two locally attracting fixed points.* In the ODE, the limiting behavior depends on the initial densities. However, in the stochastic spatial model, there is one stronger equilibrium that is the winner starting from generic initial conditions. To determine the stronger equilibrium, one can start with one half plane in each equilibrium and watch the direction of movement of the front that separates the two equilibria.

Case 3. *Periodic orbits in the ODE.* In the spatial model densities fluctuate wildly on small length scales, oscillate smoothly on moderate length scales, and after an initial transient are almost constant on large scales. That is, there is an equilibrium state with an interesting spatial structure.

These principles are a heuristic, designed to allow one to guess the behavior of the system under consideration, but there is a growing list of examples where the conclusions have been demonstrated by simulation or proved mathematically. In the last four sections of these notes, we will explain some of the results that have been obtained in support of this picture.

I. The Versatile Voter Model

Our first two lectures are concerned with the voter model. It is one of the simplest interacting particle systems, so it has been used as a model in many different contexts. We begin by defining the model and describing the basic results of Holley and Liggett (1975) then proceed to more recently developments concerning species-area curves and species-abundance distributions

1. Basic Results

The state of the voter model at time t is given by a function $\xi_t : \mathbf{Z}^d \to S$, where S is the set of possible types. The dynamics of the voter model are simple, perhaps the simplest possible:

> The voter at x at times of a rate one Poisson process $\{T_n^x, n \geq 1\}$, decides to change its opinion and adopts the opinion of the voter at y with probability $p(x,y) = \varphi(y-x)$.

To avoid complications in proofs, we will assume that the neighbor choice function φ has the following properties:

(A1) *finite range:* $\varphi(z) \neq 0$ for only finitely many z.

(A2) $\mathbf{Z}^d$*-symmetric:* $\varphi(z) = \psi(|z|)$ where $|z| = (z_1^2 + \cdots + z_d^2)^{1/2}$.

It is sensible to require also that:

(A3) *irreducible*: for any $x, y \in \mathbf{Z}^d$ the opinion at y can by some chain of effects get to x. Formally, there is a sequence of sites $x_0 = x, x_1, \ldots, x_n = y$ so that $\varphi(x_k - x_{k-1}) \neq 0$.

A common concrete example is the *nearest neighbor case*:

$$\varphi(z) = \begin{cases} 1/2d & \text{if } |z| = 1 \\ 0 & \text{otherwise} \end{cases}$$

However, with applications in mind, we will be interested in allowing more general distributions and obtaining an understanding of how the quantities we compute depend on the underlying distribution. We assume finite range out of laziness. That way we do not worry about the minimal moment conditions needed for the results to be hold or investigate exotic cases that occur e.g., transient two dimensional random walks.

a. Construction and Duality

For $x \in \mathbf{Z}^d$ let $\{T_n^x, n \geq 1\}$ be independent rate one Poisson processes and let $\{Y_n^x, n \geq 1\}$ be independent random variables taking values in $\mathbf{Z}^d$ with $P(Y_n^x = y) = p(x, y)$. Intuitively, at time T_n^x the voter at x imitates the one at Y_n^x. To implement this in a graphical way we draw an arrow from (x, T_n^x) to (Y_n^x, T_n^x). Since there are infinitely many Possion processes and hence no first arrival, we have to do a little work to show that this recipe gives rise to a well defined process. To accomplish this, but more importantly, to introduce the concept that will be the key to our analysis of the voter model, we will define a family of *dual random walks* $\{S_s^{x,t}, 0 \leq s \leq t\}$ that trace the origin of the opinion at x at time t. These will have the property

$$(1.1) \qquad \xi_t(x) = \xi_{t-s}(S_s^{x,t})$$

In words, the opinion at x at time t is the same as the one at $S_s^{x,t}$ at time $t - s$.

The paths $S_s^{x,t}$ are easy to describe in words: we work our way down the Poisson processes from time t to time 0, jumping to the head of an arrow whenever we encounter its tail. For an example, see Figure 1.1, where the thick lines indicate the duals $S_s^{x,t}$ and $S_s^{y,t}$. Formally, the dual random walks are defined by the requirement that $S_s^{x,t} = y$ if and only if there is a sequence of sites $x_0 = x, x_1, \ldots x_n = y$ and times $0 = r_0 < r_1 \ldots < r_n \leq r_{n+1} = s$ so that

(i) for $1 \leq i \leq n$ there is an arrow from x_{i-1} to x_i at time $t - r_i$

(ii) for $1 \leq i \leq n$ there is no Poisson arrival at x_{i-1} in $[t - r_{i-1}, t - r_i)$

(iii) there is no Poisson arrival at x_n in $[t - r_n, t - r_{n+1}]$.

Note. The closed interval in (iii) is there so that the state will change at the time a Poisson arrival occurs, and we will have a traditional right continuous Markov process.

To analyze the voter model, it is convenient to extend the definition of the dual process to subsets $A \subset \mathbf{Z}^d$ by

$$\zeta_s^{A,t} = \{S_s^{x,t} : x \in A\} \quad \text{for } 0 \leq s \leq t$$

To see how this system behaves we note that

(a) If $S_r^{x,t} = S_r^{y,t}$ then $S_s^{x,t} = S_s^{y,t}$ for $r \le s \le t$.

(b) Two random walks $S_s^{x,t}$ and $S_s^{y,t}$ move independently until they hit.

Again the reader can consult Figure 1.1 for an example. Visually, we imagine that when one particle lands on another one, the two particles combine into one, so we call the process $\zeta_s^{A,t}$, $A \subset \mathbf{Z}^d$, *coalescing random walks.*

The processes $S_s^{x,t}$, $0 \le s \le t$ are nice since their properties have a direct (almost sure) relationship with corresponding properties of ξ_t, e.g.,

$$\{\xi_t(x) = i\} = \{\xi_0(S_t^{x,t}) = i\}$$

For some purposes, such as the proofs of Theorems 1.1 and 1.2 below, it is convenient to combine these processes which are defined for $0 \le s \le t$ into two that are defined for all time by requiring that for any t

$$\begin{aligned} \{S_s^x, 0 \le s \le t\} &\stackrel{d}{=} \{S_s^{x,t}, 0 \le s \le t\} \\ \{\zeta_s^A, 0 \le s \le t\} &\stackrel{d}{=} \{\zeta_s^{A,t}, 0 \le s \le t\} \end{aligned}$$

and noting that these equations give us consistent finite dimensional distributions to which Kolmogorov's theorem can be applied.

b. Basic Dichotomy

Combining the duality described in the last few paragraphs with the idea that some random walks are recurrent while others are transient, leads to a dichotomy in the behavior of the voter model between dimensions $d \le 2$ and $d > 2$ discovered by Holley and Liggett (1975).

Theorem 1.1. *Clustering occurs in $d \le 2$.* For any set of possible states S, any initial configuration ξ_0, and any sites $x, y \in \mathbf{Z}^d$ we have

$$P(\xi_t(x) \ne \xi_t(y)) \to 0 \text{ as } t \to \infty$$

Theorem 1.2. *Coexistence is possible in $d > 2$.* Let ξ_t^θ denote the process with values in $S = \{0, 1\}$ starting from an initial state in which the events $\{\xi_0^\theta(x) = 1\}$ are independent and have probability θ. In $d > 2$ as $t \to \infty$, $\xi_t^\theta \Rightarrow \xi_\infty^\theta$, a translation invariant stationary distribution in which $P(\xi_\infty^\theta(x) = 1) = \theta$.

Remark. Here $\Rightarrow$ stands for weak convergence, which in this setting is just convergence of *finite dimensional distributions*: $P(\xi_t(x_1) = i_1, \dots \xi_t(x_k) = i_k)$. By *translation invariant*, we mean that the probabilities

$$P(\xi_t(x + y_1) = i_1, \dots \xi_t(x + y_k) = i_k)$$

do not depend on x.

Proof of Theorem 1.1 If the two sites x and y trace their opinions back to the same site at time 0 then they will certainly be equal at time t so

$$P(\xi_t(x) \neq \xi_t(y)) \leq P(S_t^x \neq S_t^y)$$

Now the difference $S_s^x - S_s^y$ is a random walk stopped when it hits 0, and the random walk has jumps that have mean 0 and finite variance. Such random walks are *recurrent*, and since ours is also, by (A3), an irreducible Markov chain, it will eventually hit 0. Since 0 is an absorbing state for $S_s^x - S_s^y$ it follows that $P(S_t^x \neq S_t^y) \to 0$ and the proof is complete. □

Proof of Theorem 1.2 The inclusion-exclusion formula implies that all of the finite dimensional distributions are determined if we know $P(\xi_t(x) = 0$ for all $x \in B)$ for each B. To show the convergence of these probabilities we observe that

$$P(\xi_t(x) = 0 \text{ for all } x \in B) = E\{(1 - \theta)^{|\zeta_t^B|}\}$$

since by duality there are no 1's in B at time t if and only if all of the sites in ζ_t^B are 0 at time 0, an event with probability $(1 - \theta)^{|\zeta_t^B|}$. Since ζ_t^B is a coalescing random walk, $|\zeta_t^B|$ is a decreasing function of t and hence has a limit. Since $0 \leq (1 - \theta)^{|\zeta_t^B|} \leq 1$ it follows from the bounded convergence theorem that $\lim_{t\to\infty} E\{(1 - \theta)^{|\zeta_t^B|}\}$ exists.

At this point we have shown that $\xi_t^\theta \Rightarrow \xi_\infty^\theta$. Since the voter model is a Feller process, it follows that ξ_∞^θ is a stationary distribution. For more details see Section 2 of Durrett (1995a). Since the ξ_t^θ are translation invariant, the limits ξ_∞^θ are. Duality implies that

$$P(\xi_t^\theta(x) = 1) = P(\xi_0(S_t^{x,t}) = 1) = \theta$$

for all t so $P(\xi_\infty^\theta(x) = 1) = \theta$. □

Remark. Holley and Liggett (1975) have shown that the ξ_∞^θ are spatially ergodic and give all the stationary distributions for the voter model with $S = \{0, 1\}$. That is, all stationary distributions are a convex combination of the (distributions of the) ξ_∞^θ. For proofs of this result see the original paper by Holley and Liggett (1975) or Chapter V of Liggett (1985).

c. The voter model with mutation.

In our new process, the state at time t is given by a function $\xi_t : \mathbf{Z}^d \to (0, 1)$, with $\xi_t(x)$ being the type (or species) of the individual at x at time t. We index our types by values w in the interval $(0, 1)$, so we can pick new types at random from the set of possibilities without duplicating an existing one. As in the voter model

(i) Each site x at times of a rate one Poisson process $\{T_n^x, n \geq 1\}$, decides to change its state and imitates the state of y with probability $p(x, y) = \varphi(|y - x|)$.

The new feature here is the spontaneous appearance of new types, which can be thought of as being genetic mutations or migration of individuals from outside the system.

(ii) Each site x mutates at rate α, changing to a new type w', chosen uniformly on $(0, 1)$.

To keep our treatment of the subject as brief, we will restrict our attention here to

(iii) *two dimensional nearest neighbor case:* $d = 2$ and

$$\varphi(z) = \begin{cases} 1/4 & \text{if } |z| = 1 \\ 0 & \text{otherwise} \end{cases}$$

Results can be generalized to finite range without substantial change. Interesting new phenomena occur when we consider this system in $d > 2$, in $d = 1$, or on the complete graph, but there is not enough time to discuss them here. For details see Bramson, Cox, and Durrett (1997).

It is straightforward to modify the construction of the voter model without mutation to take care of rule (ii). We introduce independent rate α Poisson processes of "mutation events" $\{\hat{T}_n^x, n \geq 1\}$, $x \in \mathbf{Z}^d$ and independent random variables $\{U_n^x, n \geq 1\}$, $x \in \mathbf{Z}^d$ uniformly distributed on $(0, 1)$

that are the new types. That is, at time T_n^x, the type at site x is set equal to U_n^x.

Because of the last feature, when we are working backwards in time with the random walks $S_s^{x,t}$ we can stop (and kill the random walk) when we encounter a mutation event, since that will determine the state of x at time t independent of the initial conditions. This modification turns the dual starting from a finite set A, $\hat{\zeta}_s^{A,t}$, into a coalescing random walk in which each particle is killed at rate α. If $\hat{\zeta}_t^{A,t} = \emptyset$ we need no information from the initial configuration to compute the state of A at time t. Since each particle is individually killed at rate α, it is not hard to show that:

Theorem 1.3. *The multitype voter model with mutation has a unique stationary distribution ξ_∞. Furthermore, for any initial ξ_0, we have $\xi_t \Rightarrow \xi_\infty$ as $t \to \infty$.*

Proof See Section II.2 of Griffeath (1978).

In most situations of interest in biology the mutation/migration rate α will be small, so our aim here will be to study the spatial structure of ξ_∞ in the limit $\alpha \to 0$.

2. Species-area curves

Since almost the beginning of the subject, see Watson (1835), ecologists have been interested in the relationship between species and area. The exact dependence of species number S on area A has been the subject of much debate. Early studies (e.g., Hopkins (1955)) fitted the curve $S = a \ln(1 + bA)$, a relationship that would be expected if the individuals in a quadrant were a random sample from a larger population (Preston (1969)). The most accepted relationship, however (Kilburn (1966), MacArthur and Wilson (1967), May (1975), Connor and McCoy (1979), Coleman (1981), Sugihara (1981)), takes the logarithm of S to be proportional to log area.

Hubbell (1993) was the first to suggest that a stochastic spatial model could be used to investigate species area curves. Some of his results are reported in Hubbell (1995). Hubbell's model is somewhat complicated because he allows each site to be occupied by more than one species and these species interact via specified rules of competition. Here, we follow the approach in

Bramson, Cox, and Durrett (1997), and Durrett and Levin (1996), and replace his model by a much simpler one, the voter model with mutation, so that we can obtain analytical results about the structure of the equilibrium state.

To define species-area curves for the voter model with mutation, we let $B(K)$ be a box with side K centered at the origin, and for $0 \le r < \infty$ let N_r be the number of different types in $B(L^r)$ in the stationary distribution ξ_∞. To plot log-species vs. log-area, we let

$$\varphi_L(r) = \frac{\log N_r}{\log(L^2)}$$

To see what to expect, we let $\bar{\alpha} = 1/\alpha$ and note that a typical $S_s^{x,t}$ will survive for time $O(\bar{\alpha})$ without being hit by a mutation and by the central limit theorem will move a distance about $L = \bar{\alpha}^{1/2}$ in that amount of time. Thus we expect sites that are separated by a large multiple of L to be distinct with high probability. This motivates part of the following result.

Theorem 2.1. *Let $L = \bar{\alpha}^{1/2}$. If $r \ge 1$, then as $\alpha \to 0$,*

$$\frac{N_r}{L^{2r-2}(\log L)^2} \to \frac{2}{\pi} \quad \text{in probability.} \tag{2.1}$$

Readers should note that (2.1) gives a very boring limit for the species area curve:

$$\varphi_L(r) \to (r-1)^+$$

i.e., a segment of slope 0 followed by a ray of slope 1. The ray with slope 1 is easy to understand: sites separated by $L^{1+\epsilon} = \bar{\alpha}^{(1+\epsilon)/2}$ will with high probability experience mutations before their random walks coalesce. Thus when $r \ge 1$, the number of species N_r increases in proportion to the area.

The segment of slope 0 at the beginning of the limiting curve $(r-1)^+$ was initially a major disappointment for us. However, it turned out to be a blessing: although the limiting slope is 0, when α is positive one gets slopes that agree reasonably well with data. To explain this we note that $N_1 \approx (2/\pi)(\log L)^2$ so

$$\frac{\log N_1}{2\log L} \approx \frac{2\log\log L + \log(2/\pi)}{2\log L}$$

The right hand side converges to 0 but only very slowly.

α	10^{-6}	10^{-8}	10^{-10}	10^{-12}
slope	.264	.229	.202	.182

These slopes are the range of values given in Figure 2.1. We are not able to prove that the species area curve will be roughly a straight line over $0 \le r \le 1$ but one can demonstrate this using computer simulation. Figure 2.2 gives the results of three simulations of the system with $\alpha = 10^{-6}$ and hence $L = 1/\sqrt{\alpha} = 1000$. Since we are only interested in the behavior on boxes of size L^r with $0 \le r \le 1$ we have performed the simulation on a 1000×1000 grid with *periodic boundary conditions.* That is, sites on the left edge are neighbors of those on the right on the same row; those on the top edge are neighbors of those in the bottom row in the same column.

Returning to the content of Figure 2.2, the reader should notice that the three simulated curves are fairly straight and end close to the value of .264 predicted by the table above. Having a slope that depends on the mutation rate provides a new explanation of the wide variety of slopes found in species area curves, i.e., mutation/migration rates vary considerably. For more on this see Durrett and Levin (1996).

3. Species abundance distributions

In addition to the spatial arrangements of the types it is interesting to look at the distribution of the abundance (i.e., number of representatives) of the types found in a sample. Again, this question has a long history in ecology. In an influential early paper, Fisher, Corbet, and Williams (1943) considered the distribution of moth and butterfly species caught in a light trap, making the interesting observation that while there are a huge number of individuals from a few species, the majority of species were represented by a few individuals.

Preston (1948) was one of the first to suggest the use of the lognormal distribution of species abundances. The theoretical explanation for the lognormal given on pages 88-89 of May (1975) is typical. Define $r_i(t)$ to be the per capita instantaneous growth rate of the ith species at time t, that is,

$$r_i(t) = \frac{1}{N_i(t)}\frac{dN_i(t)}{dt} = \frac{d}{dt}\ln N_i(t). \tag{3.1}$$

The last equation integrates to

$$\ln N_i(t) = \ln N_i(0) + \int_0^t r_i(s)\, ds. \tag{3.2}$$

If, as May (1975) says, "the ever-changing hazards of a randomly fluctuating environment are all important in determining populations," then one might reason that the integral is a sum of random variables to which the central limit theorem can be applied, and the distribution of abundances should follow a lognormal law.

While the last argument is simple, and possibly convincing, there are a number of data sets that do not fit the lognormal distribution very well. The tropical rain forest data in Hubbell (1995) is an example (see Figure 3.1). In the rain forest, the abundances of various plant species are not independent since individuals compete for a limiting resource, light. A static approach to this competition is provided by *MacArthur's broken stick distribution* (see his (1957) and (1960) papers). He imagines that the proportions $(p_1, p_2, \ldots, p_n)$ of the area occupied by n given species to be chosen at random from the set of all possible vectors of proportions, i.e., those with nonnegative coordinates that sum to one. For this reason, Webb (1974) calls this the *proportionality space model.* A simple way of generating such p_i's is to put $n-1$ independent uniform random variables on (0,1) and look at the lengths of the intervals that result, hence, the name "broken stick distribution." Quoting May's (1975) survey again, "This distribution of relative abundance is to be expected whenever an ecologically homogeneous group of species apportion randomly among themselves a fixed amount of some governing resource." Broken stick abundance patterns have been found in data for birds by MacArthur (1960), Tramer (1969), and Longuet-Higgins (1971).

One of the weaknesses of the "broken stick" approach is that it simply chooses a nice distribution based on symmetry, without a direct consideration of the underlying mechanisms. Engen and Lande (1996) have recently (see their pages 174–175) introduced a dynamic model in which new species enter the community at times given by a Poisson process, and where the log abundances of the species $Y_t^i = \log(X_t^i)$ evolve according to the independent diffusion processes

$$dY_t^i = (r - g(\exp(Y_t^i)))\, dt + \sigma(\exp(Y_t^i))\, dB_t^i. \tag{3.3}$$

Here, $r > 0$ is a fixed growth rate, $g(x)$ is a "density regulation function", and $\sigma(x) = \sigma_e^2 + \sigma_d^2 e^{-x}$, with σ_e being the environmental and σ_d the demographic stochasticity. Engen and Lande then showed that, if $g(x) = \gamma \ln(x + \nu)$, with $\nu = \sigma_e^2/\sigma_d^2$, the species abundances in equilibrium are given by the lognormal distribution. Although the last approach is dynamic, the reader should note that the sizes of the different species there (as well as in May's derivation of the lognormal) are independent. That is, there is no competition between the species, as there is, at least implicitly, in the broken stick model.

Here we will use the voter model with mutation to derive a new species abundance distribution. To explain precisely what we will study requires several definitions. We define the *patch size in A* for the type i at time t to be the number of sites y in A with $\xi_t(y) = i$. Let $N(A,k)$ be the number of types in ξ_∞ with patch size in A equal to k, and, for $I \subset [0,\infty)$, let $N(A,I) = \sum_{k \in I} N(A,k)$ be the number of types with patch sizes in A that lie in the interval I.

Here, we only consider the case in which the viewing window A is $B(L)$, the square of side L centered at the origin. Let $|B(L)|$ be the number of points of the integer lattice $\mathbf{Z}^2$ that are in the square $B(L)$. It is convenient to divide the number of species observed by the number of sites, $|B(L)|$, to obtain the species abundance per site. That is, we consider the frequency of types with patch sizes in the interval I:

$$N^L(I) = \frac{N(B(L),I)}{|B(L)|}.$$

One immediate advantage of computing densities per site is that by invoking an appropriate law of large numbers, (see e.g., Theorem 9 on page 679 of Dunford and Schwarz (1957)), we can conclude that as the observation window, $B(L)$, gets large our estimate becomes close to the underlying mean.

Proposition 3.1. *For all sets I,*

$$N^\infty(I) = \lim_{L \to \infty} N^L(I)$$

exists and is a nonrandom constant.

We refer to $N^\infty(I)$ as the underlying *theoretical abundance distribution.*

Proposition 3.1 implies that when L is large, the observed species abundance frequencies in the square $B(L)$ will be close to the theoretical frequency, so we next inquire how large L needs to be so that $N^L(I) \approx N^\infty(I)$. It is easy to see that some L are too small. The time until mutation along the line of descent of a given individual is an exponential random variable with mean $\bar{\alpha} = 1/\alpha$. Since offspring are displaced by independent and identically distributed amounts from their parents, the family tree of an individual behaves like a random walk and will move a distance of order $\bar{\alpha}^{1/2}$ in time $\bar{\alpha}$.

The arguments in the last paragraph indicate that we will need an observation window whose length is at least of order $\bar{\alpha}^{1/2}$ to get an accurate idea of the distribution of sizes. In our first result, we will we will take our observation window to have this smallest possible size. To be precise, we will let $\beta > 0$ and suppose that $L \geq \beta\bar{\alpha}^{1/2}$. This will turn out to be large enough to look at patch sizes on a logarithmic scale. That is, we will consider $N^L([1, \bar{\alpha}^y])$, the number of species (per site) with sizes between 1 and $\bar{\alpha}^y$.

When the mutation rate gets small then the individual species have a large number of individuals and the number of species per site is small. Thus to get a sensible asymptotic statement when $\alpha \to 0$ we have to divide $N^L([1, \bar{\alpha}^y])$ by something that tends to 0.

$$F_\alpha^L(y) = \frac{N^L([1, \bar{\alpha}^y])}{\alpha(\log \bar{\alpha})^2/2\pi} \tag{3.4}$$

The exact form of the denominator may look mysterious, but our results given below will show that it is the right choice. Readers curious about why this is the right normalization can consult Bramson, Cox, and Durrett (1997) for a more detailed explanation.

To state our result about the asymptotic behavior of the distribution of sizes $F_\alpha^L(y)$, we need to define the limiting distribution function V. Let

$$V(y) = \begin{cases} 0, & y \leq 0, \\ y^2, & 0 \leq y \leq 1, \\ 1, & y \geq 1. \end{cases}$$

The nature of this distribution is clearer if we look at its density function, which is 0 unless $0 < y < 1$ in which case $v(y) = V'(y) = 2y$. The graph of $v(y)$ is a right triangle. The next result is then a "log-triangular" limit theorem for species abundances in the two dimensional voter model with mutation.

Theorem 3.1. *Let $\beta > 0$, and assume that $L = L(\alpha) \geq \beta\bar{\alpha}^{1/2}$. Then, for any $\varepsilon > 0$,*

$$P\left(\sup_y |F_\alpha^L(y) - V(y)| > \varepsilon\right) \to 0 \quad \text{as } \alpha \to 0. \tag{3.5}$$

An obvious first reaction to the triangular limiting density $v(y)$ is that it does not look very much like Hubbell's data given in Figure 3.1. We will return to this point when we discuss the simulations below. The key to connecting our result with the Hubbell's data lies in a more refined result than Theorem 3.1 which we will now introduce.

Theorem 3.1 does not apply directly to the histograms of abundance counts reported in the literature. For example, in Preston (1948), abundance counts are grouped into "octaves," 1–2, 2–4, 4–8, 8–16, 16–32, ..., splitting in half the observations that are exactly powers of 2. To avoid trouble with the boundaries, some later investigators (see e.g., Chapter 3 of Whittaker (1972)) viewed the 1 cell as an interval [0.5,1.5], and then multiplied by 3 to get disjoint classes [1.5,4.5], [4.5,13.5], etc.

To treat such histograms, as well as other possibilities, we could fix an $r > 1$ to be the width of the cells, and look at the area-normalized abundance of species, $N^L([ar^k, ar^{k+1}))$ where a is some constant, which can always be chosen so that $r^{-1} < a \leq 1$. For example, in this setting Whittaker's cells have $a = 0.5$, $r = 3$, and Preston's correspond roughly to $a = 1$, $r = 2$. One can easily generalize what we are about to do too $a \neq 1$. However, our formulas are already too messy, so we will try to suppress clutter by restricting our attention to the case $a = 1$.

Theorem 3.2 below provides the refinement of Theorem 3.1 needed to analyze the histograms that are used to estimate the densities. To see what form the result should take, we note that $r^k = \bar{\alpha}^y$ where $y = \log(r^k)/\log(\bar{\alpha})$ so if in Theorem 3.1 convergence of the underlying density functions also holds, then we would have

$$\frac{N^L([r^k, r^{k+1}))}{\alpha(\log\bar{\alpha})^2/2\pi} \approx \int_{\log(r^k)/\log\bar{\alpha}}^{\log(r^{k+1})/\log\bar{\alpha}} 2y\,dy$$

Rearranging and then evaluating the integral, we have

$$\begin{aligned} N^L([r^k, r^{k+1})) &\approx \frac{\alpha(\log\bar{\alpha})^2}{2\pi}\left\{\left(\frac{\log(r^{k+1})}{\log\bar{\alpha}}\right)^2 - \left(\frac{\log(r^k)}{\log\bar{\alpha}}\right)^2\right\} \\ &= \frac{\alpha}{2\pi}\cdot(2k+1)(\log r)^2 \approx \frac{\alpha k}{\pi}(\log r)^2 \end{aligned} \tag{3.6}$$

Our next result, Theorem 3.2, shows that the counts $N^L([r^k, r^{k+1}))$ are simultaneously well approximated by the formula just derived over a wide range as $\alpha \to 0$. Fix $r > 1$ and $\varepsilon > 0$, and let $E_{L,\varepsilon}(k)$ be the event that our approximation in the kth cell, $[r^k, r^{k+1})$, is off by at least a factor of ε from what we expect. That is, $E_{L,\varepsilon}(k)$ is the event

$$\left| N^L([r^k, r^{k+1})) - \frac{\alpha k}{\pi}(\log r)^2 \right| > \varepsilon \alpha k \,. \tag{3.7}$$

Note that we have simplified the right hand side by dispensing with $(\log r)^2/2\pi$ which for fixed r is a constant.

Unfortunately, the approximation in (3.6) does not apply to large patch sizes. To say how large is too large, we introduce the cutoff size $\hat{\alpha} = \pi\bar{\alpha}/\log\bar{\alpha}$.

Theorem 3.2. *Let $r > 1$, $\beta > 0$, and assume that $L \geq \beta\bar{\alpha}^{1/2}(\log\bar{\alpha})^2$. Then, for any $\varepsilon > 0$ we can pick δ small enough so that*

$$\limsup_{\alpha \to 0} P\left(E_{L,\varepsilon}(k) \text{ for some } k \text{ with } r^k \in [\delta^{-1}, \delta\hat{\alpha})\right) \leq \varepsilon$$

The form of the conclusion of Theorem 3.2 is dictated by the fact that the approximation given in (3.6) does not apply well when k is small or k is of order $\hat{\alpha}$. Thus, we have to pick δ small to restrict the range of values considered, in order to get a small error in the limit as $\alpha \to 0$.

The additional restriction $L \geq \beta\bar{\alpha}^{1/2}(\log\bar{\alpha})^2$ in Theorem 3.2, compared to the requirement $L \geq \beta\bar{\alpha}^{1/2}$ in Theorem 3.1, comes from the fact that we are considering abundance sizes rather than their logarithms, so losing that fraction of the mass of a patch which is outside of the observation window can have a significant effect. Thus, to have the sampled distribution agree with the underlying theoretical distribution, we need to choose L substantially larger than $\bar{\alpha}^{1/2}$. A second complication is that we are making a statement simultaneously for about $\log\bar{\alpha}$ histogram cells, so we need $L/\bar{\alpha}^{1/2} \to \infty$ with at least a certain rate to be able to control the errors for all of the cells simultaneously.

The restriction to $r^k \geq 1/\delta$, in Theorem 3.2, is needed in order to use the asymptotics employed in our proof. Small patches are formed due to rare events, and require a separate analysis. The largest patch size covered by Theorem 3.2 is $[r^k, r^{k+1})$, where r^k is small relative to $\hat{\alpha}$. The distribution of larger patch sizes, i.e., those of the form $[a\hat{\alpha}, b\hat{\alpha})$, differs in a fundamental way from the distribution of the smaller patch sizes.

Theorem 3.3. *Let $\beta > 0$, and assume that $L = L(\alpha) \geq \beta\bar{\alpha}^{1/2}(\log\bar{\alpha})^2$. Then, for any $\varepsilon > 0$, and a, b with $0 < a < b$,*

$$\lim_{\alpha\to 0} P\left(\left|\hat{\alpha}N^L([a\hat{\alpha}, b\hat{\alpha}]) - \int_a^b u^{-1}e^{-u}\,du\right| > \varepsilon\right) = 0.$$

Some readers may be disturbed by the limiting density $u^{-1}e^{-u}$ above, which has an infinite integral over $(0, \infty)$. This behavior must, in fact, occur, since the scale in Theorem 3.3 for the frequency of patch sizes of interest is $1/\hat{\alpha} \approx \alpha \log\bar{\alpha}$, while Theorem 3.1 tells us that the scale for the total frequency is $\alpha(\log\bar{\alpha})^2$, which is of greater order of magnitude for small α.

Theorem 3.3 is a close relative of Theorem 3.4 which is taken from Sawyer (1979), Theorem 1.2.

Theorem 3.4. *Let $\nu(O)$ be the number of sites with the same type as the origin. As the mutation rate $\alpha \to 0$,*

$$P(\nu(O) \leq b\hat{\alpha}) \to \int_0^b e^{-u}\,du, \qquad \textit{for all } b \geq 0.$$

The same result obviously holds for any other fixed site x replacing the origin O, or for a site chosen at random from $B(L)$. Now, when a site is chosen at random, a patch has probability of being chosen which is proportional to its size. Removing this "size-bias" from Sawyer's result introduces the factor u^{-1} into the limiting density in Theorem 3.

Simulation results. To help to understand the results and to what extent the results apply when α is only moderately small, we will simulate the system with $\alpha = 10^{-4}$. By the heuristics above, the time for a typical line of descent to encounter a mutation will be of order $\bar{\alpha} = 10^4$, and it will move a distance of order $\bar{\alpha}^{1/2} = 100$ units over this time. Multiplying this distance by 10, we choose our experimental universe to be a 1000×1000 grid. To avoid edge effects we will again use *periodic boundary conditions*.

The first statistic we investigate is one that comes from the "size-biased" viewpoint. Introduce histogram bins $[1, 500]$, $[501, 1000]$, ..., and then, for a given bin, throw in all of the individuals that belong to species with patch sizes in that range. Each patch of size k is counted k times, so Theorem 3.4 implies that the distribution we observe will be approximately an exponential with mean $\hat{\alpha}$. To compute this mean, we note that

$$\hat{\alpha} = \frac{\pi\bar{\alpha}}{\log\bar{\alpha}} \approx \frac{10,000\pi}{9.21034} \approx 3411. \tag{3.8}$$

Figure 3.2 shows the average of 10 histograms for our parameters. The agreement with the exponential distribution with mean 3411 predicted by Theorem 3.4 (the curve of diamonds in Figure 3.2) is fairly good.

To investigate the ordinary (not size-biased) species abundances, we introduce histogram bins $[1,2)$, $[2,4)$, $[4,8)$, $\ldots$, and count the number of species with patch sizes in the indicated ranges. Figure 3.3 displays the average of 20 replications of the experiment (which are independent of the 10 given above). To compare with theory, we begin with Theorem 3.3. If, for example, we want to know the number of species with patch sizes in $[2^k, 2^{k+1})$, then our estimate, based on Theorem 3.3, will be

$$\frac{L^2}{\hat{\alpha}} \int_{2^k/\hat{\alpha}}^{2^{k+1}/\hat{\alpha}} u^{-1} e^{-u} \, du \,. \tag{3.9}$$

Evaluating the integral numerically for the cells of interest gives the line of diamonds in Figure 3.3. The fit is good for cells $k \geq 7$. From the remarks after Theorem 3.3, we should not necessarily expect this approximation to work well for cells too far to the left where Theorem 3.2 gives a different answer.

We now turn to Theorem 3.2. In (3.7), we employed the estimate

$$N^L([r^k, r^{k+1})) \approx \alpha k (\log r)^2 / \pi$$

for the abundance counts. Recalling that N^L is the number of species per site, we multiply by $|B(L)| \approx L^2$ to get

$$N(B(10^3), [2^k, 2^{k+1})) \approx 10^6 \cdot 10^{-4} \cdot \frac{(\log 2)^2}{\pi} k \approx 15.29\, k \tag{3.10}$$

This formula, the line of squares in Figure 3.3, provides a poor fit for the simulation data. One of the problems with the prediction from Theorem

3.2 is that the proof of (3.7) already contains approximations based on the assumption that k is large. The error can be reduced by going back into the proof of Theorem 3.2 in Bramson, Cox and Durrett (1997), and pulling out the following formula for the number of species per site:

$$N^L([2^k, 2^{k+1})) \approx \alpha \int_0^\infty p_t(e^{-2^k p_t} - e^{-2^{k+1} p_t})\, dt. \tag{3.11}$$

See the discussion at the end of Section 6 there.

Here, p_t is the density at time t of a system of coalescing random walks started from all sites occupied. A complete description of coalescing random walks can be found in Bramson, Cox, and Durrett (1997). There, we employed the asymptotics $p_t \approx (\log t)/\pi t$, as $t \to \infty$, from Bramson and Griffeath (1980), to reduce the right side of (3.11) to $\alpha k(\log 2)^2/\pi$. One can instead use simulations to estimate p_t for small t, and then using the asymptotic formula after that to numerically evaluate the integral in (3.11).

In order to estimate $N(B(1000), [2^k, 2^{k+1}))$, we use simulations on a $1000\times$ 1000 grid to estimate p_t for times up to time 5000, and then the asymptotic formula after that. The result is given by the circular symbols in Figure 3.3. This results in a drastic improvement over Theorem 3.2. The value predicted by (3.11) for the $k = 0$ cell is only about 60% of the observed value, but the fit at the other cells $1 \leq k \leq 9$ is now good. Note that if we combine the formulas in (3.11) and (3.9) by taking the minimum of the two expressions, the result is accurate for all cells except $k = 0$.

The shapes of the distributions for the simulation data in Figure 3 and the field data in Figure 1 from Hubbell (1995) are quite similar. Besides being on the same scale, they exhibit related departures from lognormality — the distributions are not symmetric about their greatest value, and they have an over-abundance of species with small numbers. Such similarity is not necessarily evidence of a common underlying cause, of course. However, there are reasons to suggest that this agreement is not an accident. In the rain forest and in our model, species compete for a fixed amount of a limiting resource (e.g., light or area).

II. Stochastic Spatial Models vs. ODEs

In the last four sections we will describe a number of examples to illustrate Durrett and Levin's (1994b) idea that the behavior of stochastic spatial models can be determined from properties of the mean field ODE, which is obtained by pretending that neighboring sites are independent.

4. Case 1. Attracting Fixed Point

We begin with a simple but fundamentally important example. Our treatment will be brief. The reader can find the facts we quote and much more in Liggett (1985) or Durrett (1988).

Example 4.1. Contact process. This system was introduced by Harris (1974). Each site can be in state 0 = vacant or 1 = occupied. The system evolves according to the following rules:

(i) An occupied site becomes vacant at a rate δ.

(ii) A occupied site gives birth at rate β. A particle born at x is sent to y with probability $p(x,y) = \varphi(y-x)$.

(iii) The site y becomes occupied if it was vacant, and stays occupied if it was occupied.

One of the simplest and most studied cases is the two dimensional nearest neighbor model which has $\varphi(1,0) = \varphi(0,1) = \varphi(-1,0) = \varphi(0,-1) = 1/4$. In words, offspring are sent to one of the four nearest neighbors chosen at random.

The contact process as formulated above has two parameters but only needs one. By scaling time we can and will suppose that $\beta = 1$. In this case, particles die at rate δ and give birth at rate at most 1 since births onto occupied sites are lost. From this it is easy to see that if we start with a finite number of occupied sites and $\delta > 1$ then the contact process will *die out*, i.e., reach the all 0 configuration with probability 1. We define the critical value δ_f, for survival from finite sets, to be the supremum of all of the values of δ so that dying out has a probability < 1 for some finite initial state.

There is a second slightly more sophisticated notion of "survival" for the contact and other process: the existence of a stationary distribution for the Markov chain which does not concentrate on the absorbing state in which all sites are vacant. To see when such a stationary distribution will exist, we start with the observation that the contact process is *attractive*: i.e., increasing the number of 1's increases the birth rate and decreases the death rate. Then we quote

Lemma 4.1. *Let ξ_t^1 denote the process starting from all 1's. If the process ξ_t is attractive then as $t \to \infty$, $\xi_t^1 \Rightarrow \xi_\infty^1$, a stationary distribution.*

Here $\Rightarrow$ is short for *converges in distribution*, which means that for any sites $x_1, \ldots, x_n$ and possible states $i_1, \ldots, i_n$ we have convergence of the *finite dimensional distributions*

$$P(\xi_t^1(x_1) = i_1, \ldots \xi_t^1(x_n) = i_n) \to P(\xi_\infty^1(x_1) = i_1, \ldots \xi_\infty^1(x_n) = i_n)$$

This result and all the others we cite for the contact process can be found in any of the four books on the subject: Liggett (1985), Griffeath (1978), Durrett (1988) and (1995b).

Of course the limit in Lemma 4.1 could assign probability 1 to the all 0 configuration, and it will if δ is too large, e.g., $\delta > 1$. Let δ_e be the supremum of the values of δ for which the limit is not all 0's. For the quadratic contact process, Example 6.1, we will have

$$0 = \delta_f < \delta_e < \infty$$

However, for the contact process these two critical values coincide. To explain why this is true, we note that by using an explicit construction and working backwards in time, much as we did for the voter model one can show:

Lemma 4.2. *Let $p_t(A, B)$ be the probability some site in B is occupied at time t when we start with 1's on A (and 0's elsewhere) at time 0. Then $p_t(A, B) = p_t(B, A)$.*

Taking A = all sites and B = a single point we see that the density of occupied sites at time t is the same as the probability that the process survives until time t starting from a single occupied site. Thus $\delta_e = \delta_f$ and we denote their common value by δ_c, where the c stands for *critical value*.

Mean Field Theory. If we consider the contact process on a grid with n sites and modify the rules so that all sites are neighbors then the number of occupied sites at time t is a Markov chain $N(t) \in \{0, 1, \ldots, n\}$ with transition rates:

$N(t) \to N(t) - 1$ at rate $\delta N(t)$.

$N(t) \to N(t) + 1$ at rate $N(t)\left(1 - \frac{N(t)}{n}\right)$.

If we let $u_n(t) = N(t)/n$ be the fraction of occupied sites and let $n \to \infty$ then it is not hard to show (e.g., using (7.1) of Chapter 8 of Durrett (1996)) that the u_n converge to the solution of the *"mean field" ordinary differential equation.*

$$\frac{du}{dt} = -\delta u + u(1-u) \tag{4.1}$$

The term in quotes comes from the physics. It refers to the fact that each site only feels the density of occupied sites. Writing 0 for vacant and 1 for occupied the density is then the mean value of the occupancy variables. The mean field equation can also be obtained from the spatial model by letting $u(t)$ be the fraction of sites occupied at time t and assuming that adjacent sites are independent. Under this assumption the rate at which new particles are produced is $u(1-u)$, while particles disappear at rate $-\delta u$. Since the second recipe is simpler we will use it for all of the other computations.

The mean field ODE for the contact process predicts that δ_c is 1 and for $\delta < \delta_c$ the equilibrium density of occupied sites is $1 - \delta$. In the nearest neighbor contact process there is a significant positive correlation between the states of neighboring sites (see Harris (1977)) so this overestimates the critical value. Numerical results tell us the critical value of the two dimensional nearest neighbor contact process is about 0.607. See Brower, Furman and Moshe (178) and Grassberger and de la Torre (1979).

Although the nearest neighbor case has been the most studied, it turns out that the contact process gets simpler when we consider

(4.2) **Long Range Limits.** Let $\psi \geq 0$ be a continuous function that is integrable and not identically 0, and define the dispersal kernel in part (ii) of the definition of the contact process to be $\varphi(z) = c_r\psi(z/r)$ where c_r chosen to make the probabilities add up to one.

Bramson, Durrett, and Swindle (1989), have shown (see also Durrett (1992) for the weaker version stated here)

Theorem 4.1. *As $r \to \infty$, $\delta_c(r) \to 1$ and*

$$P(\xi^1_\infty(x) = 1) \to 1 - \delta$$

Note that as $r \to \infty$ the critical value and equilibrium densities converge to those predicted by mean field theory.

Example 4.2. Grass Bushes Trees. In our second model, the possible states are $0 =$ grass, $1 =$ bushes, $2 =$ trees. 0's are thought of as vacant sites. Types 1 and 2 behave like contact processes subject to the rule that 2's can give birth onto sites occupied by 1's but not vice versa. In formulating the dynamics, we are thinking of the various types as species that are part of a successional sequence. With Tilman's (1994) work in mind we define the model for an arbitrary number of species.

(i) If $i > 0$, type i individuals die at a constant rate δ_i and give birth at rate β_i.

(ii) A particle of type i born at x is sent to y with probability $p_i(x, y) = \varphi_i(y - x)$.

(iii) If number of the invading type is larger it takes over the site.

Starting our analysis with the case of two types, we note that 2's don't feel the presence of 1's, so they are a contact process and will survive if $\delta_2/\beta_2 < \delta_c$. The main question then is: when can the 1's survive in the space that is left to them?

To investigate this question Durrett and Swindle (1991) considered what happens when long range limits are taken as described in (4.2) above. As in the case of the long range contact process, the motivation is that in this case the densities will behave like solutions to the mean field ODE, which is obtained by pretending that adjacent sites are always independent

$$\begin{aligned} du_1/dt &= u_1 \{ \beta_1(1 - u_1 - u_2) - \delta_1 - \beta_2 u_2 \} \\ du_2/dt &= u_2 \{ \beta_2(1 - u_2) - \delta_2 \} \end{aligned} \tag{4.3}$$

For example in the du_1/dt equation the first term represents births of 1's onto sites in state 0 (vacant), the second term represents constant deaths, and the third births of 2's onto sites occupied by 1's.

From the second equation in (4.3) the equilibrium density of 2's will be

$$\bar{u}_2 = \frac{\beta_2 - \delta_2}{\beta_2}$$

Inserting this into the first equation and solving one finds there is an equilibrium with $\bar{u}_1 > 0$ if

$$\beta_1 \cdot \frac{\delta_2}{\beta_2} - \delta_1 - \{\beta_2 - \delta_2\} > 0 \tag{4.4}$$

As written, this condition can be derived by asking the question: "Can the 1's invade the 2's when they are in equilibrium?" That is, will u_1 increase when it is small enough. To see this note that in the absence of 1's, the 2's have an equilibrium density of $\bar{u}_2 = (\beta_2 - \delta_2)/\beta_2$. Plugging this into the first equation and ignoring the $-\beta_1 u_1^2$ term gives

$$\frac{du_1}{dt} = u_1 \left(\beta_1 \frac{\delta_2}{\beta_2} - \delta_1 - (\beta_2 - \delta_2) \right)$$

The next two results say that when the range r is large enough the spatial model behaves like the ODE. First we need to define the behaviors that we will observe. We say that *coexistence occurs* if there is a stationary distribution that concentrates on configurations with infinitely many sites in each of the possible states. We say that *1's die out*, if whenever there are infinitely many 2's in the initial configuration $P(\xi_t(x) = 1) \to 0$ as $t \to \infty$.

Durrett and Swindle (1991). *If (4.4) holds then coexistence occurs for large range.*

Durrett and Schinzai (1993). *Suppose that the quantity on the left-hand side of (4.4) is < 0. If the range is large 1's die out.*

Remarks. The results in Durrett and Schinazi (1993) also apply to the Crawley and May's (1987) model of the competition between annuals and perennials. In this case the perennials are a nearest neighbor contact process but annuals have a long dispersal distance. For another competition model that has been analyzed using long range limits, see Durrett and Neuhauser (1997).

5. Rapid Stirring Limits

In the previous section we saw that stochastic spatial models simplify considerably when the range is large. Our next goal is to explain that this also occurs when the particles are subject to fast stirring. Formally, a stirring event involving x and y will change the state of the process from ξ to $\xi^{x,y}$ where

$$\xi^{x,y}(y) = \xi(x) \qquad \xi^{x,y}(x) = \xi(y) \qquad \xi^{x,y}(z) = \xi(z) \quad z \neq x, y$$

In words, stirring exchanges the values found at x and y.

The stirring mechanism has product measures as its stationary distributions. See Griffeath (1978), Section II.10. So when it acts at a rapid rate we expect that nearby sites will be almost independent. To keep the particles from flying out of our field of vision as the stirring rate is increased, we scale space by multiplying by $\epsilon = \nu^{-1/2}$. Since this is the usual diffusion scaling, it should not be surprising that the particle system converges to the solution of a reaction diffusion equation.

To state a general result, we consider processes $\xi_t^\epsilon : \epsilon \mathbf{Z}^d \to \{0, 1, \ldots, \kappa - 1\}$ that have

(i) *translation invariant finite range flip rates.* That is, there are sites $y_1, \ldots y_N$ and for each state i a function h_i so that

$$c_i(x, \xi) = h_i(\xi(x), \xi(x + \epsilon y_1), \ldots, \xi(x + \epsilon y_N))$$

(ii) *rapid stirring:* for each $x, y \in \epsilon \mathbf{Z}^d$ with $\|x - y\|_1 = \epsilon$, we exchange the values at x and y at rate ϵ^{-2}.

With these assumptions we get the following *mean field limit theorem* of De Masi, Ferrari, and Lebowitz (1986). (For the version given here, see Durrett and Neuhauser (1994).)

Theorem 5.1. *Suppose $\xi_0^\epsilon(x)$ are independent and let $u_i^\epsilon(t, x) = P(\xi_t^\epsilon(x) = i)$. If $u_i^\epsilon(0, x) = g_i(x)$ where $g_i(x)$ is a continuous function of x then as $\epsilon \to 0$, $u_i^\epsilon(t, x) \to u_i(t, x)$ the bounded solution of*

$$\partial u_i / \partial t = \Delta u_i + f_i(u) \qquad u_i(0, x) = g_i(x) \tag{5.1}$$

where

$$f_i(u) = < c_i(0,\xi)1_{(\xi(0)\neq i)} >_u - \sum_{j\neq i} < c_j(0,\xi)1_{(\xi(0)=i)} >_u \tag{5.2}$$

and $<\varphi(\xi)>_u$ *denotes the expected value of* $\varphi(\xi)$ *under the product measure in which state* j *has density* u_j*, i.e., when* $\xi(x)$ *are i.i.d. with* $P(\xi(x) = j) = u_j$.

To explain the form of the reaction term, we note that when ϵ is small, stirring operates at a fast rate and keeps the system close to a product measure. The rate of change of the densities can then be computed assuming adjacent sites are independent.

Theorem 5.1 only concerns expected values, but once it is established we can easily demonstrate the next result which says that in the fast stirring limit on a suitably rescaled lattice, the particle system becomes deterministic and looks like solutions of the PDE.

Theorem 5.2. *If* $f(x,t)$ *is a continuous function with compact support in* $\mathbf{R}^d \times [0,\infty)$ *then*

$$\epsilon^d \sum_{x\in\epsilon\mathbf{Z}^d} \int_0^\infty f(x,t)1_{(\xi_t^\epsilon(x)=i)}\,dt \to \iint_0^\infty f(x,t)u(t,x)\,dt\,dx \quad \text{in probability}$$

Our main interest in the PDE limit described in Theorems 5.1 and 5.2 is to obtain information about the particle system with fast (but finite) stirring rate. To do this we need one more result. The main assumption may look strange. Its form is dictated by the "block construction" technique we use to prove things. A complete discussion of this technique can be found in Durrett (1995b). Here we content ourselves to simply state one useful result.

($\star$) There are constants $A_i < a_i < b_i < B_i$, L, and T so that if $u_i(0,x) \in (A_i, B_i)$ when $x \in [-L,L]^d$ then $u_i(T,x) \in (a_i,b_i)$ when $x \in [-3L,3L]^d$.

Durrett and Neuhauser (1994) have shown:

Theorem 5.3. *If* ($\star$) *holds for the PDE then there is coexistence for the particle system with fast stirring.*

At this point we have reduced the task of proving theorems for particle systems to proving a specific type of result ($\star$) for the associated PDE. Leaving the reader to meditate on whether or not this is progress, we turn to the first of several concrete examples that can be treated by this method.

Example 5.1. Predator Prey Systems. Each site can be in state 0 = vacant, 1 = fish, or 2 = shark. If we let f_i be the fraction of the nearest neighbors of x (i.e., y with $\|y-x\|_1 = \epsilon$) that are in state i then we can write the flip rates as follows:

$$\begin{array}{llll} 0 \to 1 & \beta_1 f_1 & 1 \to 2 & \beta_2 f_2 \\ 1 \to 0 & \delta_1 & 2 \to 0 & \delta_2 + \gamma f_2 \end{array}$$

Here we have shifted our perspective from occupied sites giving birth, to vacant sites receiving particles from their neighbors. After this translation is made, the two rates on the left say that in the absence of sharks, the fish are a contact process.

The third rate says that sharks can reproduce by giving birth onto sites occupied by fish, an event which kills the fish. This transition is more than a little strange from a biological point of view, but it has the desirable property that the density of sharks will decrease when the density of fish is too small. The final rate says that sharks die at rate δ_2 when they are isolated and the rate increases linearly with crowding.

To be able to use our results about rapid stirring limits we also of course have to suppose that the sharks and fish swim around. That is, for each pair of nearest neighbors x and y stirring occurs at rate ϵ^{-2}. Applying Theorem 5.1 we see that if $\xi_0^\epsilon(x)$, $x \in \epsilon \mathbf{Z}^d$ are independent and $u_i^\epsilon(t,x) = P(\xi_t^\epsilon(x) = i)$ for $i = 1,2$ then as $\epsilon \to 0$, $u_i^\epsilon(t,x) \to u_i(t,x)$ as $\epsilon \to 0$, the bounded solution of

$$(5.3) \qquad \begin{aligned} \frac{\partial u_1}{\partial t} &= \Delta u_1 + \beta_1 u_1(1-u_1-u_2) - \beta_2 u_1 u_2 - \delta_1 u_1 \\ \frac{\partial u_2}{\partial t} &= \Delta u_2 + \beta_2 u_1 u_2 - u_2(\delta_2 + \gamma u_2) \end{aligned}$$

with $u_i(0,x) = f_i(x)$. To check the right-hand side, we note that if x is vacant and neighbor y is occupied by a fish, an event of probability $(1-u_1-u_2)u_1$ when sites are independent, births from y to x occur at rate $\beta_1/2d$ and there are $2d$ such pairs. The $-\beta_2 u_1 u_2$ in the first equation and the $\beta_2 u_1 u_2$ in the second come from sharks giving birth onto fish. The last term in each equation comes from the death events.

When the initial functions $f_i(x)$ do not depend on x, we have $u_i(t,x) = v_i(t)$ where the v_i's satisfy the ODE

$$\begin{aligned} \frac{dv_1}{dt} &= v_1\{(\beta_1 - \delta_1) - \beta_1 v_1 - (\beta_1+\beta_2)v_2\} \\ \frac{dv_2}{dt} &= v_2\{-\delta_2 + \beta_2 v_1 - \gamma v_2\} \end{aligned} \tag{5.4}$$

Here we have re-arranged the right hand side to show that the system is an example of the standard predator-prey equations for species with limited growth. See e.g., page 263 of Hirsch and Smale (1974).

The first step in understanding (5.3) is to look at (5.4) and ask: "What are the fixed points, i.e., solutions of the form $v_i(t) \equiv \rho_i$?" It is easy to solve for the ρ_i. There is always the trivial solution $\rho_1 = \rho_2 = 0$. In the absence of sharks the fish are a contact process, so if $\beta_1 > \delta_1$ there is a solution $\rho_1 = (\beta_1 - \delta_1)/\beta_1$, $\rho_2 = 0$. Finally, if we assume that the $\rho_1, \rho_2 \neq 0$, we can solve two equations in two unknowns to get

$$\rho_1 = \frac{(\beta_1-\delta_1)\gamma + \delta_2(\beta_1+\beta_2)}{\beta_1\gamma + \beta_2(\beta_1+\beta_2)} \qquad \rho_2 = \frac{(\beta_1-\delta_1)\beta_2 - \delta_2\beta_1}{\beta_1\gamma + (\beta_1+\beta_2)\beta_2}$$

which has $\rho_2 > 0$ if

$$(\beta_1 - \delta_1)/\beta_1 > \delta_2/\beta_2 \tag{5.5}$$

To understand this condition we note that if the fish are in equilibrium and the sharks have small density, then neglecting the $-\gamma v_2$ term and inserting the equilibrium value for v_1, the second equation in (5.4) becomes

$$\frac{dv_2}{dt} = v_2\left\{-\delta_2 + \beta_2 \cdot \frac{\beta_1 - \delta_1}{\beta_1}\right\} \tag{5.6}$$

The condition (5.5) says that the quantity in braces is positive, i.e., the density of sharks will increase when it is small.

Having found conditions that guarantee the existence of an interior fixed point, the next step is to check that it is attracting. Figure 5.1 shows an example of the ODE, which confirms this in the special case considered there. However, one does not need to use a computer to see that this will occur. To prove this, one begins with the easy to check fact that

$$H(v_1,v_2) = \beta_2(v_1 - \rho_1 \log v_1) + (\beta_1+\beta_2)(v_2 - \rho_2 \log v_2)$$

is a Lyapunov function, i.e., it is decreasing along solutions of the ODE (5.4). A simple argument by contradiction then shows that all orbits starting at points with each density $v_i > 0$ converge to (ρ_1, ρ_2). The presence of a globally attracting fixed point leads us to guess that

Theorem 5.4. *Suppose that $(\beta_1 - \delta_1)/\beta_1 > \delta_2/\beta_2$. If ϵ is small there is a nontrivial translation invariant stationary distribution in which the density of sites of type i is close to ρ_i.*

In view of Theorem 5.3 it suffices to prove $(\star)$. For details see Durrett (1993). The proof involves Brownian motion in a minor role, but is otherwise an analytic proof built on results of Redheffer, Redlinger, and Walter (1988) who considered the problem in a bounded domain with Neumann boundary conditions.

Example 5.2. Predator Mediated Coexistence. Here the possible states of a site are 0 = vacant, 1, 2 = two prey species, 3 = predator. Types $i = 1, 2$ behave like a contact process, dying at a constant rate δ_i and being born at vacant sites at rate β_i times the fraction of neighbors in state i. 3's die at a constant rate δ_3, are born at sites occupied by 1's at rate β_3 times the fraction of neighbors in state 3, and are born at sites occupied by 1's at rate β_4 times the fraction of neighbors in state 3. Finally, of course, there is stirring at rate ν: for each pair of nearest neighbors x and y we exchange the values at x and at y at rate ν.

In the absence of predators, this system reduces to the competing contact process, where the stronger species, identified by the larger of the two ratios β_i/δ_i, will competitively exclude the other. (See Neuhauser (1992) where the result is proved under the assumption that $\delta_1 = \delta_2$.) However, if the predators feeding rate on the stronger species is larger, its presence may stabilize the competition between the two species.

One way of seeing this is to consider the mean field ODE:

$$\begin{aligned} du_1/dt &= u_1\{\beta_1 u_0 - \delta_1 - \beta_3 u_3\} \\ du_2/dt &= u_2\{\beta_2 u_0 - \delta_2 - \beta_4 u_3\} \\ du_3/dt &= u_3\{\beta_3 u_1 + \beta_4 u_4 - \delta_3\} \end{aligned}$$

Here one can solve three equations in three unknowns to find conditions for an interior fixed point but a more fruitful approach is to derive conditions from an *invadability analysis*. Half of this may be described as follows.

By results for predator prey systems above, 2's and 3's can coexist if

$$\frac{\beta_2 - \delta_2}{\beta_2} > \frac{\delta_3}{\beta_4}$$

and when this holds their equilibrium densities will be $v_2 = \delta_3/\beta_4$ and

$$v_3 = \frac{(\beta_2 - \delta_2)\beta_4 - \beta_2}{(\beta_2 + \beta_4)\beta_4}$$

Examining the behavior of the ODE near $(0, v_2, v_3)$ we see that 1's can invade the (2,3) equilibrium if

$$\beta_1 - \delta_1 - \beta_1 v_2 - (\beta_1 + \beta_3)v_3 > 0$$

In a similar way one can derive conditions for the (1,3) equilibrium to exist and for the 2's to be able to invade it. When both sets of conditions hold we say there is ***mutual invadability***. It is easy to prove that in this case that the ODE has an interior fixed point. By considerably extending the methods of Durrett (1993), Shah (1997) has shown

Theorem 5.5. *If mutual invadability holds for the ODE then coexistence occurs for the stochastic spatial model with fast stirring.*

To get a feel for the resulting phase diagram, set $\beta_3 = 4$, $\beta_4 = 3/2$, all the $\delta_i = 1$, and vary β_1 and β_2. The formulas above imply that 1 and 3 coexist if $\beta_1 > 4/3$, 2 and 3 coexist if $\beta_2 > 3$, and finally all three species can coexist inside the region bounded by the equations.

$$\beta_1 > \beta_2, \quad \beta_2 < \frac{17}{32}\beta_1 + \frac{5}{8}, \quad \beta_2 > \frac{9}{14}\beta_1 + \frac{15}{14}$$

The last few lines are summarized in Figure 5.3. Note that there is a region where all three species can coexist but 2's and 3's cannot. Upon reflection this is not surprising: it simply says the 2's are not a sufficiently good food source to maintain the predator by themselves.

6. Case 2. Two Locally Attracting Fixed Points

As in our consideration of Case 1, we will begin with an example that has two states: 0 = vacant and 1 = occupied. The rules are like the contact process but now it takes two particles to make a new one. For this reason many of the early papers refer to this as the sexual reproduction process. However to emphasize that here the birth rate is quadratic instead of linear we will use the more modest name.

Example 6.1. Quadratic contact process. This system is also sometimes called Schlogl's second model. See Schlogl (1972) and Grassberger (1982).

(i) An occupied becomes vacant at a rate δ.

(ii) A vacant site becomes occupied at a rate equal to $k(k-1)/6$ where k is number of occupied neighbors.

Note that as in the contact process we have scaled time to make the maximum possible birth rate = 1.

The critical value for survival of this process starting from a finite set $\delta_f = 0$. To see this note that if the initial configuration starts inside a rectangle it can never give birth outside of the rectangle and hence is doomed to die out whenever δ is positive. Somewhat surprisingly, the critical value for the existence of a stationary distribution $\delta_e > 0$. Bramson and Gray (1991) have shown

Theorem 6.1. *There is a $\delta_0 > 0$ so that if $\delta \le \delta_0$ then the limit starting from all 1's is a nontrivial stationary distribution.*

The numerical value of δ_0 produced in the proof of Theorem 6.1 is very small. To obtain quantitative estimates we can turn to simulation to conclude that $\delta_e \approx 0.1$. Or we can take a fast stirring limit and use Theorems 5.1 and 5.2 that with rapid stirring the system behaves like the following PDE.

$$\frac{du}{dt} = \Delta u - \delta u + (1-u)u^2 \tag{6.1}$$

As in the study of the predator-prey model, we begin with the mean field ODE.

$$\frac{du}{dt} = -\delta u + (1-u)u^2 \tag{6.2}$$

When $\delta > 1/4$, $-\delta + u(1-u) < 0$ for all $u \in (0,1)$ so 0 is a globally attracting fixed point. When $\delta \in (0, 1/4)$ the quadratic equation $\delta = u(1-u)$ has two roots

$$0 < \rho_1 = \frac{1-\sqrt{1-4\delta}}{2} < \rho_2 = \frac{1+\sqrt{1-4\delta}}{2} < 1$$

This might suggest that as stirring becomes more rapid the critical value for a nontrivial equilibrium δ_c approaches 1/4. However, results of Noble (1992) and Durrett and Neuhauser (1994) show

Theorem 6.2. *As $\epsilon \to 0$, the critical value converges $\delta_c(\epsilon) \to 2/9$. Furthermore, if $\delta < 2/9$ then the equilibrium density $P(\xi^1_\infty(x) = 1) \to \rho_2$.*

To explain the value 2/9's we recall that in one dimension the limiting reaction diffusion equation has *traveling wave solutions*

$$u(x,t) = w(x - ct) \tag{6.3}$$

that keep their shape but move at velocity c. This and the other PDE results we will quote for this example can be found in Fife and McLeod (1977).

Setting $f(u) = -\delta u + (1-u)u^2$, since it will be clearer to do things for a general reaction term, it is easy to check that the recipe in (6.3) will lead to a solution of (6.1) if and only if

$$-cw'(y) = w''(y) + f(w) \tag{6.4}$$

Suppose to fix an orientation of the wave that: w tends to ρ_2 as $y \to -\infty$ and $w \to 0$ as $y \to \infty$. Multiplying by $w'(y)$ and integrating we have

$$\begin{aligned} -c \int w'(y)^2\,dy &= \int w''(y)w'(y)\,dy + \int f(w(y))w'(y)\,dy \\ &= 0 - \int_0^{\rho_2} f(z)\,dz \end{aligned} \tag{6.5}$$

Here to get the 0 we observed that the antiderivative of $w''w'$ is $(w')^2/2$ which vanishes at infinity, and in the second integral we have changed variables $z = w(y)$, and reversed the order of the limits.

(6.5) does not allow us to compute the value of c but since $\int w'(y)^2\,dy > 0$ it does tell us that the sign of c is the same as that of $\int_0^{\rho_2} f(z)\,dz$. A little calculus now confirms that the speed is positive for $\delta < 2/9$ and negative for $\delta > 2/9$. To check this easily, note that when $\delta = 2/9$ the three roots of the cubic are 0, 1/3, and 2/3, so symmetry dictates that the positive and negative areas must cancel and the speed is 0.

Sketch of the proof of Theorem 6.2. To prove that if $\delta < 2/9$ coexistence occurs for rapid stirring it suffices to check ($\star$) in Section 5 and apply Theorem 5.3. This can be done easily with the help of results in Fife and McLeod (1977), and was done Noble's (1992) Ph.D. thesis.

The other direction is a little more tricky since one must show that if $\delta > 2/9$ and stirring is rapid, the 1's die out, not just that their density in equilibrium is close to 0. Durrett and Neuhauser (1994) do this by using the PDE result to drive the density of 1's to a low level and then use auxiliary arguments to check that the 1's will then die out. □

Up to this point we have concentrated only on the critical value for a nontrivial equilibrium. From the proof of Theorem 6.2 one gets easily that

Theorem 6.3. As $\epsilon \to 0$, the critical value for survival from a finite set, $\delta_f(\epsilon) \to 2/9$.

It is known in general that $\delta_f \geq \delta_e$. This may sound obvious but it is difficult to prove. See Bezuidenhout and Gray (1994). Once there is stirring at a positive rate $\delta_f > 0$. In fact we

Conjecture. If the stirring rate $\nu > 0$ then $\delta_f = \delta_e$.

Remark. The techniques described above have been used by Durrett and Swindle (1994) to prove results for a catalytic surface. Keeping to biological models, we will continue with

Example 6.2. Colicin. The inspiration for this model came from Chao (1979) and Chao and Levin (1981). Bacteria may produce toxic substances, known collectively as bacteriocins, that kill or inhibit the growth of competing bacteria of different genotypes. In general, bacteria that are capable of producing such chemicals are immune to their action. The colicins, the most extensively studied class of bacteriocins, are produced by the bacterium *Escherica coli* and other members of the family Enterobacteriaceae. For more about the biology, and an alternative approach to the modeling, see Frank (1994).

To model the competition we will use a spatial model with three states: 0 = vacant, 1 = occupied by a colicin producer, 2 = occupied by a colicin sensitive bacterium. If we let f_i be the fraction of the four nearest neighbors in state i, we can formulate the transition rates as follows:

birth	rate	death	rate
$0 \to 1$	$\beta_1 f_1$	$1 \to 0$	δ_1
$0 \to 2$	$\beta_2 f_2$	$2 \to 0$	$\delta_2 + \gamma f_1$

In words, each type is born at empty sites at a rate proportional to the fraction of neighbors of that type. The colicin producing strain dies at a constant rate δ_1, while the colicin sensitive strain experiences deaths at rate δ_2 plus γ times the fraction of colicin producing neighbors.

To see what behavior to expect from the spatial model, we begin by writing down the mean field ODE. Let u_1 be the density of colicin-producing and let u_2 be the density of the ordinary, colicin-sensitive bacteria. Assuming that all sites are independent we have

$$(6.6) \qquad \begin{aligned} \frac{du_1}{dt} &= \beta_1 u_1(1 - u_1 - u_2) - \delta_1 u_1 \\ \frac{du_2}{dt} &= \beta_2 u_2(1 - u_1 - u_2) - \delta_2 u_2 - \gamma u_1 u_2 \end{aligned}$$

The system (6.6) has locally stable boundary equilibria at

$$(1 - \delta_1/\beta_1, 0) \quad \text{and} \quad (0, 1 - \delta_2/\beta_2)$$

provided

$$(6.7) \qquad \delta_i < \beta_i \qquad \frac{\delta_2}{\beta_2} < \frac{\delta_1}{\beta_1} < \frac{\delta_2 + \gamma}{\beta_2 + \gamma}$$

There is moreover an interior saddle point $(\bar{u}_1, \bar{u}_2)$ in this case. See Figure 6.1 for a picture of what happens when $\delta_1 = \delta_2 = 1$, $\beta_1 = 3$, $\beta_2 = 4$ and $\gamma = 3$. The interpretation of the inequalities in order from left to right is

(i) the birth rate exceeds the death rate so either type can maintain a population in isolation from the other;

(ii) there is a cost to colicin production, reflected in a lower carrying capacity in isolation

(iii) the competitive benefit of colicin production is sufficiently large so that an established colicin-producing community can repel invasion by the wild type.

The implication of this analysis is that colicin production is an evolutionarily stable strategy, but so is nonproduction. In the dynamical system pictured in Figure 6.1, if the density of the colicin sensitive bacteria is near the equilibrium value then the colicin producing bacteria cannot invade. That is, if they are introduced at a low level then their density will shrink to 0. On the other hand, if the colicin producers are introduced at a large enough level, their density will increase to 1 and the density of the colicin sensitive strain will approach 0. In words, selection will only favor genotypes when they are common, rare species cannot invade, and genetic diversity will not be maintained. This situation is "disruptive frequency dependent selection" (see B.R. Levin (1988), Thoday (1959–64)).

The last paragraph identifies the colicin system as belonging to Case 2, so we expect that there is one stronger type that is the winner starting from generic initial conditions, i.e., configurations in which there are infinitely many sites in each of the possible states. In the case of colicin, it is not natural to introduce rapid stirring and even if we did we would not know how to analyze the PDE's that result. Thus we turn to the computer to confirm our theoretical predictions.

Computer simulations. Figure 6.2 shows the density of colicin producers and colicin sensitive bacteria in a simulation of the spatial model with parameters: $\delta_1 = \delta_2 = 1$, $\beta_1 = 3$, $\beta_2 = 4$ and $\gamma = 3$. Here the lattice is 100×100 and to avoid edge effects we have used ***periodic boundary conditions.*** That is, sites on the bottom row are neighbors of those on the top row; sites on the left edge are neighbors of those on the right edge. We start at time 0 from *product measure.* That is, the states of the sites at time 0 are assigned independently, i.e., by making repeated calls to a random number generator. We started the simulation with colicin producers (1's) at density 0.01 and the colicin sensitive strain (2's) at density 0.50; but as the graph shows, the colicin producers gradually increase to their equilibrium level while the density of colicin sensitive bacteria drops to 0. The colicin producers first establish themselves in clumps that grow linearly in radius and take over the system.

The victory of the colicin producers in the last example is due to the fact that the colicin induced death rate $\gamma = 3$ is large enough to compensate for the fact that the colicin producing strain has birth rate $\beta_1 = 3$ versus $\beta_2 = 4$ for the colicin sensitive strain. If we reduce γ to 1 the situation reverses and the colicin sensitive strain is victorious even when it starts from a low density. For values of γ between 1 and 3 coexistence might be possible but this does not occur: there is a critical value γ_c so that 2's take over when $\gamma < \gamma_c$, while 1's take over when $\gamma > \gamma_c$.

More generally if we fix $\delta_1 = \delta_2 = 1$, $\beta_2 = 4$, and vary β_1 and γ then we get the phase diagram drawn in Figure 6.3. The figure is a free-hand sketch that emphasizes the generic qualitative properties but is not exact. For each fixed value of β_1 there is a critical value $\gamma_c(\beta_1)$ so that 2's take over when $\gamma < \gamma_c(\beta_1)$ while 1's take over when $\gamma > \gamma_c(\beta_1)$. When $\beta_1 = 4 = \beta_2$, $\gamma_c(\beta_1) = 0$. Decreasing β_1 increases $\gamma_c(\beta_1)$ until it reaches ∞ at a point we have labeled β_c. β_c, which is ≈ 1.65 for the neighborhood $\mathcal{N}_0$, is the minimum value of the birth rate needed for a single strain to survive in the absence of the other. When there is a single strain the model reduces to the basic contact process, see Durrett and Levin (1994a).

Example 6.3. A Three Species Colicin System. In the two examples above, the ODE and the spatial model sometimes disagreed on who would win the competition, but both approaches agreed that one type would always competitively exclude the other. We will now describe a system in which three species coexist in the spatial model, but in the ODE there is always only one winner.

To describe the system in words, we assume 1's and 2's both produce colicin, to which they are immune, and to which 3 is sensitive. The rates for this system are:

birth	rate	death	rate
$0 \to 1$	$\beta_1 f_1$	$1 \to 0$	δ_1
$0 \to 2$	$\beta_2 f_2$	$2 \to 0$	δ_2
$0 \to 3$	$\beta_3 f_3$	$3 \to 0$	$\delta_3 + \gamma_1 f_1 + \gamma_2 f_2$

Here, f_i is the fraction of the four nearest neighbors in state i. In our concrete example we will set all the $\delta_i = 1$ and

$$\beta_1 = 3 \quad \beta_2 = 3.2 \quad \beta_3 = 4 \quad \gamma_1 = 3 \quad \gamma_2 = 0.5$$

Here we imagine that species 1 produces more colicin than 2 does but has the lowest birth rate. The parameters are chosen so that 1's win against 3's while 3's win against 2's. When only 1's and 2's are present the system reduces to the multitype contact process studied in Neuhauser (1992). Since $\beta_2 > \beta_1$, the 2's win against the 1's in this case.

If we write u_i for the fraction of sites in state i then the mean field ODE is:

$$
\begin{aligned}
\frac{du_1}{dt} &= \beta_1 u_1 u_0 - \delta_1 u_1 \\
\frac{du_2}{dt} &= \beta_2 u_2 u_0 - \delta_2 u_2 \\
\frac{du_3}{dt} &= \beta_3 u_3 u_0 - u_3(\delta_3 + \gamma_1 u_1 + \gamma_2 u_2)
\end{aligned} \tag{6.8}
$$

If we insert the values for the concrete example then the picture in Figure 6.4 results. In the $u_1 u_2$ plane all trajectories starting with u_1 and u_2 positive are attracted to $(0, \hat{u}_2, 0)$ where $\hat{u}_i = (\beta_i - \delta_i)/\beta_i$. In the three dimensional ODE there is a surface which connects the two separatrices in the $u_1 u_3$ and $u_2 u_3$ planes, so that above the surface trajectories converge to $(0, 0, \hat{u}_3)$ while those below converge to $(0, \hat{u}_2, 0)$. These conclusions are true whenever $\beta_1 < \beta_2$ and equilibria exist in the interior of the $u_1 u_3$ and $u_2 u_3$ planes. (Conditions for this can be derived from (6.7).)

In contrast to the behavior of the ODE, the spatial model shows coexistence, at least for a long time. See Figure 6.5 for a simulation of the process on a 200×200 grid with periodic boundary conditions. Here we started in an initial product measure in which the states $i = 1, 2, 3$ each have density $1/3$ and plotted the observed density of the three species every one thousand units of time out to time 50,000. The densities fluctuate but none of them seems in danger of hitting 0.

7. Case 3. Periodic Orbits

Our first example was introduced by Silvertown et al. (1992) to investigate the competitive interaction of five grass species. We have given it a new name to place it in context in the theory of interacting particle systems.

Example 7.1. The Multitype Biased Voter Model. Each site will always be occupied by exactly one of the species $1, 2, \ldots, K$. The process is described by declaring that:

(i) An individual of species i produces new offspring of its type at rate β_i.

(ii) An offspring of type i produced at x is sent to y with probability $q_i(x, y) = \varphi_i(|y - x|)$ where $|y - x|$ is the distance from x to y. To avoid unnecessary complications, we will suppose that $\varphi_i(1) > 0$ and that there is an $R < \infty$ so that $\varphi_i(r) = 0$ when the distance $r > R$. In other words, there is a finite dispersal range, but nearest neighbors are always accessible.

(iii) If site y is occupied by type j, and type i disperses to that site, a successful invasion occurs (i.e., the state of y changes from j to i) with probability p_{ij}; if invasion does not occur, the site y is unchanged.

To explain the name we note that if there are only two types then the model reduces to the biased voter model introduced by Williams and Bjerknes (1972) and studied by Griffeath (1978) and Bramson and Griffeath (1980), (1981). For a summary of their results see Chapter 3 of Durrett (1988).

If we were to ignore space and assume that the states of the sites in the grid are always independent, then the fraction of sites occupied by species i, u_i, would satisfy

$$\frac{du_i}{dt} = \sum_j u_i u_j \{\beta_i p_{ij} - \beta_j p_{ij}\} \tag{7.1}$$

In the model (and of course also in the ODE) only the value of $\lambda_{ij} = \beta_i p_{ij}$ is important, so we can describe the concrete example investigated by Silvertown et al (1992) by giving the matrix λ_{ij}.

i	$j \rightarrow$	1	2	3	4	5
1	Agrostis	0	0.09	0.32	0.23	0.37
2	Holcus	0.08	0	0.16	0.06	0.09
3	Poa	0.06	0.06	0	0.44	0.11
4	Lolium	0.02	0.06	0.05	0	0.03
5	Cynosurus	0.02	0.03	0.05	0.03	0

Simulations of this process from a randomly chosen initial state were not very interesting to watch. "Three of the five species went extinct very rapidly. The two survivors *Agrostis* and *Holcus* were the same as the species that survived the longest in the aggregated models." To explain why this occurs, we say that species i *dominates* species j and we write $i \geq j$ if $a_{ij} = \lambda_{ij} - \lambda_{ji} \geq 0$. When the difference is > 0, we say i *strictly dominates* j and write $i > j$. In the concrete case given above, *Agrostis* strictly dominates all other species but beats *Holcus* by only 0.01, so it should not be surprising that *Agrostis* takes over the system with *Holcus* offering the most resistance. Indeed, Durrett and Levin (1997) have shown

Theorem 7.1. *Assume that the dispersal distribution φ_i does not depend on i and that type 1 is strictly dominant over type i for $2 \le i \le K$. If we let A_t^1 denote the event that type 1 is still alive at time t then $P(A_t^1, \xi_t(x) \neq 1) \to 0$ as $t \to \infty$.*

This result says simply that if all species disperse equally, a competitive dominant type will almost certainly outcompete all others. To explain the mathematical statement, note that if we start with infinitely many sites in state 1 then $P(A_t^1) = 1$ for all $t > 0$ and the theorem implies that type 1 comes to dominate at every site. If we only start with finitely many 1's then bad luck in the early stages can wipe out all the 1's. Our result says that if this does not happen then the 1's will take over the system.

The outcome in Theorem 7.1 is the one we should expect. It is also the one predicted in the mean field case by the ordinary differential equations, (7.1). To see this for the ODE, note that the domination condition implies that all the $a_{1i} > 0$, so $u_1(t)$ is increasing. Being increasing and bounded by 1, $\lim_{t\to\infty} u_1(t)$ exists; but this is only possible if $du_1/dt \to 0$, which implies $\sum_{i>1} u_i(t) \to 0$.

Cyclic Biased Voter Model. In view of the discussion just completed, the simplest system that can have interesting behavior is a three-species systems with a competitive loop: $1 < 2 < 3 < 1$. This may at first appear to be a rather special and esoteric situation, but its generality becomes clearer when it is recognized that late successional species (the competitive dominants) typically would be replaced by early successional species following a disturbance. Thus, if, for example, species 1, 2, and 3 are respectively grass, bushes, trees, or some other representation of the successional cycle, the ordering $1 < 2 < 3 < 1$ makes sense in terms of competitive replacement. Bramson and Griffeath (1989) have considered this system with $n \ge 3$ competitors in one dimension. Griffeath alone (1988) and with his co-workers Fisch and Gravner (1991a, 1991b) has studied related cellular automata. Tainaka (1993, 1995) has considered a variation on the model in which 1's mutate into 3's with the paradoxical result that this enhances the density of 1's.

In our situation, if we suppose $1 < 2 < 3 < 1$ and let

$$\beta_1 = \lambda_{13} \qquad \beta_2 = \lambda_{21} \qquad \beta_3 = \lambda_{32};$$

then the system (7.1) can be written as

$$(7.2)\qquad \begin{aligned} \frac{du_1}{dt} &= u_1(\beta_1 u_3 - \beta_2 u_2) \\ \frac{du_2}{dt} &= u_2(\beta_2 u_1 - \beta_3 u_3) \\ \frac{du_3}{dt} &= u_3(\beta_3 u_2 - \beta_1 u_1) \end{aligned}$$

If for example we take $\beta_1 = 0.3$, $\beta_2 = 0.7$, and $\beta_3 = 1.0$ then the ODE behaves as indicated in Figure 7.1. There is a family of periodic orbits around the fixed point $(0.5, 0.15, 0.35)$.

To show that in general we get pictures similar to the concrete example, we begin by dividing each equation by the product of the betas that appear in it to conclude that any fixed point ρ has

$$\frac{\rho_3}{\beta_2} = \frac{\rho_2}{\beta_1} = \frac{\rho_1}{\beta_3}$$

Recalling that the equilibrium densities must sum to one, we conclude that

$$\rho_1 = \frac{\beta_3}{\beta_1 + \beta_2 + \beta_3} \qquad \rho_2 = \frac{\beta_1}{\beta_1 + \beta_2 + \beta_3} \qquad \rho_3 = \frac{\beta_2}{\beta_1 + \beta_2 + \beta_3}$$

To see that there is a family of periodic orbits surrounding the fixed point we write $H(u) = \sum_i \rho_i \log u_i$ and note that

$$\begin{aligned} \frac{\partial H}{\partial t} &= \sum_i \frac{\rho_i}{u_i} \frac{du_i}{dt} \\ &= c\left(\frac{u_3}{\beta_2} - \frac{u_2}{\beta_1}\right) + c\left(\frac{u_1}{\beta_3} - \frac{u_3}{\beta_2}\right) + c\left(\frac{u_2}{\beta_1} - \frac{u_1}{\beta_3}\right) = 0, \end{aligned}$$

where $c = \beta_1\beta_2\beta_3/(\beta_1 + \beta_2 + \beta_3)$. Thus H is constant along solutions of the ODE.

The situation described above is similar to that of May and Leonard (1975) and Gilpin (1975) who considered a system in which there were invariant sets of the form $\sum_i \log u_i = K$. Gilpin (1975) observed that the "system is neutrally stable on the plane $u_1 + u_2 + u_3 = 1$, therefore stochastic effects (environmental noise) will cause it to decay to a single species system." This conclusion does not apply to the stochastic spatial model. Well separated regions oscillate out of phase, and the result is a stable equilibrium density for each of the three types.

Figure 7.2 gives the percentage of sites occupied by species 1 for the first 500 units of time when we look at the system in windows of size 30×30

or 120 × 120, or average over the whole 480 × 480 system (which again has periodic boundary conditions). Note that in the smallest viewing window, the densities fluctuate wildly; but when the averages are taken over the largest length scale, the oscillations are confined to the initial period when the system is converging to equilibrium.

Between the two extremes mentioned in the last paragraph is a moderate length scale that physicists would call the correlation length. This is the "most interesting" scale on which to view the system. Densities computed in boxes with sides of the correlation length vary smoothly in time, but undergo substantial changes. As we mentioned earlier, Rand and Wilson (1995), and Keeling et al. (1997) have considered the problem of precisely defining this length scale in terms of the variance of box averages. Pascual and Levin (1998) have recently taken a different approach by identifying the length scale at which to aggregate to achieve a maximum amount of determinism in the evolution of the local densities.

Example 7.2. Hawks-Doves. Our next model is a spatial version of Maynard Smith's (1982) evolutionary games. Our formulation follows Brown and Hansell (1987). Others have studied this system using cellular automata: Nowak and May (1992), (1993), Hubermann and Glance (1993), Nowak, Bonhoeffer, and May (1994), and May (1994), (1995).

In our model (and most of the others that have been considered), there are two types of individuals, called Hawks and Doves, whose interaction is described by a game matrix. The three examples we will be interested in are given by:

#1	H	D
H	.4	.8
D	.6	.3

#2	H	D
H	.7	.4
D	.4	.8

#3	H	D
H	−.6	.9
D	−.9	.7

Finally we list the general case, which serves to define notation we will use

	H	D
H	a	b
D	c	d

To explain the general game matrix we note that b is, for example, the payoff to a hawk when interacting with a dove. When the population consists of a fraction p of hawks and $1-p$ of doves then the payoff for hawks is $ap+b(1-p)$.

We interpret $ap + b(1-p)$, which may be positive or negative (see e.g., game #3), as the net birth rate of hawks in this situation.

Once we have decided on a game matrix then following Brown and Hansell (1987), we can let $\eta_t(x)$ and $\zeta_t(x)$ be the number of hawks and doves at x at time t and formulate the dynamics as follows:

(i) *migration.* Each individual changes its spatial location at rate ν and when it moves, it moves to a randomly chosen nearest neighbor of x, i.e., it picks with equal probability one of the four points $x+(1,0)$, $x-(1,0)$, $x+(0,1)$, $x-(0,1)$ that differ from x by 1 in one of the coordinates.

(ii) *deaths due to crowding.* Each individual at x at time t dies at rate $\kappa(\eta_t(x)+\zeta_t(x))$.

(iii) *game step.* Let $\mathcal{N} = \{z \in \mathbf{Z}^2 : |z_1|, |z_2| \le 2\}$ be a 5×5 square centered at (0,0). Let

$$\hat{\eta}_t(x) = \sum_{z \in \mathcal{N}} \eta_t(x+z) \qquad \hat{\zeta}_t(x) = \sum_{z \in \mathcal{N}} \zeta_t(x+z)$$

be the number of hawks and doves in the interaction neighborhood of x at time t, and let

$$p_t(x) = \hat{\eta}_t(x)/(\hat{\eta}_t(x) + \hat{\zeta}_t(x))$$

be the fraction of hawks. Each hawk experiences a birth (or death) rate of $ap_t(x) + b(1-p_t(x))$ while each dove experiences a birth (or death) rate of $cp_t(x) + d(1-p_t(x))$.

If we assume that all sites remain independent then we arrive at the following mean field ODE for the densities of hawks (u) and doves (v):

$$\text{(7.3)} \qquad \begin{aligned} \frac{du}{dt} &= u\left\{a\frac{u}{u+v} + b\frac{v}{u+v} - \kappa(u+v)\right\} \\ \frac{dv}{dt} &= v\left\{c\frac{u}{u+v} + d\frac{v}{u+v} - \kappa(u+v)\right\} \end{aligned}$$

Note that a species specific linear term in the net birth (death) rate, r, is easily accommodated within this framework as part of a and b or c and d since $u/(u+v) + v/(u+v) = 1$.

The Hawks and Doves model quite naturally divides itself into three cases, which were our original motivation for formulating the three cases announced in the introduction. To motivate the division into cases, we change variables

$p = u/(u+v)$, $s = u+v$ in the dynamical system to get

$$(7.4) \qquad \begin{aligned} \frac{dp}{dt} &= (a-b-c+d)p(1-p)(p-p_0) \\ \frac{ds}{dt} &= s\{\alpha p^2 + \beta p + \gamma\} - \kappa s^2 \end{aligned}$$

where

$$p_0 = \frac{b-d}{b-d+c-a} \qquad \alpha = a-b-c+d \qquad \beta = b+c-2d \qquad \gamma = d-\kappa s$$

The equation for dp/dt is identical to the usual equation from population genetics for weak selection with selection coefficient $a-b-c+d$. If the hawk strategy is never worse than the dove strategy, that is, $a \geq c$ and $b \geq d$, then $p_0 \geq 1$ or $p_0 \leq 0$ (ignoring the trivial case $a = c$, $b = d$). The same conclusion holds if the dove strategy dominates the hawk strategy; but if neither strategy dominates the other, p_0 represents a mixed strategy equilibrium. That is, if a fraction p_0 of the players play the hawk strategy and a fraction $1-p_0$ play the dove strategy, both strategies have the same payoff. To check this note that

$$p_0 a + (1-p_0)b = p_0 c + (1-p_0)d \quad \text{if and only if} \quad p_0 = \frac{b-d}{b-d+c-a}.$$

When $p_0 \in (0,1)$, it may be (Case 1) an attracting or (Case 2) a repelling fixed point. Matrices #1 and #2 above are examples of Case 1 and Case 2 respectively. Since we have discussed these situations at length, we turn now to the case in which $p_0 = 1$, i.e., the hawk strategy always dominates the dove strategy. If $a > 0$ the system is boring since the hawks will take over the world. However if $a < 0$ and $d > 0$, as in matrix #3, things are quite interesting.

This case is often called Prisoner's Dilemma after the two person non-zero sum game in which two individuals have a choice to cooperate (C) or defect (D). The payoffs to the first and second player for their actions are given as follows:

$$\begin{array}{ccc} & C & D \\ C & (R,R) & (S,T) \\ D & (T,S) & (P,P) \end{array}$$

Here $T > R > P > S$ so the defector strategy dominates cooperation, but double defection leads to less happiness than cooperation of each player. See Luce and Raiffa (1957) or Owen (1968).

It is easy to see that in Case 3, the ODE dies out. Figure 7.3 gives a picture of the ODE for matrix #3. From the picture it should be clear that the fraction of individuals that are hawks increases in time. This observation leads easily to:

Theorem 7.2. *If the initial condition for the dynamical system has $u(0) > 0$ then $(u(t), v(t)) \to (0,0)$.*

Proof. From (7.4) it follows that $p(t) = u(t)/(u(t) + v(t))$ converges to 1 as $t \to \infty$. Once $p(t)$ gets close enough to 1, both growth rates are negative and the populations decay to 0 exponentially fast. □

In contrast, the hawks and doves coexist in the our stochastic spatial model. A typical simulation of the interacting particle system in Case 3 begins with a period in which the hawk population grows faster than the dove population until the fraction of hawks is too large and both species start to die out. When the density gets low we have a few doves who are completely isolated and give birth at rate $d = 0.7$. These doves start colonies that grow and would fill up the space to the doves preferred equilibrium density, except for the fact that along the way they encounter a few hawks that managed to escape extinction. These hawks reproduce faster than the doves, the fraction of hawks grows, and the cycle begins again.

Figure 7.4 gives a graph of the density of hawks and doves vs. time for a simulation on a 50×50 grid, while Figure 7.5 shows the same statistics for a 150×150 grid. As the system size increases the oscillations decrease. The explanation for this is simple: if we look at a 150×150 grid then the cycle of growth of the hawks fraction, decrease of the population, and regrowth from isolated doves in any 50×50 subsquare is much like that of the simulation on the 50×50 grid. However, the 150×150 system consists of nine 50×50 subsquares which do not oscillate in a synchronized fashion, so the cycles cancel each other out to some extent.

Example 7.3. Epidemics with regrowth of susceptibles. In this model the states are 0 = susceptible, 1 = infected, and 2 = removed. Writing f_i for the fraction of the four nearest neighbors in state i we can write the rates as

$$0 \to 1 \quad \beta_1 f_1 \qquad 1 \to 2 \quad \delta \qquad 2 \to 0 \quad \alpha$$

Durrett and Neuhauser (1991) have shown that if the epidemic without regrowth (i.e., when $\alpha = 0$) does not die out, then whenever $\alpha > 0$ there is a nontrivial translation invariant stationary distribution.

If α is small and we make the correspondence: infecteds = Hawks, suceptibles = Doves, and removed = vacant, then the behavior of the model (when viewed in windows of size $1/\alpha$) is much like the Hawks-Doves system.

(i) Epidemics sweep through the system wiping out most susceptibles,

(ii) When susceptibles are scarce, the epidemic becomes subcritical and the density of infecteds then drops to a low level.

(iii) When infecteds are scarce, susceptibles increase. When the density of susceptibles is large enough, one of the few surviving infecteds starts another epidemic.

For simulation results on this phenomenon, see Durrett (1995c). The conference proceedings, Mollison (1995), in which that paper appears is an excellent source for information on all sorts of epidemic models.

Example 7.4. WATOR. The name is short for WAter TORus, a system considered in A.K. Dewdney's Computer Recreations column in Scientific American in December 1984. Each site can be in state 0 = vacant, 1 = occupied by a prey (fish), or 2 = occupied by a predator (shark). The original model was defined in discrete time, but we reformulate it in continuous time as follows:

(i) Fish are born at vacant sites at rate β_1 times the fraction of neighbors occupied by fish.

(ii) Each shark at rate 1 inspects q neighboring sites, chosen without replacement from the neighbor set. It moves to the first fish it finds and eats it. A shark that has just eaten gives birth with probability β_2. A shark that finds no fish dies with probability δ.

(iii) There is *stirring* (also called *swimming*) at rate ν: for each pair of nearest neighbor sites x and y we exchange the values at x and at y at rate ν.

The stirring mechanism automatically preserves the restriction of at most one individual per site and has the mathematical advantage that the trajectory of any single particle is just a continuous time random walk. Of course, if one watches the movements of two particles there is a (very small) correlation between their locations due to the occasional stirring steps that affect both particles at the same time.

Letting $u_i(t)$ be the fraction of sites in state i at time t, and computing the rate of change by supposing that adjacent sites are always independent we see that the mean field ODE in this case is:

$$\begin{aligned} du_1/dt &= \beta_1 u_1(1-u_1-u_2) - u_2\{1-(1-u_1)^q\} \\ du_2/dt &= \beta_2 u_2\{1-(1-u_1)^q\} - \delta u_2(1-u_1)^q \end{aligned} \tag{7.4}$$

Here, the first term on the right represents the birth of fish onto vacant sites. To explain the second and third terms, we note that $u_2\{1-(1-u_1)^q\}$ gives the fraction of sites occupied by sharks times the probability a given shark will find at least one fish when it inspects q neighbors, so β_2 times this gives the rate at which new sharks are produced. For similar reasons the fourth term represents the sharks who find no fish to eat, got a bad coin flip, and were told to die.

To begin to understand the ODE we note that in the absence of fish, sharks can't breed and their density drops to 0. Conversely, in the absence of sharks, fish don't die and will fill up the space. The last two results give the direction of motion of the ODE on two sides of the right triangle that we use for the possible states of the system: $\Gamma = \{(u_1,u_2) : u_1, u_2 \geq 0, u_1+u_2 \leq 1\}$.

Since fish do not die in the absence of sharks, there is a boundary equilibrium at (1,0). Considering the second equation in (7.4) and setting $u_1 = 1-\epsilon_1$ and $u_2 = \epsilon_2$ where the ϵ_i are small shows that (1,0) is always a saddle point. This behavior suggests the presence of a fixed point $(\bar{u}_1, \bar{u}_2)$ with both components positive, a fact which can easily be confirmed by algebraic manipulation. To do this neatly, and to pave the way for later calculations, we will first rewrite the system in (7.4) as

$$\begin{aligned} du_1/dt &= A(u_1) - u_2 B(u_1) \\ du_2/dt &= u_2 C(u_1) \end{aligned} \tag{7.5}$$

where $A(u_1) = \beta_1 u_1(1-u_1)$,

$$B(u_1) = \beta_1 u_1 + \{1-(1-u_1)^q\}$$

and $C(u_1) = \beta_2 - (\beta_2+\delta)(1-u_1)^q$. In order for $du_2/dt = 0$ we must have

$$C(\bar{u}_1) = 0 \qquad \text{or} \qquad \bar{u}_1 = 1 - \left(\frac{\beta_2}{\beta_2+\delta}\right)^{1/q} \tag{7.6}$$

Having found $\bar{u}_1$ we can now set $du_1/dt = 0$ to find

$$\bar{u}_2 = A(\bar{u}_1)/B(\bar{u}_1) \tag{7.7}$$

To investigate the nature of the fixed point at $(\bar{u}_1, \bar{u}_2)$ we let $v_i = u_i - \bar{u}_i$ be the displacement from it in the ith component. Assuming the v_i are small and using (7.6) and (7.7) we arrive at the linearized equation

$$\begin{aligned} dv_1/dt &= Fv_1 + Gv_2 \\ dv_2/dt &= Hv_1 \end{aligned} \tag{7.8}$$

where $F = A'(\bar{u}_1) - \bar{u}_2 B'(\bar{u}_1)$, $G = -B(\bar{u}_1)$, and $H = u_2 C'(\bar{u}_1)$. This ODE is analyzed in the Appendix of Durrett and Levin (1998) with the following result.

Theorem 7.4. *The interior fixed point is always locally attracting when $q \le 3$. Conversely, if $q > 3$ and the values of β_2 and δ are held constant, decreasing β_1 leads to a Hopf bifurcation that produces a limit cycle.*

Figure 7.5 gives a picture of a case of the ODE with a limit cycle: $\beta_1 = 1/3$ $\beta_2 = 0.1$, $\delta = 1$, and $q = 4$.

To make connections between our model and reaction diffusion equations, we use Theorems 5.1 and 5.2 to conclude that if we let the stirring rate $\nu \to \infty$ and consider our process on a scaled version of the square lattice in which the spacing between sites is reduced to $\nu^{-1/2}$ then the densities of fish and sharks converge to the solution of the partial differential equation:

$$\begin{aligned} \partial u_1/\partial t &= \Delta u_1 + g_1(u_1, u_2) \\ \partial u_2/\partial t &= \Delta u_2 + g_2(u_1, u_2) \end{aligned} \tag{7.9}$$

where the g_i are the right-hand sides of the equations in (7.4).

The next result, proved in Durrett and Levin (1998), says that sharks and fish coexist in the reaction-diffusion equation.

Theorem 7.5. *Suppose that the initial conditions $u_i(x, 0)$ are continuous, always in the set Γ of sensible values, and each u_i is not identically 0. Then there are positive constants ρ, ϵ_1 and ϵ_2 so that for large t, $u_i(x,t) \ge \epsilon_i$ whenever $|x| \le \rho t$.*

In words, the densities stay bounded away from zero on a linearly growing set. Using methods of Durrett (1993), and Durrett and Neuhauser (1994), it is not hard to convert Theorem 2 into a conclusion about the particle system. See also Sections 4 and 8 of Durrett (1995), or the more recent work of Shah (1997).

Theorem 7.6. *When the stirring rate is large there is coexistence, i.e., there is a stationary distribution for the particle system that concentrates on configurations with infinitely many sites in each of the possible states.*

Simulations. The last result proves the existence of the stationary distribution but does not yield much information about its spatial structure. To understand that, we turn to simulation. Immediately, however, we run into the difficulty that while fast stirring is convenient for making connections with reaction-diffusion equations, it is painful to implement on the computer, since most of the computational effort is spent moving the particles around.

To find a variant of the WATOR model that we can more easily simulate, we note that at any moment when a fish or shark at x inspects its neighbors, it sees a set of sites that have been subject to stirring at rate ν since the previous time site x decided to try to change. Since the flip rates stay constant as $\nu \to \infty$ this time is of order 1, and the neighbors will move a distance of order $\nu^{1/2}$. With this mind, we will replace stirring by choose our neighbors at random (with replacement) from a square of radius $r = \nu^{1/2}$ centered at the point of interest.

Figures 7.7 and 7.8 show results of computer simulations when $r = 5$, i.e., neighbors are chosen at random from an 11×11 square centered at the point. Note that densities oscillate wildly when measured in a 50×50 window but are much smoother in time in a 200×200 window.

REFERENCES

Bezuidenhout, C. and Gray, L. (1994) Critical attractive spin systems. *Ann. Probbab.* **22**, 1160–1194

Bramson, M., Cox, J.T., and Durrett, R. (1996) Spatial models for species area curves. *Ann. Prob.* **24**, 1727–1751

Bramson, M., Cox, J.T., and Durrett, R. (1997) A spatial model for the abundance of species. *Ann. Prob.*, to appear

Bramson, M., Durrett, R. and Swindle, G. (1989) Statistical mechanics of Crabgrass. *Ann. Prob.* **17**, 444–481

Bramson, M. and Griffeath, D. (1980). Asymptotics for some interacting particle systems on $\mathbf{Z}^d$. *Z. Warsch. verw. Gebiete* **53**, 183–196.

Bramson, M. and Griffeath, D. (1989). Flux and fixation in cyclic particle systems. *Ann. Probab.* **17**, 26-45.

Bramson, M,, Durrett, R. and Swindle, S. (1989) Statistical mechanics of crabgrass. *Ann. Probab.* **17**, 444-481

Bramson, M. and Gray, L. (1991) A useful renormalization argument. *Random Walks, Brownian Motion, and Intracting Brownian Motion.* Edited by R. Durett and H. Kesten. Birkhauser, Boston.

Bramson, M. and Griffeath, D. (1980) On the Williams-Bjerknes tumor growth model, II. *Math. Proc. Camb. Phil. Soc.* **88**, 339–357

Bramson, M. and Griffeath, D. (1981) On the Williams-Bjerknes tumor growth model, I. *Ann. Probab.* **9**, 173–185

Brower, R.C., Furman, M.A., and Moshe, M. (1978) Critical exponents for the Reggeon quantum spin model. *Phys. Lett. B.* **76**, 213-219

Brown, D.B. and Hansel, R.I.C. (1987) Convergence to an evolutionary stable strategy in the two-policy game. *Am. Nat.* **130**, 929–940

Chao, L. (1979) The population of colicinogenic bacteria: a model for the evolution of allelopathy. Ph.D. dissertation, U. of Massachusetts.

Chao, L. and Levin, B.R. (1981) Structured habitats and the evolution of anti-competitor toxins in bacteria. *Proc. Nat. Acad. Sci.* **78**, 6324–6328

Connor, E.F. and McCoy, E.D. (1979). The statistics and biology of the species-area realtionship. *Amer. Nat.* **113**, 791–833.

Cox, J.T. and Durrett, R. (1995) Hybrid zones and voter model interfaces. *Bernoulli* **1**, 343–370

Crawley, M.J. and R.M. May (1987) Population dynamics and plant community structure: competition between annuals and perennials. *J. Theor. Biol.* **125**, 475-489

DeMasi, A., Ferrari, P. and Lebowitz, J. (1986) Reaction diffusion equations for interacting particle systems. *J. Stat. Phys.* **44**, 589-644

Dunford, N. and Schwarz, J.T. (1957) *Linear Operators*, Vol. 1. Interscience Publishers, John Wiley and Sons, New York

Durrett, R. (1988) *Lecture Notes on Particle Systems and Percolation.* Wadsworth Pub. Co. Belmont, CA

Durrett, R. (1992) A new method for proving the existence of phase transitions. Pages 141-170 in *Spatial Stochastic Processes*, edited by K.S. Alexander and J.C. Watkins, Birkhauser, Boston

Durrett, R. (1993) Predator-prey systems. Pages 37–58 in *Asymptotic problems in probability theory: stochastic models and diffusions on fractals.* Edited by K.D. Elworthy and N. Ikeda, Pitman Research Notes 83, Longman Scientific, Essex, England

Durrett, R. (1995a). *Probability: Theory and Examples.* Duxbury Press, Belmont, CA.

Durrett, R. (1995b) Ten Lectures on Particle Systems. *Ecole d'Eté de Probabilités de Saint Flour, 1993.* Lecture Notes in Math 1608, Springer, New York

Durrett, R. (1995c) Spatial epidemic models. Pages 187–201 in *Epidemic Models: Their Structure and Relation to Data.* Edited by D. Mollison. Cambridge U. Press

Durrett, R. (1996) *Stochastic Calculus.* CRC Press, Boca Raton, FL

Durrett, R. and Levin, S.A. (1994a) Stochastic spatial models: A user's guide for ecological applications. *Phil. Trans. Roy. Soc. B* **343**, 1047–1066

Durrett, R. and Levin, S. (1994b) The importance of being discrete (and spatial). *Theoret. Pop. Biol.* **46**, 363-394

Durrett, R. and Levin, S. (1996) Spatial models for species area curves. *J. Theor. Biol.*, **179**, 119–127

Durrett, R. and Levin, S.A. (1997) Spatial aspects of interspecific competition. *Preprint.*

Durrett, R. and Levin, S.A. (1998) Pattern formation on planet WATOR. *In preparation.*

Durrett, R. and Neuhauser, C. (1991) Epidemics with recovery in $d = 2$. *Ann. Applied Prob.* **1**, 189–206

Durrett, R. and Neuhauser, C. (1994) Particle systems and reaction diffusion equations. *Ann. Probab.* **22**, 289-333

Durrett, R. and Neuhauser, C. (1997) Coexistence results for some competition models. *Ann. Appl. Prob.* **7**, 10–45

Durrett, R. and Schinazi, R. (1993) Asymptotic critical value for a competition model. *Ann. Applied. Prob.* **3**, 1047-1066

Durrett, R. and Swindle, G. (1991) Are there bushes in a forest? *Stoch. Proc. Appl.* **37**, 19-31

Durrett, R., and Swindle, G. (1994) Coexistence results for catalysts. *Prob. Th. Rel. Fields* **98**, 489–515

Engen, S. and R. Lande (1996) Population dynamic models generating the lognormal species abundance distribution. *Math. Biosci.* **132**, 169–183

Fife, P.C. and McLeod, J.B. (1977) The approach of solutions of nonlinear diffusion equations to travelling front solutions. *Arch. Rat. Mech. Anal.* **65**, 335–361

Fisch, R., Gravner, J. and Griffeath, D. (1991a). Cyclic cellular automata in two dimensions. Pages 171-185 in *Spatial Stochastic Processes.* edited by K. Alexander and J. Watkins. Birkhauser, Boston.

Fisch, R., Gravner, J. and Griffeath, D. (1991b). Threshold-range scaling of excitable cellular automata. *Statistics and Computing.* **1**, 23–39.

Fisher, R.A., Corbet, A.S., and Williams, C.B. (1943). The relation between the number of species and the number of individuals in a random sample of an animal population. *J. Animal Ecol.* **12**, 42–58.

Frank, S.A. (1994) Spatial polymorphism of bacteriocins and other allelopathic traits. *Evolutionary Ecology*

Gilpin, M.E. (1975). Limit cycles in competition communities. *Am. Nat.* **109**, 51–60.

Grassberger, P. (1982) On phase transitions in Schlogl's second model. *Z. Phys. B.* **47**, 365–376

Grassberger, P. and de la Torre, A. (1979) Reggeon field theory (Schlogl's first model) on a lattice: Monte Carlo calculation of critical behavior. *Ann. Phys.* **122**, 373-396

Griffeath, D. (1979). *Additive and Cancellative Interacting Particle Systems.* Springer Lecture Notes in Mathematics, **724**.

Griffeath, D. (1988). Cyclic random competition: a case history in experimental mathematics. *Notices of the Amer. Math. Soc.* 1472–1480

Harada, Y., Ezoe, H., Iwasa, Y., Matsuda, H., and Sato, K. Population persistence and spatially limited local interaction. *Theor. Pop. Biol.* **48**, 65–91

Harris, T.E. (1974) Contact interactions on a lattice. *Ann. Prob.* **2**, 969-988

Harris, T.E. (1977) A correlation inequality for Markov processes in partially ordered state spaces. *Ann. Probab.* **6**, 355-378

Holley, R.A. and Liggett, T.M. (1975). Ergodic theorems for weakly interacting systems and the voter model. *Ann. Prob.* **3**, 643–663.

Hirsch, M.W. and S. Smale (1974) *Differential Equations, Dynamical Systems, and Linear Algebra*, Academic Press, New York

Hubbell, S.P. (1992). Speciation, dispersal, and extinction: An equilibrium theory of species-area relationships. *Preprint.*

Hubbell, S.P. (1995) Towards a theory of biodiversity and biogeograph on continuous landscapes. Pages 173–201 in *Preparing for Global Change: A Midwestern Perspective.* Edited by G.R. Carmichael, G.E. Folk, and J.L. Schnoor. SPB Academic Publishing, Amsterdam.

Huberman, B.A. and Glance, N.S. (1993) Evolutionary games and computer simulations. *Proc. Nat. Acad. Sci., USA*. **90**, 7712-7715

Keeling, M.J., Mezic, I., Hendry, R.J., McGlade, J., and Rand, D.A. (1997) Characteristic length scales of spatial models in ecology. *Phil. Trans. R. Soc. London B* **352**, 1589–1601

Levin, B.R. (1988) Frequency dependent selection in bacterial populations. *Phil. Trans. R. Soc. London B* **319**, 459–472

Liggett, T.M. (1985) *Interacting Particle Systems*. Springer-Verlag, New York

Longuet-Higgins, M.S. (1971) On the Shannon-Weaver index of diversity, in relation to the distribution of species in bird censuses. *Theor. Pop. Biol.* **2**, 271–289

Luce, R.D., and Raiffa, H. (1968) *Games and Decisions*. John Wiley and Sons, New York

MacArthur, R.H. (1957) On the relative abundance of bird species. *Proc. Nat. Acad. Sci. USA*. **43**, 293–295

MacArthur, R.H. (1960) On the relative abundance of species. *Am. Nat.* **94**, 25–36

MacArthur, R.H. and Wilson, E.O. (1967). *The Theory of Island Biogeography*. Princeton Monographs in Population Biology.

MacCauley, E., Wilson, W.G., and de Roos, A.M. (1993) Dynamics of age-structured and spatially structured predator-prey interactions: Individual bassed models and population-level formulations. *Amer, Natur.* **142**, 412–442

Matsuda, H., Ogita, N., Sasaki, A., and Sato, K. (1992) Statistical mechanics of population: the lattice Lotka-Volterra model. *Prog. Theor. Phys.* **88**, 1035–1049

Matsuda, H., Tamachi, N., Sasaki, A., and Ogita, N. (1987) A lattice model for population biology. Pages 154–161 in *Mathematical Topics in Biology* edited by E. Teramoto and M. Yamaguchi, Springer Lecture Notes in Biomathematics.

May, R.M. (1975) Patterns of species abundance and diversity. Pages 81–120 in *Ecology and Evolution of Communities*. Edited by M.L. Coday and J.M. Diamond, Belknap Press, Cambridge, MA

May, R.M. (1994) Spatial chaos and its role in ecology and evolution. Pages 326–344 in *Frontiers in Mathematical Biology*. Lecture Notes in Biomathematics 100, Springer, New York

May, R.M. (1995) Necessity and chance: Deterministic chaos in ecology and evolution. *Bulletin of the AMS.*, New Series, **32**, 291–308

May, R.M. and Leonard, W.J. (1975). Nonlinear aspects of competition between species. *SIAM J. of Applied Math* **29**, 243–253.

Maynard Smith, J. (1982) *Evolution and the Theory of Games*. Cambridge U. Press, Cambridge, England

Mollison, D. (1995) *Epidemic Models: Their Structure and Relation to Data.* Cambridge U. Press

Neubert, M.G., Kot, M., and Lewis, M.A. (1995) Dispersal and pattern formation in a discrete-time predator-prey model. *Theoret. Pop. Biol.* **48**, 7–43

Neuhauser, C. (1992) Ergodic theorems for the multi-type contact process. *Prob. Theor. Rel. Fields.* **91**, 467–506

Noble, C. (1992) Equilibrium behavior of the sexual reproduction process with rapid diffusion. *Ann. Probab.* **20**, 724-745

Nowak, M.A., and May, R.M. (1992). Evolutionary games and spatial chaos. *Nature* **359**, 826–829

Nowak, M.A., and May, R.M. (1993). The spatial dilemmas of evolution. *Int. J. Bifurcation and Chaos.* **3**, 35–78

Nowak, M.A., Bonhoeffer, S., and May, R.M. (1994). More spatial games. *Int. J. Bifurcation and Chaos.* **4**, 33-56

Owen, G. (1968) *Game Theory*. W.B. Saunders Co., Philadelphia

Pacala, S.W. and Levin, S.A. (1996). Biologically generated spatial pattern and the coexistence of competing species In: (D. Tilman and P. Kareiva, eds.) Spatial Ecology: The Role of Space in Population Dynamics and Interspecific Interactions. Princeton University Press, Princeton, NJ. *To appear.*

Pascual, M. and Levin, S.A. (1998) From individuals to population densities: searching for the intermediate scale of nontrivial determinism. *Preprint*

Preston, F.W. (1948) The commonness, and rarity, of species. *Ecology* **29**, 254–283

Preston, F.W. (1962). The canonical distribution of commonness and rarity. *Ecology* **43**, I. 185–215, II. 410–432.

Rand, D.A., Keeling, M., and Wilson, H.B. (1995). Invasion, stability, and evolution to criticality in spatially extended, artificial host-pathogen ecologies. *Proc. Roy. Soc. London B.* **259**, 55-63

Rand, D.A. and Wilson, H.B. (1995) Using spatio-temporal chaos and intermediate scale determinism in artificial ecologies to quantify spatially-extended systems. *Proc. Roy. Soc. London.* **259**, 111-117

Redheffer, R., Redlinger, R. and Walter, W. (1988) A theorem of La Salle-Lyapunov type for parabolic systems. *SIAM J. Math. Anal.* **19**, 121–132

Sawyer, S. (1979) A limit theorem for patch sizes in a selectively-neutral migration model. *J. Appl. Prob.* **16**, 482–495

Schlogl, F. (1972) Chemical reaction models for non-equilibrium phase transitions. *Z. Physik* **253**, 147–161

Shah, N. (1997) Predator-mediated coexistence. *Ph.D. Thesis Cornell U.*

Silvertown, J., Holtier, S., Johnson, J. and Dale, P. (1992). Cellular automaton models of interspecific competition for space – the effect of pattern on process. *J. Ecol.* **80**, 527–534.

Tainaka, K. (1993). Paradoxical effect in a three candidate voter model. *Physics Letters A* **176**, 303–306.

Tainaka, K. (1995). Indirect effects in cyclic voter models. *Physics Letters A* **207**, 53–57

Tramer, E.J. (1969) Bird species diversity; components of Shannon's formula. *Ecology*, **50**, 927–929

Thoday, J.M. et al (1959–64) Effects of disruptive selection. I-IX. *Heredity.* **13**, 187–203, 205–218; **14**, 35–49; **15**, 119-217; **16**, 219–223; **17**, 1–27; **18**, 513–524; **19**, 125–130

Tilman, D. (1994) Competition and bio-diversity in spatially structured habits. *Ecology.* **75**, 2-16

Watson, H. (1835) *Remarks on the Geographical Distribution of British Plants.* Longman, London.

Webb, D.J. (1974) The statistics of relative abundance and diversity. *J. Theor. Biol.* **43**, 277–292

Whittaker, R.H. (1970) *Communities and Ecosystems.* MacMillan, New York.

Williams, T. and Bjerknes, R. (1972) Stochastic model for abnormal clone spread through epithelial basal layer. *Nature.* **236**, 19–21

Williamson, M. (1988). Relationship of species number to area, distance and other variables. Chapter 4 in *Analytical Biogeography* edited by A.A. Myers and P.S. Giller, Chapman and Hall, London.

Wilson, W.G. (1996) Lotka's game in predator-prey theory: linking populations to individuals. *Theoret. Pop. Biol.* **50**, 368–393

Wilson, W.G., de Roos, A.M., and MacCauley, E. (1993) Spatial instabilities with the diffusive Lotka-Volterra system: Individual-based simulation results. *Theoret. Pop. Biol.* **43**, 91–127

Reaction Transport Systems in Biological Modelling

K.P.Hadeler,
Universität Tubingen
auf der Morgenstelle 10
D-72076 Tubingen

Contents

1 Spreading in space

Interaction with the environment and motion in space are two fundamental features of living organisms. Also non-living particles like molecules react and move – driven by diffusion. In most experiments and in industrial production chemists do not rely on diffusion alone to make particles move, they try to suppress spatial effects by stirring, i.e., by introducing some additional motion. Unstirred chemical systems show a wider range of behaviors as compared to stirred reactions. At least on a chemical level, there is no conceptual distinction between interaction and motion, e.g. membrane transport processes are chemical reactions. In contrast to chemical species living organisms show various "intelligent" strategies for spatial spread. Whereas most animals move, we consider plants as sedentary. Nevertheless they spread in various ways, by creeping or by spreading seeds.

The most detailed descriptions trace individuals in the form of stochastic processes with large state spaces: each individual carries a type, i.e., an array of characters, in particular its position, then, e.g. its age, size, etc. The process describes birth, death, and change of type. There are some problems with this approach. First of all it may be difficult to model individual behavior for lack of insight into what happens in detail (like in many chemical reactions) or lack of experimental data. Even if one can give detailed descriptions (say of the people in a small village) it is difficult to present the results of the modeling effort in a communicable way. One either produces simulations, e.g. graphic representations of single runs, or sampled and averaged quantities. On the other hand one can rightaway take to averaged quantities such as densities and derive differential or integral equations for these. Then the connection between the stochastic (or discrete) approach and the continuous model is an important and mathematically interesting problem: to what extent can differential equations be obtained by taking limits in stochastic processes? We will address this question at several instants.

It is often observed that, while moving and interacting among themselves and with the environment, individuals produce spatial patterns. Typical examples are wormholes in wood, the traces foraging snails leave on the walls of an aquarium covered by algae, witches' circles of fungi, large scale patterns in mussel colonies or forests (larvae and seeds moving), skin and coat patterns in animals or leaf patterns in plants that are also attributed to reactions and motion of substances or cells.

We see the patterns and we want to know or to explain how they are produced. In some cases – worm holes – it seems obvious: by the individual worms (which actually are not worms but larvae of beetles). Still, we want to know how the worms avoid each others tracks and how rather regular random patterns result from individual actions. In other cases like in the Turing phenomenon (see section 13) the underlying mechanism is a collective phenomenon and it is less obvious how it comes about.

The classical model for spatial spread is the diffusion equation (the second Fickian law, in thermodynamic context the heat equation)

$$u_t = D\Delta u \tag{1.1}$$

for a density $u = u(t, x)$, $x \in \mathbb{R}^n$. Derivatives are indicated by subscripts: $u_t = \partial u/\partial t$

a.s.o. The operator Δ is the Laplacian, $\Delta u = \sum_{i=1}^{n} u_{x_i x_i}$, and D is the diffusion constant. In dimension 1 the equation reads $u_t = Du_{xx}$. The equation (1.1) can be derived analytically from a material law (first Fickian law, see below) or from a stochastic description, i.e., from Brownian motion or a Wiener process.

The role of diffusion and spatial spread in population dynamics has been discussed in great detail by Okubo [77], with reference to various concrete examples. Probably Skellam [93] was the first who systematically investigated how animals and plants spread, migrate, and diffuse. A standard reference to diffusion in general is [13].

It is worthwhile to study the ideas underlying equation (1.1). Take a one-dimensional grid $\mathbb{Z}$ and a particle moving on the grid in discrete time steps. At time $k = 0$ the particle starts at the grid point $i = 0$ and then, in successive time steps, moves either right or left, in each case with probability 1/2. Let X_k be the random variable giving the position of the particle at time k. Then X_k is even for k even and odd for k odd. Apart from this complication X_k follows the binomial distribution $B_{n,p}$ with $n = k$ and $p = 1/2$. Of course we know how to get from the binomial distribution to continuous limits by scaling. We introduce a sequence of grids $h\mathbb{Z}$ for space and $\tau\mathbb{N}_0$ for time such that $x = ih$ and $t = k\tau$. Then we let $h \to 0$, $\tau \to 0$ such that τ/h^2 tends to a constant D. One arrives at a process X_t on the real line with continuous paths. This process is called Brownian motion or the Wiener process. The probability density for this process is the normal distribution with expectation 0 and variance $\sqrt{2Dt}$. It is a solution of equation (1.1), for $n = 1$.

The diffusion equation (1.1) preserves positivity and total mass ($u(0, x) \geq 0$ for all $x \in \mathbb{R}$ implies $u(t, x) \geq 0$, and $\int_{-\infty}^{\infty} u(t, x)dx = \int_{-\infty}^{\infty} u(t, 0)dx$ holds). Brownian motion would be a good model for spatial spread were it not for the unwanted phenomenon of infinitely fast propagation. Indeed, assume the particle starts surely at $x = 0$. Then the initial datum is the Dirac delta function $u(0, x) = \delta_0(x)$, and the solution $u(t, x)$ (the normal distribution) is positive everywhere for arbitrarily small $t > 0$. In other words, even for very small $t > 0$ and very large $|x|$ there is some positive probability that the particle arrives at x. This behavior clearly contradicts basic laws of physics and thus restricts the validity of the diffusion equation (or the heat equation) as a model. This fact has been discussed by Maxwell (1867), Einstein (1906) [17] and many others. In present-day thermodynamic theory Fourier's "theory of heat" has been replaced by models based on more fundamental physical laws. The question remains: whether there is an error in the derivation of Brownian motion from the random walk on a grid, or, since the limiting process is mathematically correct, why the process fails as a physical model. Let us return to the walk on the grid. For any k the random variable assumes only finitely many values, the maximal displacement of the particle from the origin is finite. Even in the rescaled process with positive h and τ, for a fixed time horizon $t > 0$, in the time interval $[0, t]$ the particle performs only finitely many moves (roughly t/τ) and thus can arrive only at about $x = ht/\tau$. Such an argument does not apply in the limit: in Brownian motion the particle can make arbitrarily many moves in a given time interval $[0, t]$ and thus it can get arbitrarily far. Thus infinitely fast propagation results from the fact that the particle can make any number of independent (and thus uncorrelated) moves in a given time. The model particle has no inertia or relaxation time.

For more realistic models we should use different scalings in transition from a random walk on a grid to a continuous process. Such scalings have been considered in detail by Goldstein [29] and earlier by Fürth [25] and Taylor [98]. At present it is easier to describe the continuous process directly. Whereas in Brownian motion the particle is characterized by its position in space alone, here the particle has (almost surely) a well-defined velocity. Thus consider a particle on the real line that starts at the origin with speed γ. Assume the particle moves initially to the right, its velocity being $+\gamma$. Then it turns and moves with velocity $-\gamma$, turns again a.s.o. Assume that the turning events are given by a Poisson process with parameter $\mu/2$. Then the holding time with respect to the direction of motion is exponentially distributed with parameter $\mu/2$. This process is called a correlated random walk with parameters γ and $\mu/2$. In the (x,t) plane the path of the particle appears as a zig-zag curve whereas the path of Brownian motion looks highly irregular.

We can easily see what the partial differential equation of the correlated random walk should look like. The probability density is a function of time t, space x and velocity s. We write $u^s(t,x)$ with $s=\pm$ rather than $u(t,x,\pm\gamma)$. If the particle moves to the right with speed γ and does not turn at all then its density $u^+(t,x)$ is a simple wave and thus satisfies $u_t^+ + \gamma u_x^+ = 0$, similarly $u_t^- - \gamma u_x^- = 0$ if it moves to the left. If the particle also turns we get

$$\begin{aligned} u_t^+ + \gamma u_x^+ &= \tfrac{1}{2}\mu(u^- - u^+) \\ u_t^- - \gamma u_x^- &= \tfrac{1}{2}\mu(u^+ - u^-). \end{aligned} \tag{1.2}$$

This is the Goldstein-Kac system [56] system for a correlated random walk on the real line (see also [105]). Equation (1.2) is a hyperbolic system of differential equations. The system (1.2) preserves positivity (jointly for u^+, u^-) and it preserves total mass. This behavior is somewhat strange for a hyperbolic system, since, in general, the solutions of such systems, in agreement with their usual physical interpretations, are oscillating and do not preserve positivity.

We indicate some notational confusion. In most papers on correlated random walks (including previous papers by the author) the coefficient in the differential equation is μ rather than $\mu/2$. Then μ is the rate at which the Poisson process makes the particle stop and choose a velocity (which may be the previous one). If there are just two velocities then $\mu/2$ is the rate of actual turns.

If we introduce in (1.2) the total density $u = u^+ + u^-$ and the "probability flow" $v = \gamma(u^+ - u^-)$, then these two variables satisfy

$$\begin{aligned} u_t + v_x &= 0 \\ v_t + \gamma^2 u_x &= -\mu v. \end{aligned} \tag{1.3}$$

Of course u^+ and u^- can be recovered as $u^+ = (u+v/\gamma)/2$ and $u^- = (u^+ - v/\gamma)/2$. It is hard to see whether the system (1.3) has any positivity properties unless we go back to (1.2).

The following transition is known as the Kac trick [56]: differentiate the first equation (1.3) with respect to t and the second with respect to x. By eliminating v_{xt} from both equations and v_x from the first equation (1.3), we arrive at

$$u_{tt} + \mu u_t = \gamma^2 u_{xx}. \tag{1.4}$$

Thus for every C^2 solution of (1.3), the first component satisfies a telegraph equation or a damped wave equation. In the transition from (1.3) to (1.4) the solutions of (1.3) of the form $(u,v) = (0, v_0 e^{-\mu t})$, v_0 a constant, are mapped into the zero solution. These solutions can be seen as "flows without mass". It is easy to see that also the second component v satisfies equation (1.4).

We rewrite equation (1.4) in the form

$$\frac{1}{\mu} u_{tt} + u_t = \frac{\gamma^2}{\mu} u_{xx}. \tag{1.5}$$

We can formally arrive at a diffusion equation if we let μ tend to ∞. We should also let $\gamma \to \infty$ in such a way that γ^2/μ has a limit, or we can just assume that γ^2/μ is a constant,

$$D = \frac{\gamma^2}{\mu}. \tag{1.6}$$

Thus, if particles move fast and change their directions frequently, in such a way that the relation (1.6) holds, the hyperbolic equation (1.3) becomes asymptotically the parabolic system (1.1). This transition is in fact the diffusion approximation for the correlated random walk. So Brownian motion can be seen as a limiting case of the correlated random walk in which the speed and the turning frequency are very large, and both are scaled in such a way that γ^2/μ remains constant. Equation (1.6) gives the "physically correct" relation between the parameters in (1.2) and (1.1).

It will turn out that each of the three descriptions of the system, namely equations (1.2) in terms of u^+, u^-, equations (1.3) in terms of u, v, and the telegraph equation (1.4) for u, can be useful, depending on the problem to be studied.

In the theory of electromagnetism, Heaviside has derived the telegraph equation from Maxwell's equations in connection with the problem of fading in transatlantic cables (see [68]). Poincaré [84] found a representation formula for the solutions and explained their dependence on initial data. Consider a transmission line of two parallel wires. Let R be the Ohm resistance, L the self induction, C the capacity, and A the loss of isolation (leakage), each per unit of length. Let v be the voltage and i the cross current. These variables satisfy the equations

$$\begin{aligned} v_x + Li_t + Ri &= 0 \\ i_x + Cv_t + Av &= 0. \end{aligned}$$

Differentiate the first equation with respect to x and the second with respect to t, then eliminate the mixed derivatives and obtain the second order equation

$$LC\, v_{tt} + (AL + RC)\, v_t = v_{xx} - RAv.$$

This equation is the telegraph equation. One can check easily that the function i satisfies the same equation. The connection between the first order system and the telegraph equation is the same in the electromagnetic problem and the random walk problem. Positivity of solutions is important only in the latter.

Much knowledge had been accumulated on this equation before it was connected to stochastic processes.

One can ask how important the assumption of a Poisson process is. The Poisson process is the only "Markovian" process in the sense that the mere existence of the particle is sufficient to describe its state. In all other processes we have to keep track of the history of the particle, i.e., the state of the moving particle depends, in addition to its position in space, on the time since the last turn. If we denote this time by τ then we look for a system that describes the evolution of the functions $u^s(t,x,\tau)$ (see [1]). This system reads

$$\begin{aligned} u_t^+ + u_\tau^+ + \gamma u_x^+ &= -\mu(\tau)u^+ \\ u_t^- + u_\tau^- - \gamma u_x^- &= -\mu(\tau)u^- \\ u^s(t,x,0) &= \int_0^\infty \mu(\tau)u^s(t,x,\tau)d\tau, \quad s = \pm. \end{aligned} \tag{1.7}$$

Here $\mu(\tau)$ is the turning rate depending on the time since the last turn. In order to start the process one must know $u^s(0,x,\tau)$, i.e., the positions and "ages" of the existing particles. Indeed, if u is independent of direction s and position x, the system (1.7) reduces to the McKendrick model for a population structured by age, for the special case where the birth rate equals the death rate.

The system (1.3) is the simplest hyperbolic analogue of the diffusion equation (1.1). It does not show the phenomenon of infinitely fast propagation. Quite on the contrary, all particles move with speed γ, and the support of any solution with compact support broadens only like γt.

Before entering the problem of several space dimensions let us consider a slightly more general system. Let us assume that the particle speed is not constant but that particles can choose their speeds from some interval $[\gamma_0, \gamma_1]$ with $\gamma_0 > 0$. Then the velocities range over the set $V = [-\gamma_1, -\gamma_0] \cup [\gamma_0, \gamma_1]$ and the desired density is a function $u = u(t,x,s)$ with $s \in V$. If some particles just move with their given velocities and do not turn then we have a family of simple waves $u_t + s\,u_x = 0$. If the particles turn then we have to specify when and how. Again we assume that the turning events are governed by a Poisson process with a parameter μ. At a turning event, the new direction is given by a probability distribution on V, depending on the previous direction as a parameter. Thus the equation reads

$$u_t + s\,u_x + \mu u = \int_V K(s,\tilde{s})u(t,x,\tilde{s})d\tilde{s} \tag{1.8}$$

where $u = u(t,x,s)$ on the left hand side. The kernel K is nonnegative, satisfies $\int_V K(s,\tilde{s})ds = \mu$ and has appropriate symmetry properties to be discussed later in a broader context. Equation (1.2) is recovered if $\gamma_0 = \gamma_1 = \gamma$.

Equation (1.8) is called a one-dimensional transport equation (or velocity jump process [79]). The correlated random walk is a transport equation with constant particle speed. For the formulation of the problem we could also have assumed that V is the interval $[-\gamma_1, \gamma_1]$. However, then we would have arbitrarily slow particles and as a result the solution operator of equation (1.8) has poor smoothing properties. Notice that the "free stream operator" $\partial_t + s\,\partial_x$ contains the independent variable s explicitly (this is not so obvious in equation (1.3)). Nevertheless equation (1.8) is homogeneous in x and s, and also isotropic, if K is appropriately chosen.

Equation (1.8) should be distinguished from systems with long-range interaction (position jump processes [79]) where the integral is taken over space rather than velocity. In one space dimension, let $u(t,x)$ be the density of particles at time t. Assume a particle sits at position x_1, waits, and then takes a jump to a position x_2. Thus the path of the particle is not continuous. We assume that waiting time is governed by a Poisson process and that waiting time and new position are independent random variables. We further assume that the distribution depends only on $|x_2 - x_1|$. Then we arrive at an equation

$$u_t(t,x) = \mu \int_{\mathbb{R}} k(x+y)u(t,y)dy - \mu u(t,x) \tag{1.9}$$

where $k \in C(\mathbb{R})$ has the properties $k \geq 0$, $\int_{\mathbb{R}} k(y)dy = 1$, $k(-y) = k(y)$. In contrast to the Laplacian in (1.1), the integral operator in (1.9) is not a local operator: in order to evaluate $u_t(t,x)$ it is not sufficient to know u on a small neighborhood of x. Also the system (1.9) shows infinitely fast propagation.

It is easy to derive a formal diffusion approximation for the system (1.9). We rewrite the integral and apply Taylor expansion to the function u (we omit the variable t),

$$\begin{aligned} &\int k(x+y)u(y)dy = \int k(y)u(x+y)dy \\ &= \int k(y)u(x)dy + \int k(y)u_x(x)y\,dy + \int k(y)u_{xx}(x)\tfrac{y^2}{2}dy + o(||y||). \end{aligned} \tag{1.10}$$

The second term on the right vanishes. If we insert this expression into equation (1.9) then the first term cancels. If we discard the $o(||y||)$ term then we arrive at the equation (1.1) with

$$D = \frac{\mu}{2} \int_{\mathbb{R}} k(y)y^2 dy. \tag{1.11}$$

Thus the effective diffusion coefficient is proportional to the parameter μ of the Poisson process and the "variance" of the kernel k. Also in this model the state of the particle is its position. Particles do not have an individual velocity.

After this detour to position jump processes we return to the problem of generalizing (1.3) or even (1.8) by considering the case of space dimension $n \geq 2$. We introduce the free stream operator in several dimensions. This operator $\partial_t + \nabla_x \cdot s$ comes in different notations which we explain for convenience. In coordinate notation, the free stream equation reads

$$\frac{\partial u}{\partial t} + \sum_{j=1}^{n} s_j \frac{\partial u}{\partial x_j} = 0. \tag{1.12}$$

It says that each particle moves with "its" velocity along a straight line. In coordinate-free notation (1.12) becomes

$$\frac{\partial u}{\partial t} + (\nabla_x \cdot s)u = 0 \tag{1.13}$$

(sometimes the brackets are omitted) or

$$\frac{\partial u}{\partial t} + s \cdot \nabla_x u = 0 \tag{1.14}$$

The first formula says that $\nabla_x \cdot s$ is a formal inner product of the velocity vector s and the vector $\nabla_x = (\partial_{x_1}, \ldots, \partial_{x_n})$ and that the resulting "scalar" operator is applied to the function u. The second notation says that $s \cdot \nabla_x u$ is the inner product of the velocity vector and the gradient of the function u.

Now we introduce the feature that particles change their velocity, in particular their direction of motion. Let $V \subset \mathbb{R}^n$ be the set of admissible velocities. Mostly V is a spherical shell $V = \{s \in \mathbb{R}^n : \gamma_0 \leq \|s\| \leq \gamma_1\}$ with $\gamma_0 > 0$. Particles "die" according to a Poisson process and "are born" at the same position with a new speed. The new speed can be correlated with the previous speed. Let $K(\cdot, \tilde{s})$ be the probability density of new velocities, given the previous velocity *tildes*. Then

$$K(s, \tilde{s}) \geq 0, \quad \int_V K(s, \tilde{s}) ds = \mu. \tag{1.15}$$

The equation of motion becomes

$$\frac{\partial u}{\partial t} + \nabla_x \cdot su = -\mu u + \int_V K(s, \tilde{s}) u(t, x, \tilde{s}) d\tilde{s}. \tag{1.16}$$

This equation describes pure motion: no particle is produced or deleted. The total number of particles

$$\int_{\mathbb{R}^n} \int_V u(t, x, s) ds dx \tag{1.17}$$

(in case it is bounded) is an invariant of motion. It is obvious that the system preserves positivity. With respect to the space variable x the system is fully symmetric (i.e., it is invariant under the full symmetry group of $\mathbb{R}^n$.) So far no symmetry with respect to the variable s has been assumed.

We can find system (1.16) in different contexts. It has been introduced into biological modeling by Othmer et al. [79] as the velocity jump process (as opposed to a position jump process). Earlier it had appeared in transport problems in physics, e.g. in the form of neutron transport equations. Still earlier it has shown up as a linearization (at some spatially constant equilibrium) of the Boltzmann equation and there it has been called the Boltzmann-Lorentz equation (see [11], [12]). It is no wonder that the equation shows up in various fields. Like the Lotka-Sharpe-McKendrick model for age structured populations, the equation just expresses careful book-keeping. We arrive at equation (1.16) once we agree that particles are characterized by their individual positions and speeds, that turning events are governed by a Poisson process and that the turning distribution is constant in time.

The velocity jump process rather closely describes the motion of some bacteria. Apparently because bacteria are too small to measure gradients (of useful or toxic substances) in space they measure them over time. They run on a straight line, stop, "tumble", and choose a new direction. The speed in this process is more or less constant and the choice of direction is not far from uniform. It is an interesting phenomenon that some bacteria get up (or down, whatever they wish) the gradient by adapting the turning rate μ to the observed changes in concentration. This mechanism is explored in chemotaxis models (see section 12).

The simplest example of a system (1.16) in dimension two is the following system with four types of particles which we call "NEWS" (north, east, west, south)

$$\begin{aligned} E_t + \gamma E_x &= -\mu E + \mu M \\ W_t - \gamma W_x &= -\mu W + \mu M \\ N_t + \gamma N_y &= -\mu N + \mu M \\ S_t - \gamma S_y &= -\mu S + \mu M \\ M &= \tfrac{1}{4}(E + W + N + S). \end{aligned} \tag{1.18}$$

The system is modeled after the Broadwell system [9]. However, the turning events are determined by an autonomous Poisson process and not by collisions.

In (1.16) let us consider the special case of constant speed, i.e., $V = \gamma S^{n-1}$, where S^{n-1} is the unit sphere in $\mathbb{R}^n$. If $n = 2$, then we can parametrize S^1 by an angle φ. If we further assume isotropy then we obtain an equation

$$\begin{aligned} u_t(t,x,\varphi) + {} & \gamma(\cos\varphi\, u_{x_1} + \sin\varphi\, u_{x_2}) \\ & = -\mu u + \textstyle\int_0^{2\pi} K(\varphi - \psi) u(t,x,\psi) d\psi, \end{aligned} \tag{1.19}$$

where the function K of one variable has the properties

$$K(\varphi) \geq 0, \quad K(-\varphi) = K(\varphi), \quad K(\varphi + 2\pi) = K(\varphi), \quad \int_0^{2\pi} K(\varphi) d\varphi = \mu. \tag{1.20}$$

In the special case where the new speed has uniform distribution we get

$$u_t + \gamma(\cos\varphi\, u_{x_1} + \sin\varphi\, u_{x_2}) = -\mu u + \frac{\mu}{2\pi}\int_0^{2\pi} u(t,x,\psi) d\psi. \tag{1.21}$$

This problem is called the "Pearson walk". It has been studied by K.Pearson [83] and Lord Rayleigh (see [103]) and, in particular, in [95]. In equation (1.19) the change of velocity is decribed by an integral operator. In (1.10) we have seen how to replace an integral operator by a diffusion operator (with respect to the space variable). Here we apply a similar idea to the velocity variable and arrive at the equation

$$u_t + s \cdot \nabla_x u = \mu \Delta_s u, \tag{1.22}$$

where Δ_s is the Laplace-Beltrami operator (i.e., the Laplace operator in appropriate coordinates) on the manifold V of velocities. If V has a boundary ∂V (like the spherical shell) then boundary conditions have to be specified on ∂V. Bartlett [4] has proposed to study such a system on γS^{n-1}. Then there is no boundary. In two dimensions, $n = 2$, one can parametrize S^1 by an angle $\varphi \bmod 2\pi$ and write $u = u(t, x, y, \varphi)$,

$$u_t + \gamma\cos\varphi u_x + \gamma\sin\varphi u_y = \mu u_{\varphi\varphi}. \tag{1.23}$$

There is a rather different approach to a generalization of the correlated random walk system (1.2) to arbitrary space dimension n. The equations (1.3) can be interpreted as follows: The first equation connects a density u to a vector field v. Since in dimension 1 the quantity v_x can be interpreted as the divergence $\operatorname{div} v$, the first equation of (1.3) is a conservation law. The second equation of (1.3) can be seen as a material law which connects the flow v to the gradient of u. This view can be carried over to several dimensions. Assume a particle density $u : \mathbb{R}^n \to \mathbb{R}$ and a particle flux $v : \mathbb{R}^n \to \mathbb{R}^n$ are connected by a conservation law

$$u_t + \operatorname{div} v = 0. \tag{1.24}$$

Following the ideas of Fourier (1822) on the spread of heat, Fick (1855) stated the first Fickian law

$$v = -D\operatorname{grad} u, \tag{1.25}$$

which says that the flux is proportional to the negative gradient, the coefficient being the diffusion coefficient. If v is inserted from (1.25) into (1.24), then the diffusion equation (1.1) or the second Fickian law results: $u_t = \operatorname{div} D\operatorname{grad} u$. Cattaneo [10] has argued that an instantaneous adaptation of the flux to the gradient as stated in (1.25) is physically unrealistic. He assumed that the flux adapts to the gradient by a negative feedback loop with a small time constant τ,

$$\begin{aligned} u_t + \operatorname{div} v &= 0 \\ \tau v_t + D\operatorname{grad} u + v &= 0. \end{aligned} \tag{1.26}$$

Cattaneo derived the system in a thermodynamic setting. Apparently Maxwell had similar thoughts, see [55] for the history of the problem, [32], [73], [88] for more general thermodynamic models, and the latter two on the role of Cattaneo's work in this field.

We call (1.26) a "Cattaneo system". From (1.26) we obtain, with $w = -\operatorname{div} v$, the system

$$\begin{aligned} u_t - w &= 0 \\ \tau w_t - D\Delta u + w &= 0 \end{aligned} \tag{1.27}$$

and then, by eliminating w as in (1.4), a telegraph equation or damped wave equation

$$\tau u_{tt} + u_t = D\Delta u. \tag{1.28}$$

Similar to the one-dimensional case, in the transition from (1.27) to (1.28) the solutions $(0, v_0 e^{-\mu t})$, v_0 any vector field (constant in time) with $\operatorname{div} v_0 = 0$, are mapped into the zero solution.

The Cattaneo system seems to be quite realistic and it has useful mathematical properties. The same is true for the damped wave equation for which there is quite an amount of analytic results. However, we have seen that the system (1.3) does not preserve positivity in the naive sense: for initial data $(u(0,x), v(0,x))$ with $u(0,x) \geq 0$ the function $u(t,x)$ will in general assume also negative values. On the other hand, the system preserves positivity with respect to the cone $u^+ \geq 0$, $u^- \geq 0$. For the Cattaneo system in dimension $n \geq 2$ no concept of positivity (invariant cone) has been found so far. It seems that this negative result is connected to the fact that no microscopic description for the Cattaneo system with $n = 2$ has been found. Several stochastic processes in the plane have been proposed for the telegraph equation in dimension two, but close inspection of the proofs reveals that the resulting equation is the one-dimensional telegraph equation in disguise. In other words, the processes constructed are not really acting in the plane, but each particle moves back and forth on some straight line (see [78], [92]). It is interesting to see why these approaches have failed. In the Brownian motion interpretation a particle is just given by its position. If we would attribute to each particle an individual position and velocity then we would arrive at a transport equation. Thus particles governed by the Cattaneo system carry a little less information, say, the position and some "tendency" rather than a direction. The approach fails because the verbal description of what the particles are supposed to do does not agree with the mathematical description.

So far we have interpreted the functions $u(t,x)$, $u^s(t,x)$, $u(t,x,s)$, etc. as probability densities for the state of one particle. Then u is normalized, the integral is equal to 1. Since the equations are linear, also αu is a solution for any constant $\alpha > 0$. Non-normalized solutions will be interpreted as particle densities.

2 Introducing reactions

Consider some population of individuals which interact. Let u be the population size. The simplest population model is an ordinary differential equation $\dot{u} = f(u)$, where the function f comprises all demographic effects in a simplistic fashion. We can think of the Verhulst equation $\dot{u} = au(1 - u/K)$, where a is the rate of exponential growth at low densities and K is the carrying capacity. Now consider particles which react according to this law and at the same time undergo diffusion as in equation (1.1). What is the form of the joint system modeling reaction and diffusion? Before answering this question consider a simpler problem. Let A and B be two matrices of the same order and consider the differential equations $\dot{u} = Au$ and $\dot{v} = Bv$. We know the solutions are $u(t) = e^{tA}u(0)$ and $v(t) = e^{tB}v(0)$, respectively. Can we express the solutions of $\dot{w} = (A+B)w$ in terms of u and v? In other words, can we express e^{A+B} in terms of e^A and e^B? If A and B commute (as in the case $n = 1$), then $e^{A+B} = e^A e^B$ but this formula is wrong otherwise. The "Trotter formula"

$$e^{A+B} = \lim_{n\to\infty} (e^{(1/n)A} e^{(1/n)B})^n$$

shows that the differential equation $\dot{w} = (A+B)w$ describes a process where A and B act simultaneously. Now we reverse this argument. Let us assume we are interested in the evolution of the reaction-diffusion process during some interval $[0,T]$. Divide the interval into n uniform time steps of length $h = T/n$ and consider the following process on $2n$ time steps numbered $l = 1, 2, \ldots, 2n$. If l is odd, let the reaction act, if l is even, the diffusion. If we call the solution operator of the reaction system $F(h)$ and that of the diffusion system $G(h)$, then we get a product $(G(h)F(h))^n$ acting on the initial data. This operator is very different from $G(T)F(T)$. Under rather mild conditions one can show that $\lim_{n\to\infty}(G(h)F(h))^n$ exists and that it is the solution operator of the equation

$$u_t = D\Delta u + f(u). \tag{2.1}$$

The argument appeals to biologists: both reaction and diffusion act the full length of time T, and in the limit they act simultaneously. Equation (2.1) is called a reaction diffusion equation. Like equation (1.1) it is a parabolic equation, and it also shows the phenomenon of infinitely fast propagation. There is an énormous literature on reaction diffusion equations (e.g. Henry [45], Rothe [87], Smoller [94]) and their role in biological modeling (e.g. Fife [19], Britton [8], Murray [74]). More or less every type of biological model (predator-prey, competitive, epidemic, chemostat, age structure) has been studied together with spatial spread in the form of reaction diffusion equations. We will recall some results in due course.

Reaction diffusion equations of the form (2.1), in one space dimension,

$$u_t = Du_{xx} + f(u) \tag{2.2}$$

have been introduced into biology in 1937 by R.A.Fisher [21] and by Kolmogorov, Petrovskij, and Piscunov [63]. Therefore the equation (2.2) is frequently called Fisher's equation or KPP equation. Equation (2.1) usually appears with a nonlinearity $f(u)$ of a "quadratic" ("hump") $f(u) = u(1-u)$ or a "cubic" ("threshold") $f(u) = u(1-u)(u-\alpha)$ with $0 < \alpha < 1$ (the range of u is $0 \le u \le 1$). Fisher studied the quadratic case in a population genetic setting, KPP used an ecological interpretation. The threshold case can be interpreted in ecological terms as an "Allee" effect or in neurobiological terms as a "Huxley pulse".

For future reference we formally define the terms "hump" type and "threshold" type for functions $f \in C^1[0,1]$. In the hump case we have $f(u) > 0$ for $0 < u < 1$, and $f(0) = f(1) = 0$, $f'(0) > 0 > f'(1)$. In the threshold case there is an $\alpha \in (0,1)$ such that $f(u) < 0$ for $0 < u < \alpha$ and $f(u) > 0$ for $\alpha < u < 1$. Furthermore $f(0) = f(\alpha) = f(1) = 0$ and $f'(0) < 0$, $f'(\alpha) > 0$, $f'(1) < 0$.

Since in reaction diffusion equations the reaction term and the diffusion term are just added, one can think of adding a reaction term $f(u)$ to the systems (1.2), (1.8), or (1.26). The simplest case is the correlated random walk (1.2). Assuming that the function $f(u)$ describes "production" of particles, we distribute the amount $f(u)$ to the two types u^+, u^- and obtain what we call the isotropic reaction walk system (Holmes [53], [34])

$$\begin{aligned} u_t^+ + \gamma u_x^+ &= \tfrac{1}{2}\mu(u^- - u^+) + \tfrac{1}{2}f(u) \\ u_t^- - \gamma u_x^- &= \tfrac{1}{2}\mu(u^+ - u^-) + \tfrac{1}{2}f(u). \end{aligned} \tag{2.3}$$

Then the variables $u = u^+ + u^-$ and $v = \gamma(u^+ - u^-)$ satisfy

$$\begin{aligned} u_t + v_x &= f(u) \\ v_t + \gamma^2 u_x &= -\mu v. \end{aligned} \tag{2.4}$$

Similarly one can introduce a reaction term into the velocity jump process (1.8) and get

$$u_t + \nabla_x \cdot su = -\mu u + \int_V K(s,\bar{s})u(t,x,\bar{s})d\bar{s} + \frac{1}{|V|}f(\bar{u}), \tag{2.5}$$

where

$$\bar{u}(t,x) = \int_V u(t,x,s)ds \tag{2.6}$$

and $|V| = \int_V ds$. Finally, we can formulate the isotropic Cattaneo system as

$$\begin{aligned} u_t + \operatorname{div} v &= f(u) \\ \tau v_t + D \operatorname{grad} u + v &= 0. \end{aligned} \tag{2.7}$$

Notice that (2.3) is a special case of (2.5) and (2.4) a special case of (2.7).

It is rather strange that the Kac trick can be applied to the nonlinear equation (2.7) as well (as Holmes [53] has observed for (2.4)). The result is a nonlinear "reaction telegraph equation" or nonlinear damped wave equation

$$\tau u_{tt} + (1 - \tau f'(u)) u_t = D \Delta u + f(u). \tag{2.8}$$

The formal limit for $\tau \to 0$ is the reaction diffusion equation (2.1). In general, the damping coefficient $1 - \tau f'(u)$ will not be positive, i.e., the equation will not necessarily have a dissipative character. Indeed, the hypothesis $1 - \tau f'(u) > 0$ will show up at various occasions in the sequel.

Although the systems (2.3)-(2.8) have rather nice properties mathematically, the approach neglects the fact that in biological applications the term $f(u)$ in (2.1) describes the net production rather than the newly born particles and that (2.3)-(2.8) are inappropriate if particles are removed: in a structured population model (here the population is structured by velocity) one cannot subtract the disappearing particles uniformly from all classes, each particle has to be removed from its class. For simplicity, assume that f has the form $f(u) = m(u)u - g(u)u$, where $m(u)$ and $g(u)$ are the density-dependent fertility and mortality, respectively. Then, instead of (2.3) we have [34]

$$\begin{aligned} u_t^+ + \gamma u_x^+ &= \tfrac{1}{2}\mu(u^- - u^+) + \tfrac{1}{2} m(u) u - g(u) u^+ \\ u_t^- - \gamma u_x^- &= \tfrac{1}{2}\mu(u^+ - u^-) + \tfrac{1}{2} m(u) u - g(u) u^- \end{aligned} \tag{2.9}$$

or, equivalently (see (2.4))

$$\begin{aligned} u_t + v_x &= f(u) \\ v_t + \gamma^2 u_x &= -(\mu + g(u)) v, \end{aligned} \tag{2.10}$$

and instead of (2.5) we get

$$u_t + \nabla_x \cdot s u = -\mu u + \int_V K(s, \bar{s}) u(t, x, \bar{s}) d\bar{s} + \frac{1}{|V|} m(\bar{u}) \bar{u} - g(\bar{u}) u. \tag{2.11}$$

For the Cattaneo system (2.7), with $n \geq 2$, there is no extension of this form. Since the flow v has no direct interpretation in terms of population groups, we cannot split the nonlinearity.

Interpretation as a branching process

In Fisher's interpretation of equation (2.2) the value $u(t, x)$ is the frequency of a gene (ranging from 0 to 1) at the position x at time t. In the interpretation of KPP it is the density of matter (ranging from 0 to ∞) at time t. McKean [70] has given a totally different interpretation of the same equation in terms of branching processes. Subsequently, there have been extended discussions on the comparison of the two interpretations, the modeling assumptions and the meaning of critical parameters, e.g. the minimal wave speed, in practical situations (see [72]).

McKean designs a stochastic process of the following form. At any time $t \geq 0$ there are $\nu(t)$ particles. These have positions on the real axis $X_1(t), \ldots, X_{\nu(t)}$. Each of these particles performs a Brownian motion independent of all the other particles. At the same time the particles are subject to a branching process. Any existing particle has exponential holding time. When it splits it gives rise to finitely many daughters. The daughters start their motion at the position of the mother. Brownian motion and branching act independently of each other.

Thus, the process is characterized by the following parameters: the diffusion rate D, the Poisson parameter b and the distribution of the number of daughters given by its generating function

$$g(z) = \sum_{k=2}^{\infty} g_k z^k \tag{2.12}$$

with

$$g_k \geq 0, \quad k = 2, 3, \ldots; \quad \sum_{k=2}^{\infty} g_k = 1. \tag{2.13}$$

In the simplest case, branching into two daughters, we have $g(z) = z^2$.

The following notion is motivated by the observation that in Brownian motion both the probability density and the distribution function $Prob\{X(t) < x\}$ are solutions of the linear diffusion equation. Thus, for the present process, introduce $u(t, x)$ as the probability for the position of the most advanced particle,

$$\tilde{u}(t, x) = Prob\{X_i(t) < x, \, i = 1, \ldots, \nu(t)\} \tag{2.14}$$

and $u(t, x) = 1 - \tilde{u}(t, x)$. Then $u(t, x)$ is the probability that at least one particle has a position above x. Clearly, $u(t, x)$ is a nonincreasing function of x with $u(t, x) \to 1$ for $x \to -\infty$ and $u(t, x) \to 0$ for $x \to +\infty$. The function u is a solution to Fisher's equation (2.2) with

$$f(u) = b(g(u) - u). \tag{2.15}$$

Thus, in the case of two daughters, the standard example $u_t = Du_{xx} + u(1 - u)$ is recovered.

These ideas have been carried over to hyperbolic systems by Dunbar and Othmer [16], [15]. Consider a stochastic process as before with Brownian motion replaced by a correlated random walk with parameters γ and μ. Then the function u, as defined in (2.14), satisfies a reaction telegraph equation

$$u_{tt} + (\mu - 2f'(u))u_t = \gamma^2 u_{xx} + (\mu + b)f(u) \tag{2.16}$$

which looks rather similar to (2.8).

3 Boundary conditions

Most biological populations live in certain restricted areas, and also in models for slime molds or skin patterns the geometry of the space domain plays a role. Here we assume that the partial differential equation describes reaction and spread in some bounded domain $\Omega \subset \mathbb{R}^n$. We require that the boundary $\partial\Omega$ is smooth in such a way that a unique normal vector exists which depends continuously on the point in $\partial\Omega$. Whenever there is a boundary we have to describe boundary conditions. Actually, in the case of $\Omega = \mathbb{R}^n$ boundary conditions are implicitly imposed if we look for solutions in some function space (bounded, L^2-integrable etc.). An important observation is the following: the boundary conditions do not involve the nonlinearity. Thus the typical boundary conditions are linear. We can introduce nonlinear boundary conditions but these would reflect some particular reactions that take place at the boundary.

A typical boundary condition describes how the flux through the boundary is connected to the concentration or the concentration difference across the boundary. The flux through the boundary is the flux in normal direction. Although one can think of a great variety of boundary conditions, only few play a role in biological modeling. This seems partly due to the fact that we do not know what happens at the boundary in field experiments. On the other hand the behavior of a biological model may strongly depend on boundary conditions. It makes a difference whether we apply a reaction diffusion system for pattern formation to the surface of an animal (a somewhat deformed two-sphere) or to the flat skin as it is presented in a fur shop. In the latter case we have to design an appropriate "periodic" boundary condition.

In a Dirichlet boundary condition one assumes that the solution u is fixed on the boundary. A zero Dirichlet condition says that u vanishes on the boundary,

$$u(t,x) = 0 \quad \text{for} \quad x \in \partial\Omega. \tag{3.1}$$

If also $f(0) = 0$ then the condition says that the population on the boundary is at the equilibrium $u = 0$ with respect to the reaction.

If a Dirichlet condition is given, then there is a flux through the boundary. The flux can only be determined by solving the initial boundary value problem and then evaluating the normal derivative u_ν. (The vector ν is the outer normal unit vector at $x \in \partial\Omega$). In a stochastic interpretation a zero Dirichlet condition says that particles are absorbed at the boundary. At this point a warning is appropriate: when we say absorbed we mean that particles hitting the boundary are removed. We do not mean to say that particles are somehow piling up at the boundary such that u becomes large near the boundary.

In a Neumann condition the flux through the boundary is prescribed. In most biological models we have a zero Neumann condition

$$u_\nu(t,x) = 0 \quad \text{for} \quad x \in \partial\Omega. \tag{3.2}$$

In this case the values $u(t,x)$ on the boundary are not known (and in general different from zero). In a stochastic interpretation this boundary condition says that particles are reflected at the boundary. A Robin boundary condition connects the flux through the boundary to the concentration, $u_\nu = c\,u$, where c is a constant.

In parabolic equations like reaction diffusion equations, one can (and has to) prescribe at each boundary point the density or the flux (or a linear combination of these). In hyperbolic problems one can prescribe boundary conditions only at boundary points where characteristic directions are pointing inward. We will explain the phenomenon using the simplest possible example. Consider the equation for a simple wave $u_t + \gamma u_x = 0$ on the half-line $[0, \infty)$, with $\gamma > 0$. If we prescribe initial data $u(0, x) = u_0(x)$ for $x > 0$, then we can obtain the solution in the domain where $x > \gamma t$, namely $u(t, x) = u_0(x - \gamma t)$. There is no information on the solution in the wedge $0 < x < \gamma t$ unless we prescribe also boundary data $u(t, 0) = \varphi(t)$ for $t > 0$. Then the solution in the domain $0 < x < \gamma t$ is $u(t, x) = \varphi(\gamma t - x)$. Now consider the same problem on the interval $[0, l]$ with some $l > 0$. Then the solution for the half-line is also the solution on the interval. The solution is uniquely determined by the initial data and the data on the left-hand boundary. The values of the solution at $x = l$ are determined. We cannot prescribe any information at $x = l$. The wave transports the values of u_0 and φ along the characteristic curves $x - \gamma t = 0$ to the line $x = l$ and the arriving values would in general not match with the prescribed data.

Now consider the correlated random walk (1.2) on a bounded interval $[0, l]$. For u^+ we can impose a condition at $x = 0$, for u^- at $x = l$. The zero Dirichlet condition reads

$$u^+(t, 0) = 0, \quad u^-(t, l) = 0. \tag{3.3}$$

No particle can enter at $x = 0$. Existing u^+ particles turn or move further right. If they arrive at $x = l$ then they are removed. In this context "removed" means that these particles are not considered anymore, although they could be followed along the characteristic curve beyond $x = l$. But for $x > l$ the quantity u^+ has no meaning. In general $u^+(t, l)$ will be different from zero, in agreement with the interpretation that these particles have the velocity γ when they cross the boundary. The value $u^+(t, l)$ can be seen as the number of particles (per time) leaving the domain at $x = l$.

For the system (1.3) the zero Dirichlet condition reads

$$u(t, 0) + \frac{1}{\gamma} v(t, 0) = 0, \quad u(t, l) - \frac{1}{\gamma} v(t, l) = 0. \tag{3.4}$$

Thus the Dirichlet condition for the correlated random walk looks like a Robin condition. Only in the limit $\gamma \to \infty$ we get (formally) a Dirichlet condition.

The zero Neumann condition for the correlated random walk says that particles are reflected at the boundary. Thus u^+ particles arriving at $x = l$ become u^- particles and analogously at $x = 0$. Hence, the Neumann condition has the form

$$u^+(t, 0) = u^-(t, 0), \quad u^-(t, l) = u^+(t, l). \tag{3.5}$$

For the system (1.3) the Neumann condition reads

$$v(t, x) = 0 \quad \text{for} \quad x = 0, l. \tag{3.6}$$

It says that the flux across the boundary is zero.

Also in the case of the transport equation (1.16) we can impose boundary conditions only at ingoing characteristics. Again, the zero Dirichlet condition says that particles cannot enter,

$$u(t,x,s) = 0 \quad \text{for} \quad x \in \partial\Omega, \quad \nu^T s < 0, \tag{3.7}$$

where ν is the outer normal at x and $\nu^T s$ is the inner product between the outer normal and the velocity. The zero Neumann condition is a reflection law

$$u(t,x,\tilde{s}) = u(t,x,s) \quad \text{for} \quad x \in \partial\Omega, \quad \nu^T \tilde{s} = -\nu^T s. \tag{3.8}$$

Finally consider the Cattaneo system (1.26). The homogeneous Dirichlet condition says that no particles should enter the domain along a characteristic direction. Thus we get a Robin type condition

$$u(t,x) = \sqrt{\frac{T}{D}}\,\nu^T v(t,x) \quad \text{for} \quad x \in \partial\Omega. \tag{3.9}$$

The Neumann condition simply says that the flux through the boundary vanishes,

$$\nu^T v(t,x) = 0 \quad \text{for} \quad x \in \partial\Omega. \tag{3.10}$$

While the transition from the random walk system to the telegraph equation is rather transparent (except for some vector fields mapped into zero), the transition for the boundary conditions is rather involved. In particular for the Dirichlet condition one gets a time-dependent condition which is very different from the standard boundary conditions for the telegraph equation motivated by electromagnetic theory, see [38].

4 Linear initial boundary problems

Here we consider the initial value problems for the equations of section 1 in a naive sense, i.e., without paying attention to regularity or existence of solutions in appropriate function spaces. For the moment we consider only problems on the real line, i.e. space dimension 1. The central equation is the telegraph equation or damped wave equation. In many respects, this equation is "between" the diffusion equation and the standard wave equation. This is particularly obvious if we connect the equations by a parameter $\alpha \in [0,1]$,

$$\alpha u_{tt} + (1-\alpha)u_t = Du_{xx}. \tag{4.1}$$

For $\alpha = 0$ we have the diffusion equation, for $\alpha = 1$ the wave equation and for any $\alpha \in (0,1)$ a damped wave equation. Thus the telegraph equation arises from the diffusion equation by a singular perturbation leading from $\alpha = 0$ (parabolic) to $\alpha > 0$ (hyperbolic), whereas it becomes the wave equation if the damping vanishes.

For the diffusion equation (1.1) we need one initial datum. Thus the initial value problem reads

$$u_t = Du_{xx}, \quad u(0,x) = u_0(x). \tag{4.2}$$

The solution of this problem is given by Poisson's formula

$$u(t,x) = \frac{1}{2\sqrt{\pi D t}} \int_{-\infty}^{\infty} e^{-(x-y)^2/(4Dt)} u_0(y)dy. \tag{4.3}$$

The dissipative character of the equation becomes obvious from the behavior of certain integrals. For initial data which decay sufficiently rapidly for $x \to \pm\infty$ the total mass

$$m(t) = \int_{-\infty}^{+\infty} u(t,x)dx \tag{4.4}$$

and the Dirichlet integral

$$d(t) = \int_{-\infty}^{+\infty} u_x^2(t,x)dx \tag{4.5}$$

both exist. The total mass is constant in time and $d(t)$ is decreasing to zero.

For the standard wave equation, i.e. (4.1) with $\alpha = 1$, we need two initial data

$$u_{tt} = \gamma^2 u_{xx}, \quad u(0,x) = u_0(x), \quad u_t(0,x) = u_1(x). \tag{4.6}$$

In the standard interpretation of the vibrating string of infinite length u_0 and u_1 correspond to the initial deplacement and impulse. The solution of the initial value problem is represented by d'Alembert's formula

$$u(t,x) = \frac{1}{2}(u_0(x+\gamma t) + u_0(x-\gamma t)) + \frac{1}{2\gamma}\int_{x-\gamma t}^{x+\gamma t} u_1(s)ds. \tag{4.7}$$

The formula tells us that the solution at the point x and time t depends only on the initial data in the interval $[x-\gamma t, x+\gamma t]$. By the same argument, the initial data at the point x influence the solution at time t only at those points y, where $|y-x| \le \gamma t$. The equation has also an invariant of motion. For data which decay suffiently fast the "energy" integral

$$E(t) = \int_{-\infty}^{+\infty} [\gamma^2 u_t^2(t,x) + u_x^2(t,x)]dx \tag{4.8}$$

exists and remains constant in time.

Also, for the correlated random walk (1.2) with initial data $u^{\pm}(0,x) = u_0^{\pm}(x)$, there is a representation formula. The solution of the Cauchy problem is [38]

$$\begin{aligned}
u^+(t,x) = {} & u_0^+(x-\gamma t)e^{-(\mu/2)t} \\
& + \tfrac{\mu}{4\gamma}e^{-(\mu/2)t}\int_{x-\gamma t}^{x+\gamma t} \frac{I_0'\left(\frac{\mu}{2\gamma}\sqrt{\gamma^2t^2-(\xi-x)^2}\right)}{\sqrt{\gamma^2t^2-(\xi-x)^2}}[\gamma t-(\xi-x)]u_0^+(\xi)d\xi \\
& + \tfrac{\mu}{4\gamma}e^{-(\mu/2)t}\int_{x-\gamma t}^{x+\gamma t} I_0\left(\tfrac{\mu}{2\gamma}\sqrt{\gamma^2t^2-(\xi-x)^2}\right)u_0^-(\xi)d\xi \\
u^-(t,x) = {} & u_0^-(x+\gamma t)e^{-(\mu/2)t} \\
& + \tfrac{\mu}{4\gamma}e^{-(\mu/2)t}\int_{x-\gamma t}^{x+\gamma t} \frac{I_0'\left(\frac{\mu}{2\gamma}\sqrt{\gamma^2t^2-(\xi-x)^2}\right)}{\sqrt{\gamma^2t^2-(\xi-x)^2}}[\gamma t+(\xi-x)]u_0^-(\xi)d\xi \\
& + \tfrac{\mu}{4\gamma}e^{-(\mu/2)t}\int_{x-\gamma t}^{x+\gamma t} I_0\left(\tfrac{\mu}{2\gamma}\sqrt{\gamma^2t^2-(\xi-x)^2}\right)u_0^+(\xi)d\xi,
\end{aligned} \tag{4.9}$$

where I_0, I_0' are the modified Bessel function and its derivative. This formula is based on Poincaré's formula for the telegraph equation. In deriving this formula one starts with the representation of solutions of the three-dimensional wave equation by spherical means, then derives the formula for the two-dimensional wave equation by Hadamard's principle of descent, then, again descending in dimension, one gets a formula for the telegraph equation in dimension one. Finally, one arrives at (4.9) using the connection between the telegraph equation and the random walk system.

The formula can be seen as an intermediate between Poisson's and d'Alembert's formulae. The functions $I_0(x)$ and $I_0'(x)/x$ are both positive and even and they go to infinity for $|x| \to \infty$. The terms containing integrals are smooth and behave rather similar to the last term in (4.7). Since the integrals extend from $x - \gamma t$ to $x + \gamma t$, no signal can travel faster than with speed γ. The formula also shows the similarity with d'Alembert's formula (4.7). The initial data at x influence the solution in the whole range from $x - \gamma t$ to $x + \gamma t$. The formula shows that discontinuities in the initial data persist forever. The integral terms are smoothing, the terms $u_0^{\pm}(x \mp \gamma t)e^{-\mu t/2}$ show that the jumps are moving to $\pm\infty$ and are exponentially damped, but nevertheless persist.

Similar formulae can be derived for the Cattaneo system for $n \geq 2$. They do not look very useful. For the various boundary value problems representions of solutions can be obtained in terms of Fourier series or eigenfunction expansions, see [38]. For the transport equation (1.16), for the case of constant speed and uniform distribution of the new velocity, there are explicit formulae for the distribution of the position of a moving particle [95].

5 Random walk systems for several species

The "scalar" systems of Sections 2,3 can be generalized to the case of several dependent variables in various ways. Here we assume that particles move according to a correlated random walk. There are m types or species of particles. The densities of particles of the i-th species moving to the right or left are denoted by u_i^+, u_i^-, respectively. We introduce the column vectors $u^+ = (u_i^+)$, $u^- = (u_i^-)$ and the diagonal matrices of speeds and turning rates

$$\Gamma = (\gamma_i \delta_{ij}), \quad M = (\mu_i \delta_{ij}). \tag{5.1}$$

Thus we exclude "cross diffusion". Then the following system ([49], [51]) describes the motion of m species of particles

$$\begin{aligned} u_t^+ + \Gamma u_x^+ &= \tfrac{1}{2} M(u^- - u^+) \\ u_t^- - \Gamma u_x^- &= \tfrac{1}{2} M(u^+ - u^-). \end{aligned} \tag{5.2}$$

A reaction random walk system which allows for any interaction of the $2m$ types u_i^s has the form

$$\begin{aligned} u_t^+ + \Gamma u_x^+ &= \tfrac{1}{2} M(u^- - u^+) + \tfrac{1}{2} f_+(u^+, u^-) \\ u_t^- - \Gamma u_x^- &= \tfrac{1}{2} M(u^+ - u^-) + \tfrac{1}{2} f_-(u^+, u^-). \end{aligned} \tag{5.3}$$

In the isotropic case particles interact independently of their direction of motion. With $u = u^+ + u^-$ we have

$$\begin{aligned} u_t^+ + \Gamma u_x^+ &= \tfrac{1}{2}M(u^- - u^+) + \tfrac{1}{2}f(u) \\ u_t^- - \Gamma u_x^- &= \tfrac{1}{2}M(u^+ - u^-) + \tfrac{1}{2}f(u). \end{aligned} \tag{5.4}$$

The most general case with complete symmetry between directions is characterized by

$$f_-(u^+, u^-) = f_+(u^-, u^+). \tag{5.5}$$

In that case we can just write $f_+(u^+, u^-) = \tilde{f}(u^+, u^-)$ and we obtain the system

$$\begin{aligned} u_t^+ + \Gamma u_x^+ &= \tfrac{1}{2}M(u^- - u^+) + \tfrac{1}{2}\tilde{f}(u^+, u^-) \\ u_t^- - \Gamma u_x^- &= \tfrac{1}{2}M(u^+ - u^-) + \tfrac{1}{2}\tilde{f}(u^-, u^+). \end{aligned} \tag{5.6}$$

As in Section 3 we could introduce reactions which produce particles and others which remove particles. There are the many different ways in which such reactions can be realized.

We can rewrite the system (5.4) in the form

$$\begin{aligned} u_t + v_x &= f(u) \\ v_t + \Gamma^2 u_x &= -Mv. \end{aligned} \tag{5.7}$$

Then the u-component of any C^2 is a solution of the reaction telegraph system

$$u_{tt} + (M - f'(u))u_t = \Gamma^2 u_{xx} + Mf(u). \tag{5.8}$$

Similarly one can study vector-valued Cattaneo systems and related systems of reaction-telegraph equations in several space dimensions (see [51])

$$\begin{aligned} u_t + \operatorname{div} v &= f(u) \\ \tau v_t + D \operatorname{grad} u + v &= 0, \end{aligned} \tag{5.9}$$

$$\tau u_{tt} + (I - \tau f'(u))u_t = \operatorname{div} D \operatorname{grad} u + f(u). \tag{5.10}$$

Here u, v are vectors, D is a diagonal matrix with positive entries, f is a vector field, the operators grad and div act componentwise, $f'(u)$ is the derivative (Jacobian) of the function f. Of course multitype reaction transport systems can be formulated following the pattern of equation (2.11).

6 Local existence and spectral problems

For the solutions of the various initial value problems and initial boundary value problems local existence of the solutions can be proved in appropriate function spaces. As they can be expected, the problems of local existence, uniqueness, and regularity of solutions are rather closely connected to the corresponding problems for semilinear damped wave equations. Thus we refer, in a broad sense, to the monographs [43],

[99]. The Dirichlet and the Neumann problem for the Cattaneo system have been treated in [51]. Similarly, for existence of solutions of transport equations we refer to the monographs [11], [12]. Some work has still to be done to cover the problems which result from models in biology. Some approaches such as in [54], [5], [65], [31], [102] are sufficiently general to provide guidelines how this could be done.

Local existence of solutions is intimately connected to bounds for certain linear operators and, in particular, to properties of the spectrum. In the cases considered here the leading part of the spectrum consists of isolated eigenvalues. On the other hand these eigenvalues, or rather their real parts, determine the stability of stationary solutions. Typically the leading eigenvalues do not depend on the chosen space. Thus it makes sense to speak of eigenvalues in a formal way, without defining the function space, and get nevertheless useful information on stability and on bifurcation phenomena.

In the parabolic equation (2.1) with Dirichlet boundary conditions (3.1) we can linearize at the zero solution $u = 0$ to obtain

$$u_t = D\Delta u + au \quad \text{in}\, \Omega, \quad u(t,x) = 0 \quad \text{on}\, \partial\Omega \tag{6.1}$$

where we assume $a = f'(0) > 0$. We can obtain information about the stability of the zero solution from the spectral problem

$$D\Delta u + au = \lambda u \quad \text{in}\, \Omega, \quad u = 0 \quad \text{on}\, \partial\Omega. \tag{6.2}$$

In the one-dimensional case, $\Omega = [0, l]$, the eigenvalues are $a - \lambda_k$, $\lambda_k = k^2\pi^2/l^2$, $k = 1, 2, \ldots$. Thus the zero solution is stable if $a < \pi^2/l^2$ and unstable if $a > \pi^2/l^2$. For the corresponding Neumann problem the leading eigenvalue is just $\lambda = a$.

Now consider the problem (6.1) in several dimensions. Let λ_k be the k-th eigenvalue of the negative Laplacian. Then the zero solution is stable if $a - \lambda_1 < 0$ and unstable if $a > \lambda_1$. The physical interpretation is as follows. The term au describes production of matter at a rate a, the diffusion operator describes diffusion which eventually moves matter to the boundary where it may become absorbed in view of the zero Dirichlet condition. The zero solution is stable when the absorption on the boundary controls the production in the interior. Thus the zero solution loses its stability if the rate a is large or the domain is large in an appropriate sense.

Also in the hyperbolic case information on the stability of the zero solution can be gained from the spectrum of certain linear operators. If we linearize the system (2.10) at the zero solution then we obtain a linear system

$$\begin{aligned} u_t + v_x &= au \\ v_t + \gamma^2 u_x &= -bv \end{aligned} \tag{6.3}$$

where $a = f'(0)$, $b = \mu + g(0)$. We assume $a > 0$, $b > 0$. In the special case of (2.3) we have $b = \mu$. With zero Dirichlet boundary conditions (3.4) we get the spectral problem

$$\begin{aligned} \lambda u + v_x &= au \\ \lambda v + \gamma^2 u_x &= -bv, \end{aligned} \tag{6.4}$$

with boundary conditions (3.4). The solutions of the differential equation (6.4) satisfy

$$-\gamma^2 u_{xx} = (a-\lambda)(\lambda+b)u. \tag{6.5}$$

This problem has the characteristic equation (see [38])

$$\frac{\sinh(2\sqrt{(b+\lambda)(\lambda-a)}l/\gamma)}{2\sqrt{(b+\lambda)(\lambda-a)}l/\gamma}\cdot\frac{l}{\gamma}+\frac{2(b-a+2\lambda)}{(b+a)^2}=0. \tag{6.6}$$

Equation (6.6) has a simple scaling invariance. If b, γ, a are multiplied by the same factor then the eigenvalues are multiplied by this factor. In particular, in the case $b=\mu$, $a=0$, the characteristic equation can be written as an equation for λ/μ depending on a single parameter $\mu l/\gamma$. The left hand side of (6.6) is an entire analytic function. In [76] the existence of an analytic characteristic function has been proved for a wider class of problems.

In view of the positivity properties of the system (2.2)(3.3) the spectral bound s_0 (the supremum of the real part of any point of the spectrum) is itself an eigenvalue. Thus the zero solution is stable for $s_0 < 0$ and unstable for $s_0 > 0$. For given a, we have stability for sufficiently small values of l. There is a critical length l^* of the interval which for $b=\mu$ is given by the equation

$$\tan\left(\frac{\sqrt{\mu a}}{\gamma}l\right) = -\frac{2\sqrt{\mu a}}{\mu - a}. \tag{6.7}$$

The corresponding equation for the parabolic case is

$$\frac{l}{\gamma}=\frac{\pi}{\sqrt{\mu a}}. \tag{6.8}$$

In particular for small l, the two conditions are markedly different.

Now consider the case $b=\mu$, $a=0$ in greater detail. In the parabolic case the leading eigenvalue is $\lambda_1 = -\pi^2/l^2$. We put $\lambda=\mu\nu$ and we write the characteristic equation

$$\frac{1+\nu-\sqrt{(2+\nu)\nu}}{1+\nu-\sqrt{(2+\nu)\nu}} - e^{2\sqrt{((2+\nu)\nu}(\mu l/\gamma)} = 0. \tag{6.9}$$

One can show that the quotient $-\lambda_1/\mu$ is an increasing function of the compound parameter $\mu l/\gamma$. Certain values can be explictly computed: for $\mu l/\gamma \to 0$ we have $\lambda_1/\mu \to -\infty$, for $\mu l/\gamma = 1$ we get $\lambda_1/\mu = -2$, for $\mu l/\gamma = \pi/2$ we get $\lambda_1/\mu = -1$, and finally $\lambda_1/\mu \to 0$ for $\mu l/\gamma \to -\infty$.

The Neumann problem (2.4)(3.6) is considerably simpler. The differential equation (6.4) leads to (6.5) as before. Hence the eigenvalues satisfy

$$(\lambda-a)(\lambda+b) = -k^2\pi^2\gamma^2/l^2, \quad k=0,1,2,\ldots \tag{6.10}$$

Thus there is a leading eigenvalue, $\lambda = a$, possibly some further real eigenvalues, and an infinite family of complex eigenvalues with real parts $(a-b)/2$.

For the eigenvalue problems related to Cattaneo systems see [38]. Whereas spectral problems for the Cattaneo system are rather transparent due to the close connection to the Laplacian, the spectra of transport operators (in several dimensions) are much harder to describe in detail. Usually the existence of a leading eigenvalue (related to the Perron root of some positive operator) can be shown. The non-negative eigenfunction can be interpreted as a stationary distribution of a linear transport problem.

7 Invariance principles and Lyapunov functions

The diffusion equation (1.1) has some very important properties which reflect properties of the underlying physical models. These are preservation of positivity, conservation of mass and dissipativeness. The same is true for the damped wave equation (the telegraph equation) and thus for certain random walk and Cattaneo systems. The dissipative character of the system can be expressed in terms of certain quantities ("energy integrals") which decrease along time-dependent solutions (unless these are stationary). Many of these can be generalized to nonlinear equations and thus form a basis for the construction of Lyapunov functions for the corresponding dynamical systems.

First consider equation (1.1). Suppose $u_0(x)$ is a smooth initial datum on $\mathbb{R}^n$ which decays for $|x| \to \infty$. Let $u(t,x)$ be the corresponding solution. If $u_0(x) \geq 0$ for all x then $u(t,x) \geq 0$ for all x and $t \geq 0$. Furthermore

$$\int u\,dx = \text{const} \tag{7.1}$$

where the integral extends over $\mathbb{R}^n$. The dissipativeness follows from

$$\frac{d}{dt}\int u^2 dx = -2D\int (\nabla u)^2 dx \tag{7.2}$$

and

$$\frac{d}{dt}\int (\nabla u)^2 dx = -2D\int (\Delta u)^2 dx. \tag{7.3}$$

Thus both $\int u^2 dx$ and the Dirichlet integral decay with time.

Some of these properties carry over to correlated random walks, Cattaneo problems and transport equations. Some of these equalities and inequalities appear, in somewhat different form, in the literature on the the qualitative behavior of nonlinear damped wave equations, see, e.g., [43], [99]. Formulae for correlated random walks and Cattaneo systems have been found in [49], [51], based on the early work in [7].

First consider the correlated random walk system (1.2) or (1.3). Let $(u_0^+(x), u_0^-(x))$ be a nonnegative initial datum which decays rapidly for $|x| \to \infty$. Let $(u^+(t,x), u^-(t,x))$ be the corresponding solution. From the explicit representation (4.9) it follows that $u^+(t,x) \geq 0$, $u^-(t,x) \geq 0$ for all x and $t \geq 0$.

Furthermore preservation of mass holds

$$\int u\,dx = \int (u^+ + u^-)dx = \text{const}. \tag{7.4}$$

For the function v we find
$$\frac{d}{dt}\int v\,dx = -\mu\int v\,dx \tag{7.5}$$
which says that, on the average, the difference between u^+ and u^- decays exponentially.

Next we observe the important equality (which replaces (7.2) in some sense)
$$\frac{d}{dt}\int(\gamma^2u^2+v^2)\,dx = -2\mu\int v^2dx. \tag{7.6}$$

For smooth solutions also the functions (u_x, v_x) and (u_t, v_t) are solutions. Then (7.6) leads to the equations
$$\frac{d}{dt}\int(\gamma^2u_x^2+v_x^2)\,dx = -2\mu\int v_x^2\,dx = -2\mu\int u_t^2\,dx \tag{7.7}$$
and
$$\frac{d}{dt}\int(\gamma^2u_t^2+v_t^2)dx = -2\mu\int v_t^2\,dx. \tag{7.8}$$
Since
$$\gamma^2u_t^2+v_t^2 = \gamma^2(\gamma^2u_x^2+v_x^2)+2\mu\gamma^2u_xv+\mu^2v^2 \tag{7.9}$$
holds for any solution (u, v), equation (7.8) can also be written
$$\frac{d}{dt}\int[\gamma^2(\gamma^2u_x^2+v_x^2)+2\mu\gamma^2u_xv+\mu^2v^2]dx = -2\mu\int(\gamma^2u_x+\mu v)^2dx. \tag{7.10}$$
With (7.7) we also find
$$\frac{d}{dt}\mu\int(2\gamma^2u_xv+\mu v^2)dx = -2\mu\int(v_t^2-\gamma^2u_t^2)\,dx. \tag{7.11}$$
Combining (7.8) and (7.11) with $\kappa\in(0,\mu)$ gives
$$\frac{d}{dt}\int[(\gamma^2u_t^2+v_t^2)-\kappa(2\gamma^2u_xv+\mu v^2)]dx = -2\int(\kappa\gamma^2u_t^2+(\mu-\kappa)v_t^2)\,dx \tag{7.12}$$
or, equivalently,
$$\begin{aligned}&\tfrac{d}{dt}\textstyle\int[\gamma^2(\gamma^2u_x^2+v_x^2)+(\mu-\kappa)(2\gamma^2u_xv+\mu v^2)]dx\\ &= -2\textstyle\int(\kappa\gamma^2v_x^2+(\mu-\kappa)(\gamma^2u_x+\mu v)^2)dx.\end{aligned} \tag{7.13}$$
The integral on the left hand side is nonnegative, thus bounded below. The right hand side is nonpositive and thus tends to zero for $t\to\infty$.

The connection to the standard "energy" function for the (damped) wave equation is obvious from
$$\int(\gamma^2u_x^2+u_t^2)dx = \int(\gamma^2u_x^2+v_x^2)dx. \tag{7.14}$$
In view of (7.7) this energy is decaying which shows the dissipative character of the system.

For the Cattaneo system, $n \geq 2$, there is no obvious principle of positivity. Since the first equation in (1.26) is a conservation law, we have conservation of mass. The direct analogue of (7.5) is

$$\frac{d}{dt}\int v\,dx = -\frac{1}{\tau}\int v\,dx. \tag{7.15}$$

The equation (7.6) can be generalized to

$$\frac{d}{dt}\int (Du^2 + \tau v^2)dx = -2\int v^2 dx \tag{7.16}$$

where u^2 is just the square of u and $v^2 = v \cdot v$ is the inner product. Now one has to be somewhat careful to get the correct generalization of (7.7) by replacing u, v by derivatives. It is easier to consider the energy integral for the corresponding telegraph equation or damped wave equation (1.28) which immediately gives

$$\frac{d}{dt}\int (D(\nabla u)^2 + \tau u^2)dx = -2\int u_t^2 dx \tag{7.17}$$

again showing the dissipative character of the system.

The transport equation, by construction, preserves positivity and total mass. Its dissipative character is not so obvious, at least not for general kernels.

Now consider nonlinear problems. For scalar reaction diffusion equations (2.2) there are maximum, comparison and invariance principles. Assume $f(0) \geq 0$ and $f(1) \leq 0$. Then every initial datum $u_0(x)$ with values in $[0, 1]$ leads to a solution with values in $[0, 1]$.

In the isotropic reaction random walk system (2.3) preservation of positivity does not hold generally. Preservation of positivity holds if the turning rate μ is large compared to the derivative $f'(u)$. Thus preservation of positivity gets lost together with the dissipative character of the nonlinear system. For the more realistic random walk system (2.9) preservation of positivity holds under rather weak conditions, see [50].

Again, for the Cattaneo system (2.7), $n \geq 2$, nothing is known.

The conservation laws and related inequalities suggest how Lyapunov functions for nonlinear problems can be constructed. For the function f define the potential

$$F(u) = \int_0^u f(s)\,ds. \tag{7.18}$$

For the reaction diffusion equation (2.2) define

$$V(u) = \int (\tfrac{1}{2}D(\nabla u)^2 - F(u))\,dx. \tag{7.19}$$

Then

$$\frac{d}{dt}V(u) = -\int u_t^2 dx. \tag{7.20}$$

Thus, with boundedness assumptions on F, compact intervals and appropriate boundary conditions the function V will be a Lyapunov function.

It is not evident how one should proceed in the random walk case (2.3) or (2.4). As in (7.5), (7.6) we get

$$\frac{d}{dt}\int u\,dx = \int f(u)dx, \quad \frac{d}{dt}\int v\,dx = -\mu\int v dx,$$

$$\frac{d}{dt}\int(\gamma^2 u^2 + v^2)\,dx = -2\mu\int v^2 dx + 2\gamma^2\int u f(u)\,dx.$$

Instead of (7.7) we get

$$\frac{d}{dt}\int(\gamma^2 u_x^2 + v_x^2)dx = 2\int(\gamma^2 f'(u)u_x^2 - \mu v_x^2)dx \tag{7.21}$$

and

$$\frac{d}{dt}\int(\gamma^2 u_t^2 + v_t^2)dx = 2\int(\gamma^2 f'(u)u_t^2 - \mu v_t^2)dx. \tag{7.22}$$

A useful observation, compare (7.11), is [49]

$$\frac{d}{dt}\int(2\gamma^2 F(u) + \mu v^2 + 2\gamma^2 u_x v)dx = 2\int(\gamma^2 u_t^2 - v_t^2)dx. \tag{7.23}$$

As in (7.12) one can combine (7.22) and (7.23) with some $\kappa \in (0,\mu)$ to get

$$\begin{aligned}\tfrac{d}{dt}\textstyle\int[(\gamma^2 u_t^2 + v_t^2) \quad & -\kappa(2\gamma^2 F(u) + \mu v^2 + 2\gamma^2 u_x v)]dx \\ & = -2\textstyle\int((\kappa - f'(u))\gamma^2 u_t^2 + (\mu-\kappa)v_t^2)dx.\end{aligned} \tag{7.24}$$

To find similar expressions for the Cattaneo system (2.7) one can use (7.24) as a guideline. First compute

$$\begin{aligned}\tfrac{d}{dt}\textstyle\int[D(f(u) - \nabla v)^2 \quad & +\tfrac{1}{\tau}(D\nabla u + v)^2]dx \\ & = 2\textstyle\int[Du_t(f'(u)u_t - \nabla v_t) - v_t(D\nabla u_t + v_t)]dx \\ & = 2\textstyle\int(Df'(u)u_t^2 - v_t^2)dx - 2\int D(u_t\nabla v_t + v_t\nabla u_t)dx.\end{aligned} \tag{7.25}$$

The last integral vanishes. Then compute

$$\begin{aligned}\tfrac{d}{dt}\textstyle\int[2DF(u) + \quad & v^2 + 2D\nabla u\cdot v]dx \\ & = 2\textstyle\int[Df(u)u_t + vv_t + D\nabla u_t\cdot v + D\nabla u v_t]dx \\ & = 2\textstyle\int(Du_t^2 - \tau v_t^2)dx.\end{aligned} \tag{7.26}$$

Combining the two equations (7.25) and (7.26) with $\kappa \in (0, 1/\tau)$ gives [51]

$$\begin{aligned}\tfrac{d}{dt}\textstyle\int\{ \quad & [D(f(u) - \nabla v)^2 + \tfrac{1}{\tau}(D\nabla u + v)^2] - \kappa[2DF(u) + v^2 + 2D\nabla u\cdot v]\}dx \\ & = -2\textstyle\int[D(\kappa - f'(u))u_t^2 + (1-\kappa\tau)v_t^2]dx.\end{aligned} \tag{7.27}$$

The equations (7.24) and (7.27) show the dissipative character of random walk systems and of Cattaneo systems, respectively. If there is a $\kappa \in (0,\mu)$ (or $\kappa \in (0, 1/\tau)$ such that the function f satisfies inequality

$$f'(u) \leq \kappa \tag{7.28}$$

then the right hand sides of (7.24) and (7.27), respectively, are non-positive.

8 Qualitative behavior

Qualitative analysis is a difficult subject for ordinary differential equations, even more so for reaction diffusion equations and random walk systems. In many instants the behavior of the underlying ordinary differential equation gives some hint for what can be expected to hold for the partial differential equation. In general, however, the analogy will be misleading. Solutions of the reaction diffusion equation may blow up in finite time even if all solutions of the ordinary differential equation exist for all times [45], [87]. This happens frequently if the dimension of the independent space variable is increased above a certain threshold. This phenomenon shows, among others, that Brownian motion acts differently in different space dimensions.

There are several basic concepts in the theory of dynamical systems. Assume $\phi(t,u)$ is the flow of some dynamical system, i.e., $\phi(t,u)$ is the solution starting at time 0 with initial value $\phi(0,u) = u$. For a given initial data u the ω-limit set is the set (if it is non-empty)

$$\omega(u) = \{z : \ \exists t_k \to +\infty, \, \phi(t_k, u) \to z\}.$$

or, equivalently, $\omega(u) = \cap_{t\geq 0}\overline{\{\phi(\tau,u) : \tau \geq t\}}$. The ω-limit set gives information about the asymptotic behavior of a single trajectory: convergence to a stationary point, to a periodic orbit, or more complex behavior. Another concept is the global attractor. The global attractor is the smallest compact invariant set which attracts all bounded sets. If M is a bounded set in the state space which attracts all trajectories then under suitable compactness properties of the flow the set $\cap_{t\geq 0}\overline{\phi(t,M)}$ is the global attractor. The global attractor contains all limit sets, but also other trajectories, e.g. saddle-saddle connections.

The key tool in applying these concepts to systems of the form (2.3) or (2.7) is the equality (7.27) (or (7.24) in the case of the correlated random walk in one space dimension). The integral on the left hand side of (7.27) can be used as a Lyapunov function. Consider first the case of the homogeneous Neumann boundary condition (3.10). Suppose the function f has the property (7.28) with some $\kappa \in (0, /\tau)$. Then choose $\tilde{\kappa} \in (\kappa, 1/\tau)$ and form

$$\begin{aligned} L(u,v) = \ & \textstyle\int_\Omega \{(Df(u) - \nabla u)^2 + \frac{1}{\tau}(D\nabla u + v)^2 \\ & -\tilde{\kappa}(2DF(u) + (D\nabla u + v)^2 - (D\nabla u)^2)\} dx. \end{aligned} \tag{8.1}$$

Using (7.27) and the boundary condition one finds for smooth solutions

$$\begin{aligned} \tfrac{d}{dt} L(u,v) & = -2 \textstyle\int_\Omega [D(\tilde{\kappa} - f(u))u_t^2 + (1 - \tilde{\kappa}\tau)v_t^2] dx \\ & \leq -c \textstyle\int_\Omega [(f(u) - \nabla v)^2 + (D\nabla u + v)^2] dx \end{aligned} \tag{8.2}$$

with $c = 2\max\{\tilde{\kappa} - \kappa, \, 1 - \tau\tilde{\kappa}\}$. Hence the L_2 norms of $f(u) - \nabla u$ and of $D\nabla u + v$ decrease along solutions and thus go to zero. However, these expressions are u_t and $-v_t$.

For a Lyapunov function one needs a condition that ensures $L(u,v) \to +\infty$ for $\|(u,v)\| \to \infty$. Such a condition is

$$F(u) \to -\infty \quad \text{for} \quad |u| \to \infty. \tag{8.3}$$

If $n > 2$ then one needs a growth condition on $|f'(u)|$ to ensure the necessary compactness.

With these assumptions one can prove, in the space $L^2(\Omega)$, that the ω-limit set of each solution is a continuum of stationary points. If there are only finitely many stationary points, then every ω-limit set is a stationary point [51]. One can also show that the global attractor exists.

The Dirichlet boundary condition requires more attention [51]. In that case one uses

$$\tilde{L}(u,v) = L(u,v) + \kappa\sqrt{\tau D}\int_{\partial\Omega}(\nabla u \cdot v)^2 ds. \tag{8.4}$$

Then one proves the same results for the Dirichlet problem. The results carry over to random walk systems of the form (5.9) if the function f is a gradient field, i.e., if there is a scalar function $F : \mathbb{R}^m \to \mathbb{R}$ such that $f = F'$. The condition (7.28) assumes the form

$$\xi^T f'(u)\xi \leq \kappa \xi^T \xi \quad \text{for} \quad \xi \in \mathbb{R}^m. \tag{8.5}$$

With respect to biological models, the assumption of a gradient field is rather restrictive.

Again, such results are related to results for semilinear damped wave equations [43], [102], [99], [18], [104].

9 Stationary solutions

Stationary solutions of the Dirichlet problem (2.2), (3.1) on an interval $[0,l]$ are solutions of the boundary value problem for the ordinary differential equation

$$0 = D\ddot{u} + f(u), \tag{9.1}$$

$$u(0) = u(l) = 0. \tag{9.2}$$

The differential equation is equivalent to the first order system

$$\begin{aligned} \dot{u} &= -\tfrac{1}{D}v \\ \dot{v} &= f(u). \end{aligned} \tag{9.3}$$

This is a Hamiltonian system with Hamiltonian function

$$I(u,v) = \frac{1}{2D}v^2 + F(u), \tag{9.4}$$

where

$$F(u) = \int_0^u f(s)ds. \tag{9.5}$$

In the "hump" case, where f is positive for $0 < u < 1$, the stationary point $(0,0)$ is a center, and $(1,0)$ is a saddle point. There is an orbit homoclinic to the point $(1,0)$ which runs around the center. Solutions to the boundary value problem correspond to (pieces of) trajectories that connect the $\dot{v} = 0$ axis to itself. There is the trivial solution $u = 0$ and a one-parameter family of positive solutions parametrized by the amplitude $\bar{u} = u(l/2)$.

In general there is no one-to-one correspondence between the amplitude $\bar{u}$ and the interval length l. For $\bar{u} \to 1$ we have $l \to \infty$ (the trajectory takes a long time to pass through the neighborhood of the saddle point $(1,0)$), and for $\bar{u} \to 0$ we find $l \to l^*$ where

$$l^* = \frac{\pi}{\sqrt{Df'(0)}}. \tag{9.6}$$

There are countably many branches of nonpositive solutions corresponding to (pieces of) trajectories winding around $(0,0)$ several times. The global behavior of these branches depends on the values of the function f for $u < 0$ (that have not been specified here).

The qualitative behavior of the first branch, specifically the question under which condition on f the branch is monotone, has been investigated in [89].

In the case of the homogeneous Neumann problem we have the same Hamiltonian system (9.3), but this time we look for (pieces of) trajectories that connect the $v = 0$ axis to itself. If attention is restricted to nonnegative solutions then the only feasible solutions are $u = 0$ and $u = 1$.

From the invariant of motion (9.4) one finds that the trajectory (u, v) passing through (u_0, v_0) satisfies

$$v^2 = v_0^2 + 2D(F(u_0) - F(u)). \tag{9.7}$$

In particular, for $v_0 = 0$, $u = \bar{u}$,

$$-D\frac{du}{dt} = v = \sqrt{2D(F(\bar{u}) - F(u))}$$

and thus

$$l = \sqrt{2D}\int_0^{\bar{u}} \frac{du}{\sqrt{F(\bar{u}) - F(u)}}. \tag{9.8}$$

This formula gives the interval length explicitly as a function of the amplitude. Practical application is limited to cases where the integral can be evaluated explicitly.

Similar to the reaction diffusion case in one space dimension the problem of stationary solutions of reaction random walk systems leads to two-point boundary value problems. As an example we consider the Dirichlet problem (2.4),(3.4). Stationary solutions satisfy the differential system

$$\begin{aligned} \dot{v} &= f(u) \\ \gamma^2 \dot{u} &= -\mu v \end{aligned} \tag{9.9}$$

and the boundary condition

$$\gamma u(0) + v(0) = 0, \quad \gamma u(l) - v(l) = 0. \tag{9.10}$$

With $D = \gamma^2/\mu$ (9.9) becomes (9.3). Although the differential equation is the same as in the reaction diffusion case, there is a major difference. There we look for trajectories connecting the line $u = 0$ to itself. Here we want trajectories connecting the line $u^+ = 0$ to the line $u^- = 0$, i.e., trajectories connecting the line $\gamma u + v = 0$ to the line $\gamma u - v = 0$. Nonnegative solutions are of particular interest. There is a one-parameter family of such solutions. These are symmetric to the point $x = l/2$. This family can be parametrized by the maximum value $\bar{u} = u(l/2)$.

In the parabolic limit (1.6) the boundary condition (9.11) becomes the Dirichlet condition (9.2). For finite γ there is a marked difference in the profiles of the stationary solutions of the diffusion equation and of the random walk system. The stationary solution $u(x)$ of the diffusion equation vanishes at the boundary points $x = 0$ and $x = l$. This behavior reflects the interpretation that oncoming particles are absorbed at the boundary. For stationary solutions of the random walk system we have $u^+(0) = 0$ and $u^-(l) = 0$ but for non-trivial solutions the sum $u = u^+ + u^-$ is positive at the boundary. At $x = l$ we have $u^-(l) = 0$ but $u^+(l) > 0$ because the "absorbed" particles leave the domain with positive speed.

Stationary boundary value problems for the Cattaneo problem (2.7) can be reduced to elliptic boundary value problems for $\Delta u + f(u)$ with Robin type boundary conditions, see [38]. Not much is known about stationary solutions of general transport equations (2.11).

10 Travelling front solutions

The phenomenon of travelling fronts in the reaction-diffusion equation (2.2) has been found by Fisher, the existence of fronts and their stability properties have been investigated in the seminal paper [63]. Although Fisher [21] started from a population genetics setting, the phenomenon can be more easily explained in ecological terms.

Assume that the function f is of the "hump" type as defined in Section 2; a typical case is the Verhulst function situation $f(u) = u(1-u)$. Let $u(t, x)$ be the density of some population distributed on the real line. Consider the initial datum

$$u(0, x) = \begin{cases} 1, & x \leq 0 \\ 0, & x > 0. \end{cases} \tag{10.1}$$

Thus on the negative half-line the population has equilibrium density, the positive half-line is vacant. This situation could arise in the presence of a gate or lock in a long channel. At time $t = 0$ the lock is opened, and individuals spread from the "upper side" ($x < 0$) to the lower side. For a short time the evolution will be dominated by diffusion. The solution will look like a (reflected) normal distribution function. Then the reaction process will act at those space positions where the solution is far from equilibrium, i.e., where u is far from 0 and 1. It is evident that any population level between 0 and 1, e.g. the point where $u = 1/2$, moves to the right, and it is plausible that asymptotically a population front travels with constant shape and speed, i.e., $u(t, x)$ is asymptotic to a solution of the form $\phi(x - ct)$ with speed $c > 0$ where the shape or profile ϕ satisfies the conditions

$$\lim_{x \to -\infty} \phi(x) = 1, \quad \lim_{x \to +\infty} \phi(x) = 0, \tag{10.2}$$

$$0 < \phi(x) < 1. \tag{10.3}$$

Any shape of a front travelling with speed c satisfies the ordinary differential

$$-c\phi'(x-ct) = D\phi''(x-ct) + f(\phi(x-ct)). \tag{10.4}$$

Again we write t instead of $x - ct$ (and a dot for the derivative). With $u = \phi$, $v = \phi'$ the equation (10.4) is equivalent with the system

$$\begin{aligned} \dot{u} &= v \\ D\dot{v} &= -cv - f(u). \end{aligned} \tag{10.5}$$

The travelling front corresponds to a heteroclinic orbit of (10.5) connecting the stationary points (1, 0) and (0, 0) thereby respecting the side condition $0 < u < 1$. Here we see a general principle in the simplest (scalar) case: travelling fronts or pulses of partial differential equations are heteroclinic or homoclinic orbits of ordinary differential equations.

The stationary point (1, 0) is a saddle point for all $c > 0$, the point (0,0) is a focus for $0 < c < c^* = 2\sqrt{Df'(0)}$ and a node for $c > c^*$. One can show that there is a $c_0 \geq c^*$ such that for every $c \geq c_0$ there is a heteroclinic orbit with (10.3), and there are no such orbits for $c < c_0$. If $c_0 = c^*$ then the minimal speed is entirely determined by the behavior of f near $u = 0$. A sufficient (though not necessary) condition for $c_0 = c^*$ to hold is the "subtangential inequality" $f(u) \leq f'(0)u$. Similar results can be obtained for functions f of the "cubic" type. The main differences are that there is a unique speed of a travelling front corresponding to a saddle-saddle connection in the system (10.5). These problems have been studied by many authors, see [3], [20], [42].

We have seen that there are two hyperbolic analogues of Fisher's equation, namely the systems (2.3) and (2.8) with $f(u) = m(u)u - g(u)u$. In both cases one can ask for solutions of the form

$$u^s(t,x) = u^s(x-ct), \quad s = \pm \tag{10.6}$$

with boundary conditions

$$u^s(-\infty\ ) = \tfrac{1}{2}, \quad u^s(+\infty) = 0, \quad s = \pm \tag{10.7}$$

$$0 \leq u^s, \quad u^+ + u^- \leq 1. \tag{10.8}$$

This question leads to a system of ordinary differential equations

$$\begin{aligned} (\gamma - c)\dot{u}^+ &= \mu(u^- - u^+) + \tfrac{1}{2}f(u) \\ -(\gamma + c)\dot{u}^- &= \mu(u^+ - u^-) + \tfrac{1}{2}f(u) \end{aligned} \tag{10.9}$$

and similarly for the system (2.8). The most efficient way to treat this problem is a reduction to the parabolic case. This reduction has been proposed in [33], [34], a very short proof is given in [40]. Also in this case there is a value c^*, depending on the behavior of f and g near $u = 0$, such that no fronts exist for speeds below c^*. There is a $c_0 \in [c^*, \gamma)$ such that there is a travelling front with speed c for every $c \in [c_0, \gamma)$. In general $c_0 > c^*$. There is no direct analogue of the subtangential condition but one can show that $c_0 = c^*$ whenever the function f is concave, see [40] for a detailed discussion also of the parabolic case. Similar results can be obtained for systems (2.3) and (2.8) where the function f is of the "cubic" type, see [40].

We have seen that in the one-dimensional case the minimal speed of travelling fronts depends on the nonlinearity f in the following way. There is a number $c^* > 0$ which is the minimum of all values c for which the stable manifold of the stationary point $(0,0)$ of the system (10.5) is non-oscillatory, and there is a number $c_0 \geq c^*$ which is the minimum of all c for which travelling fronts exist. The number c^* can be obtained from a local stability analysis, i.e., it depends only on the linearization at $u = 0$, whereas c_0 depends on the global behavior of the function f. If f has certain concavity properties then we have $c_0 = c^*$. In other words, if we consider the class of functions f with $f'(0)$ fixed, then the number c^* is a lower bound for the minimal speed for any function f in that class, and this bound will be assumed for a large subset of that class.

In the reaction diffusion equation and in the Cattaneo system travelling fronts in dimension 1 immediately give rise to fronts in dimension $n \geq 1$. We have just to add dummy variables for the remaining $n-1$ coordinates to get a plane wave front in dimension n. For reaction transport equations the situation is entirely different. Even for highly symmetric kernels K a front in dimension n cannot be interpreted as a front in dimension $n+1$. This fact justifies to compute, for given $f'(0)$, the number c^* for arbitrary dimension n. In [90] the system with constant particle speed has been studied,

$$u_t + s \cdot \nabla u + \mu(u - \frac{1}{|V|}\bar{u}) = \frac{1}{|V|} f(\bar{u}) \tag{10.10}$$

where $u = u(t,x,s)$, $s \in V = \gamma S^{n-1} \subset \mathbb{R}^n$, and

$$\bar{u} = \int_V u(\cdot, s) ds. \tag{10.11}$$

Asking for a travelling front (in the x_1 direction)

$$u(t,x,s) = w(x_1 - ct, s) \tag{10.12}$$

with $0 < c < \gamma$, we get the equation (writing again t)

$$-(c - s_1)w_t + \mu w - \frac{\mu}{|V|}\bar{w} = \frac{1}{|V|} f(\bar{w}). \tag{10.13}$$

Then we linearize at $w = 0$ and introduce $w(s)e^{\lambda t}$,

$$((-c + s_1)\lambda + \mu)w = \frac{\mu + f'(0)}{|V|}\bar{w}. \tag{10.14}$$

Thus

$$w = \frac{\mu + f'(0)}{|V|}\bar{w}\frac{1}{(-c + s_1)\lambda + \mu}. \tag{10.15}$$

Integrating over s we get a "characteristic equation"

$$\frac{1}{\mu + f'(0)} = \frac{1}{|V|} \int_V \frac{ds}{\mu - c\lambda + \lambda s_1}. \tag{10.16}$$

We define c_n^* as the minimal number $c \in (0 < \gamma)$ such that (10.16) has only real roots $\lambda > 0$. With some effort one can prove the following result.

Theorem([90]): *Given $\gamma > 0$, $\mu > 0$ and $a = f'(0) > 0$, the number c_n^* is uniquely defined. The function $n \to c_n^*$ is strictly decreasing. For small n one finds*

$$c_1^* = 2\sqrt{\frac{\gamma^2}{\mu} a}\left(1 + \frac{a}{\mu}\right)^{-1} \tag{10.17}$$

$$c_2^* = 2\sqrt{\frac{\gamma^2}{2\mu} a}\left(1 + \frac{a}{2\mu}\right)^{1/2}\left(1 + \frac{a}{\mu}\right)^{-1}. \tag{10.18}$$

Expressions for c_4^*, c_6^* are also known [90].

11 Epidemic spread

The model for the advance of advantageous genes of R.A. Fisher and the mathematical investigation by Kolmogorov, Petrovskij and Piskunov as well as the Kermack-McKendrick [62] model (SIR model)

$$\begin{aligned} \dot{u} &= -\beta uv \\ \dot{v} &= \beta uv - \alpha v \\ \dot{w} &= \alpha v \end{aligned} \tag{11.1}$$

for an outbreak of an epidemic have been known since 1937. In 1965 both ideas were taken together into a model for the spread of an infectious disease in a spatially distributed population by D.Kendall [61]. Later other models for spatial spread have been designed that resemble Kendall's model superficially. In fact, the various approaches are rather distinct from a modeling point of view.

The system (11.1) describes the interaction of susceptibles u and infecteds v and the transition to the recovered (immune) state w. If one introduces diffusion into the SIR model one gets a system of three coupled reaction diffusion equations

$$\begin{aligned} u_t &= -\beta uv + D_u u_{xx} \\ v_t &= \beta uv - \alpha v + D_v v_{xx} \\ w_t &= \alpha v + D_w w_{xx}. \end{aligned} \tag{11.2}$$

The first two equations can be considered separately. Not much seems to be known about the qualitative behavior of this system. The simplest interpretation of the diffusion terms is a random motion of the three types with possibly different diffusion rates.

Kendall assumed another view. He suggested to consider the infective force to which a susceptible individual is exposed. This force is an integral over the spatially distributed population. Thus the kernel in this integral measures the contacts of a susceptible at position x with an infected at position y. Then the first two equations of the system (11.1) are replaced by

$$\begin{aligned} u_t &= -\beta u \int k(x,y)v(t,y)dy \\ v_t &= \beta u \int k(x,y)v(t,y)dy - \alpha v. \end{aligned} \tag{11.3}$$

Typically one will assume that k is a convolution kernel, $k(x,y) = k(x-y)$, $k(x) \geq 0$, $\int_{-\infty}^{\infty} k(x)dx = 1$, $k(x) = k(-x)$. Kendall replaced, in the way of a formal Taylor expansion (cf. (1.10)), the integral operator by a diffusion operator,

$$\int k(y)v(x+y)dx = v(x) + \sigma^2 v_{xx}(x) + \dots \tag{11.4}$$

Thus he arrived at the system

$$\begin{aligned} u_t &= -\beta u(v + \sigma^2 v_{xx}) \\ v_t &= \beta u(v + \sigma^2 v_{xx}) - \alpha v \ . \end{aligned} \tag{11.5}$$

Integral equations of the form (11.3) have been studied in detail in [2] and [14]. The transition from (11.3) to (11.5) introduces Brownian motion into the epidemic model. It is not obvious whether in this system it makes sense to replace Brownian motion by a correlated random walk.

On the other hand, one can consider the system (11.1) with only the infecteds spreading, thinking of infectivity spreading rather than persons, or of high migration rates induced by an infection process like in rabies. Then one obtains a degenerate parabolic system

$$\begin{aligned} u_t &= -\beta uv \\ v_t &= \beta uv - \alpha v + D v_{xx} \end{aligned} \tag{11.6}$$

Here it does make sense to study the same problem with a correlated random walk instead of Brownian motion, i.e., the system

$$\begin{aligned} u_t &= -\beta u(v^+ + v^-) \\ v_t^+ + \gamma v_x^+ &= \tfrac{1}{2}\mu(v^- - v^+) - \alpha v^+ + \tfrac{1}{2}\beta u(v^+ + v^-) \\ v_t^- - \gamma v_x^- &= \tfrac{1}{2}\mu(v^+ - v^-) - \alpha v^- + \tfrac{1}{2}\beta u(v^+ + v^-). \end{aligned} \tag{11.7}$$

In terms of the variables u, $v = v^+ + v^-$, $w = \gamma(v^+ - v^-)$ the system assumes the form

$$\begin{aligned} u_t &= -\beta uv \\ v_t + w_x &= \beta uv - \alpha v \\ w_t + \gamma^2 v_x &= -\mu w - \alpha w. \end{aligned} \tag{11.8}$$

Susceptibles u do not move, infected individuals v move according to a correlated random walk, the superscript $\pm$ indicating the direction of motion. Newly infected individuals choose the direction of motion according to a given distribution.

Similar to the travelling fronts in section 10 describing the advance of individuals or traits into a vacant territory, one can study travelling epidemic fronts which describe the advance of an epidemic into a population of susceptibles. One can normalize the problem in several ways. We take the following view. The first two equations of (11.1), seen as a system for u and v, have an invariant of motion. A trajectory connecting a stationary point $(\bar{u}, 0)$ with $\bar{u} > \alpha/\beta$ to another stationary $(\underline{u}, 0)$ with $\underline{u} < \alpha/\beta$ describes an epidemic which passes through the population thereby reducing the initial level $\bar{u}$ to the final level $\underline{u}$. The numbers $\bar{u}$ and $\underline{u}$ are connected by the equation $\bar{u} - (\alpha/\beta)\log\bar{u} = \underline{u} - (\alpha/\beta)\log\underline{u}$. We imagine the spread of an epidemic in a population distributed on the real line. At $+\infty$ the population is at the initial state, at $-\infty$ it is in the final state. Thus we look for travelling fronts with boundary conditions $u(t, +\infty) = \bar{u}$, $u(t, -\infty) = \underline{u}$, $v(t, \pm\infty) = 0$.

This problem can be studied for each of the systems (11.5), (11.6), and (11.7). It turns out that for given $\bar{u}$ (and correspondingly $\underline{u}$) there is an interval of possible speeds, where the minimal speed c_0 depends on α, β and $\bar{u}$. In the parabolic problems (11.5) and (11.6) the speeds can become arbitrarily large, in the hyperbolic case γ is the supremum of speeds.

Kendall [61] solved the problem for (11.5). In [2] it has been shown that the integral equation (11.3) has travelling front solutions and a "characteristic equation" has been derived for the minimal speed. For the parabolic problem travelling fronts have been shown by Källén [57] (see also [35]). The minimal speed is $c_0 = 2\sqrt{D(\beta\bar{u} - \alpha)}$. For the corresponding hyperbolic system (11.7) the exact conditions have been obtained in [35]. For given $\bar{u}$, the minimal speed is given by

$$c_0 = 2\frac{\sqrt{\mu + \alpha}}{\mu + \beta\bar{u}}\gamma\sqrt{\beta\bar{u} - \alpha}. \tag{11.9}$$

The general problem (11.2), where all types diffuse, seems rather difficult, even more so the corresponding hyperbolic problem. In [39] the somewhat artificial case has been studied where susceptibles can migrate and infecteds become sedentary.

More realistic models assume the form of epidemic transport systems

$$\begin{aligned} u_t(t,x,s) + s u_x(t,x,s) &= \mu \int_S K(s,\tilde{s}) u(t,x,\tilde{s}) d\tilde{s} - \mu u(t,x,s) \\ &\quad -\beta u(t,x,s) \int_S L(s,\tilde{s}) v(t,x,\tilde{s}) d\tilde{s} \\ v_t(t,x,s) + s v_x(t,x,s) &= \tilde{\mu} \int_S \tilde{K}(s,\tilde{s}) v(t,x,\tilde{s}) d\tilde{s} - \mu v(t,x,s) \\ &\quad +\beta u(t,x,s) \int_S L(s,\tilde{s}) v(t,x,\tilde{s}) d\tilde{s} - \alpha v(t,x,s). \end{aligned} \tag{11.10}$$

For this system existence of solutions on compact domain has been shown by a Kaniel-Shinbrot approach in [41].

12 Chemotaxis and aggregation

The classical model for chemotaxis has been designed as an aggregation model for slime mold amoebae by Patlak [82], Keller and Segel [59], [60]. Similar models describe bacteria attracted by a substrate. Although the views on the nature of the messenger molecule in slime molds have changed, the model has stayed valid. In fact this model is the nucleus of many systems describing chemotaxis or aggregation or it appears as a simplified version (diffusion approximation) of such systems.

The model consists of two coupled reaction diffusion equations

$$\begin{aligned} u_t &= -\mathrm{div}\,(D_1 \mathrm{grad}\, v) + \mathrm{div}\,(D_2 \mathrm{grad}\, u) \\ v_t &= g(u,v) + D_v \Delta v \end{aligned} \tag{12.1}$$

for amoebae or bacteria u and messenger or substrate v. A particular feature of the system is the triangular form of the diffusion operator. The species u diffuses freely with a diffusion rate $D_2 = D_2(u,v)$. The substance v diffuses with constant rate D_v. The species u is also driven by the gradient of v where again the rate $D_1 = D_1(u,v)$ may depend on the densities of both species. The function $g(u,v)$ describes production and depletion of the substrate v. Depending on the biological context, the substance v will be produced in an autonomous fashion (substrate) or by the species u (messenger), it will be either washed out or will be consumed by the species u.

Keller and Segel [59] and Lauffenburger [66] use $g(u,v) = f(v)u - k(v)v$, i.e., messenger is produced by amoebae and then removed. Rascle and Ziti [85] use $g(u,v) = -k(v)u$ and $D_1(u,v) = u\chi(v)$. Stevens [96] discusses the system with constant D_2, $D_1 = \chi \cdot u/v$, $D_v = 0$, $g(u,v) = \lambda u$ (and also λuv) as a model for myxobacteria which can aggregate by following slime trails. Stevens and Othmer [80] derive the system from discrete movements of organisms and diffusion laws for substances.

Depending on the nonlinearities and choices of parameters, several phenomena have been observed. Keller and Segel found destabilization of the spatially homogeneous state, others found travelling humps on a ring or fronts travelling on $\mathbb{R}$ connecting an unexploited state to a depleted state. Some of these systems show blow up of solutions in finite time. Whereas, in general, blow up of solutions can be considered as an effect of inappropriate modeling, in aggregation models blow up to a peak is a desired phenomenon. In slime mold models the appearance of a peak mimics the onset of fructification.

Real slime molds release the signalling substance cAMP by an excitable system in an oscillatory fashion. Thus one needs at least two variables to describe cAMP production. Tang and Othmer [97] have designed an excitable system of eight (then reduced to five) ordinary differential equations closely adapted to experimental data. The model shows peaked oscillations. Tang and Othmer then assume immovable cells at constant density and cAMP diffusing with constant rate. The model of four ordinary differential equations and one diffusion equation (somewhat similar to the Hodgkin-Huxley system) shows travelling waves, and in two dimensions spiral waves have been exhibited numerically. The modeling assumptions and the exhibited features are in accordance with the experimental situation at the onset of aggregation. It seems that the connection to the Keller-Segel model has not been elaborated in this case.

Since the right hand side of (12.1) has divergence form it is immediate that the system preserves total mass, i.e., $\int u dx$ remains constant (if it exists). For realistic coefficients, e.g. $D_1 = u\tilde{D}_1$, the system also preserves positivity of u. In [67] families of positive solutions have been constructed which show blow-up in finite time. In [46],

[75], [91] the blowup phenomenon for equation (12.1) has been studied in detail for radial solutions in two space dimensions.

In many cases the movements of real bacteria are not merely driven by diffusion. Moving bacteria either adhere to surfaces or they are freely swimming. In particular flagellate bacteria swim actively, thereby following gradients of substances. In view of their small size they cannot measure concentration differences in space but only in time. Thus some bacteria have developed a run-and-tumble strategy where the path in space is approximately described by a correlated random walk in three-dimensional space. Typically the speed is more or less constant whereas the turning frequency depends on external stimuli. The turning frequency is reduced if the bacterium moves in the desired direction, i.e., towards higher or lower concentrations depending on whether the substrate is attracting or repelling. These mechanisms have been investigated experimentally by Berg, Brown, Koshland and others, see [6], [69], and [1] for a detailed review. Thus the connection between chemotaxis, correlated random walks and hyperbolic systems is obvious. Most authors have used diffusion approximations rather than the hyperbolic systems. Diffusion equations have been justified by [1], [81].

Following [86] and [52] we discuss the approach in terms of correlated random walks in some detail. As before let u^+ and u^- be the particles going right or left. Within a chemotaxis model incorporating the substrate or messenger the turning rates μ^+, μ^- and the speed γ will depend on the densities and the gradients. For the discussion of the diffusion approximation it is sufficient to assume that these depend directly on space and time. Then the functions u^+ and u^- are governed by the system

$$\begin{aligned} u_t^+ + (\gamma u^+)_x &= \tfrac{1}{2}(\mu^- u^- - \mu^+ u^+) \\ u_t^- - (\gamma u^-)_x &= \tfrac{1}{2}(\mu^+ u^+ - \mu^- u^-). \end{aligned} \tag{12.2}$$

We introduce the total particle density $u = u^+ + u^-$ and the flux $v = \gamma(u^+ - u^-)$ and also parameters

$$\bar{\mu} = \tfrac{1}{2}(\mu^+ + \mu^-), \quad \delta = \tfrac{1}{2}(\mu^+ - \mu^-). \tag{12.3}$$

These variables are connected by a conservation law and a law which describes the change of the flux,

$$\begin{aligned} u_t + v_x &= 0 \\ v_t + \gamma(\gamma u)_x - (\gamma_t/\gamma)v &= -\bar{\mu}v - \delta\gamma u. \end{aligned} \tag{12.4}$$

A meaningful diffusion approximation requires careful inspection of the parameters. One should not just put $v_t = 0$. As in section 1 we assume that the speed and the turning frequency are both large. We assume that they contain a small positive parameter ϵ. We incorporate this parameter by replacing γ by γ/ϵ, $\bar{\mu}$ by $\bar{\mu}/\epsilon^2$ and δ by δ/ϵ. In biological terms: The speed and the two turning rates both depend on space and time. The speed and the average turning rate are large, the deviations of the turning rates from the average are not quite so large. With these assumptions (12.4) becomes

$$\begin{aligned} u_t + v_x &= 0 \\ \epsilon^2 v_t - \epsilon^2(\gamma_t/\gamma)v + \gamma(\gamma u)_x &= -\bar{\mu}v - \delta\gamma u. \end{aligned} \tag{12.5}$$

Now we put $\epsilon = 0$, solve for v and insert the expression for v into the conservation law. Then the variable u satisfies the parabolic equation

$$u_t = \frac{\partial}{\partial x}\left(\frac{\gamma(\gamma u)_x + \delta\gamma u}{\bar{\mu}}\right). \tag{12.6}$$

This equation can be written in a more transparent form as

$$u_t = (Du_x)_x - (qu)_x \tag{12.7}$$

with

$$D = \frac{\gamma^2}{\bar{\mu}}, \quad q = -\frac{\gamma}{\bar{\mu}}(\delta + \gamma_x). \tag{12.8}$$

Thus the asymmetry in the turning frequency leads to a drift in the diffusion approximation driving the u species in the desired direction. It is not necessary to assume differential speed or the existence of preferred directions at the turning events.

13 Turing phenomenon and pattern formation

In 1951 A.M.Turing [100] proposed a mechanism for morphogenesis which explains the formation of regular patterns in a homogeneous structure. The model is widely used in connection with the segmentation of arthropodes, skin patterns in snakes or mammals, colour patterns on marine snails and mussels, etc. (see the Maini lectures in this volume). The same mechanism is used in physics to explain certain patterns on surfaces on a submicroscopic scale.

Whereas Turing used a linear system, his idea has been formulated several times in terms of nonlinear reaction-diffusion equations, e.g. in the early papers of the Brussels school on stripe patterns and later spiral pattern in chemical reactions (Prigogine, Nicolis, Lefever and others, see [28]) and in the papers by Gierer and Meinhardt on the morphogenesis of *Hydra* ([27]). At present there is a wide variety of models and theories based on Turing's idea.

At first glance Turing's idea is surprising. Consider a reaction with two reactants converging to a stationary point. Reaction and diffusion, each by itself, lead to a stationary situation. The reaction alone, i.e., the reaction in a well-stirred reaction vessel, will lead to an equilibrium according to the mass action law. The diffusion process for possibly different types of nonreacting particles leads to spatially constant distributions of all particle types. Hence one would perhaps expect that, if reaction and diffusion act simultaneously, the spatially homogeneous chemical equilibrium is stable. Turing showed for a simple mathematical model that this expectation is unfounded, and his finding is supported by experimental evidence. If reaction and diffusion work simultaneously then the spatially homogeneous chemical equilibrium may become unstable, and a new, spatially inhomogeneous stable equilibrium arises.

In hindsight Turing's discovery is not so surprising. If two processes which both converge to equilibrium, act simultaneously, then one can exclude instability only if these processes commute, i.e., if the evolution of the joint process is independent of the order in which the individual processes are applied. The simplest example is the sum of two matrices A and B which have all their eigenvalues in the left half-plane (of the complex plane). The sum $A+B$ may nevertheless have an eigenvalue in the right half-plane. In other words, if all solutions of the differential equations $\dot{x} = Ax$, $\dot{y} = By$ go to zero, the differential equation $\dot{z} = (A+B)z$ may have unbounded solutions. However, this cannot happen if $AB = BA$. This observation on matrices is the key to Turing's theory.

Consider the reaction-diffusion equation

$$u_t = Du_{xx} + f(u). \tag{13.1}$$

Here the function $f : \mathbb{R}^m \to \mathbb{R}^m$ describes the reaction of m reactants according to the equation $\dot{u} = f(u)$, and $D = (d_j \delta_{jk})$ is a diagonal matrix with positive diffusion coefficients d_j. We consider this equation on an interval $[0, l]$ with zero Neumann boundary conditions (no-flux conditions)

$$u_x(t,0) = u_x(t,l) = 0. \tag{13.2}$$

Thus the reaction vessel is one-dimensional (think of a capillary), and no substance can leave or enter through the ends. Suppose $\bar{u} \in \mathbb{R}^m$ is a stationary point of the reaction, i.e., $f(\bar{u}) = 0$. Then $u(t,x) = \bar{u}$ is a spatially constant stationary solution of the equations (13.1) (13.2). In order to investigate its stability we linearize (13.1) (13.2), i.e., we introduce $u(t,x) = \bar{u} + v(t,x)$, expand according to Taylor's formula and neglect terms of higher order in v. From $\bar{u}_t + v_t = Du_{xx} + Dv_{xx} + f(\bar{u}) + f'(\bar{u})v + \cdots$ we obtain

$$v_t = Dv_{xx} + Av, \tag{13.3}$$

where $A = f'(\bar{u})$ is the Jacobian of f at $\bar{u}$. The function v satisfies the boundary conditions

$$v_x(t,0) = v_x(t,l) = 0. \tag{13.4}$$

Next we discuss the stability of the zero solution of the problem (13.3) (13.4). We enter the problem with a Fourier expansion of the function v,

$$v(t,x) = \sum_{k=0}^{\infty} \hat{u}_k e^{ik\pi x/l + \lambda_k t} \tag{13.5}$$

with coefficients $\hat{u}_k$. The real part of this expansion corresponds to a solution expanded into cosine functions satisfying the boundary conditions. We introduce (13.5) into (13.3), compare terms with the same k and obtain

$$\lambda_k \hat{u}_k = A\hat{u}_k - \frac{k^2\pi^2}{l^2} D\hat{u}_k, \; k = 0, 1, 2, \ldots \tag{13.6}$$

These are countably many equations in $\mathbb{R}^m$, one for each "frequency" or "mode" k. We put

$$\mu = \frac{k^2\pi^2}{l^2} \geq 0 \tag{13.7}$$

and omit the index k. Then we obtain the matrix eigenvalue problem

$$(A - \mu D)\hat{u} = \lambda \hat{u}. \tag{13.8}$$

For each value of the parameter μ the matrix eigenvalue problem (13.8) has m eigenvalues λ. The solution $v \equiv 0$ is stable if for μ as in (13.7) all these eigenvalues are located in the left half-plane. By assumption the eigenvalues of the matrix A have negative real part, and the diagonal elements of D are positive. In the case where all diffusion rates d_j are equal, i.e., if D is a multiple of the identity matrix, all eigenvalues of $A - \mu D$ have negative real part. This need not be so for more general diffusion matrices D. Now we proceed as follows. First we consider the matrix equation for a general $\mu \geq 0$, in particular we try to characterize those μ for which λ with positive real part occur. Then we check whether these μ can be interpreted as in (13.7).

This program can be completely carried through only for $m = 2$. Thus let $m = 2$ and let A, D be given as

$$A = \begin{pmatrix} a_{11} & a_{12} \\ a_{21} & a_{22} \end{pmatrix}, \quad D = \begin{pmatrix} d_1 & 0 \\ 0 & d_2 \end{pmatrix}. \tag{13.9}$$

Assume the matrix A is stable (a matrix is called stable, if all its eigenvalues have strictly negative real part), i.e.,

$$\operatorname{tr} A = a_{11} + a_{22} < 0, \quad \det A = a_{11}a_{22} - a_{12}a_{21} > 0. \tag{13.10}$$

Assume d_1, $d_2 > 0$. Then the inequalities

$$a_{11} - \mu d_1 + a_{22} - \mu d_2 < 0, \tag{13.11}$$

$$(a_{11} - \mu d_1)(a_{22} - \mu d_2) - a_{12}a_{21} > 0 \tag{13.12}$$

are the conditions for the stability of the matrix $A - \mu D$. The first condition is always satisfied. The second condition can be reformulated as

$$P(\mu) \equiv \mu^2 - T\mu - S > 0, \tag{13.13}$$

$$T = \frac{d_1 a_{22} + d_2 a_{11}}{d_1 d_2}, \quad S = \frac{a_{11}a_{22} - a_{12}a_{21}}{d_1 d_2}. \tag{13.14}$$

Here $S > 0$, while nothing is known about the sign of T. We distinguish several cases.
Case 1: $T \leq 0$. Then condition (13.12) is valid for all $\mu \geq 0$. For such a choice of A and D the zero solution (13.1) (13.2) is stable for any choice of l.
Case 2: $T > 0$. Then $P(0) > 0$, $P'(0) < 0$. The zeros of P are either real and positive or complex conjugate.
Case 2a: $T^2 < 4S$. Then P has no real roots. Hence condition (13.12) is satisfied for all $\mu \geq 0$. The stability properties are the same as in Case 1.
Case 2b: $T^2 > 4S$. Then there is an interval (μ_1, μ_2) with $\mu_2 > \mu_1 > 0$, such that $P(\mu) > 0$ for μ in this interval, and $P(\mu) \leq 0$ otherwise. For every μ from this interval the matrix $A - \mu D$ has an eigenvalue with positive real part.

We determine the conditions on the matrix elements for Case 2b to happen. Assume $T > 0$ and $T^2 > 4S$, i.e.

$$d_1 a_{22} + d_2 a_{11} > 0, \tag{13.15}$$

$$(d_1 a_{22} + d_2 \ \ a_{11})^2 > 4 d_1 d_2 (a_{11} a_{22} - a_{12} a_{21}). \tag{13.16}$$

If $a_{11} < 0$, $a_{22} < 0$ then (13.15) cannot occur. In view of (13.10) one of the two elements a_{11}, a_{22} must be positive and the other negative. We assume

$$a_{11} > 0, \ a_{22} < 0. \tag{13.17}$$

Then by (13.15) and (13.16)

$$\frac{d_2}{d_1} + \frac{a_{22}}{a_{11}} > 0, \tag{13.18}$$

$$\left(a_{22} + \frac{d_2}{d_1} a_{11}\right)^2 \ > \ 4\frac{d_2}{d_1} \cdot \det A. \tag{13.19}$$

Taking square roots and rearranging terms, we get

$$\left(\sqrt{\frac{d_2}{d_1}}\right)^2 - 2\frac{\sqrt{\det A}}{a_{11}}\sqrt{\frac{d_2}{d_1}} + \frac{a_{22}}{a_{11}} > 0. \tag{13.20}$$

This is a polynomial in $\sqrt{d_2/d_1}$ which has a single positive root. Hence

$$\sqrt{\frac{d_2}{d_1}} > \frac{\sqrt{\det A}}{a_{11}} + \frac{1}{a_{11}}\sqrt{-a_{11}a_{22}}. \tag{13.21}$$

Thus we have proved the following statement: *Let the matrix A have the properties*

$$a_{11} > 0, \quad a_{22} < 0, \quad a_{11} + a_{22} < 0, \quad a_{11}a_{22} - a_{12}a_{21} > 0. \tag{13.22}$$

(thus also $a_{12}a_{21} < 0$). Then there are diagonal matrices $D = (d_j \delta_{jk})$, $d_1 > 0$, $d_2 > 0$ such that for certain μ the matrix $A - \mu D$ has eigenvalues with positive real parts. The set of all these matrices is described by the inequality

$$\frac{d_2}{d_1} > \kappa^2, \tag{13.23}$$

$$\kappa = \frac{\sqrt{\det A}}{a_{11}} + \frac{1}{a_{11}}\sqrt{-a_{11}a_{22}}. \tag{13.24}$$

Now we assume that the matrices A and D are such that (13.22) (13.23) are satisfied. Then there is an interval (μ_1, μ_2) such that for μ in this interval the matrix $A - \mu D$ has a positive eigenvalue. In the following we vary either the continuous variable l or the discrete parameter k.

First we keep $k \in \mathbb{N}$ fixed and vary l. For this k there is a positive eigenvalue if

$$\mu_1 < (\frac{k\pi}{l})^2 < \mu_2, \tag{13.25}$$

in other words, if

$$l_2 < l < l_1, \quad l_j = \frac{k\pi}{\sqrt{\mu_j}}, \quad j = 1, 2. \tag{13.26}$$

If l runs from 0 to $+\infty$, then the k-th term ("mode") in the Fourier series becomes unstable, if l increases beyond l_1. It becomes stable again if l exceeds l_2. The mode k only fits to certain interval lengths which are not too long and not too short. At $l = l_1$ we expect a bifurcation. The spatially constant solution of the reaction diffusion equation (13.1) loses its stability, a new, stable, stationary solution bifurcates. In the neighborhood of the bifurcation point this solution has approximately the form $\cos(k\pi/l)$.

Now we fix l. Then condition (13.24) defines an interval

$$\frac{\sqrt{\mu_1}l}{\pi} < k < \frac{\sqrt{\mu_2}l}{\pi} \tag{13.27}$$

for the parameter k. There are only finitely many values of $k \in \mathbb{N}$ which satisfy this condition (or none at all). The spatially constant solution is unstable against any perturbation with these modes.

Turing's mechanism explains how stable spatial patterns may emerge. In the simplest case the pattern is a polarity: If the mode $k = 1$ becomes unstable then the interval has a "positive" and a "negative" end. The model provides some hints for the possible chemical basis of differentiation processes. At least two substances must react. The Jacobian of the reaction function at the stable equilibrium must have a special sign distribution, with the present normalization (13.22). In this situation the variable u_1 is called the "activator" and u_2 is the "inhibitor". In absence of the inhibitor the activator grows according to $\dot{u}_1 = a_{11}u_1$, whereas the inhibitor, in absence of the activator, decays as $\dot{u}_2 = a_{22}u_2$. The interaction of the two substances produces a stable equilibrium. From inequality (13.23) we infer that the diffusion rate of the d_2 must be greater than the diffusion rate d_1. With these insights the Turing phenomenon can be formulated as follows: The spatially homogeneous state can become unstable if a long range inhibitor interacts with a short range activator.

Hillen [47] has studied a similar question for the isotropic random walk system

$$\begin{aligned} u_t^+ + \Gamma u_x^+ &= \tfrac{1}{2}M(u^- - u^+) + \tfrac{1}{2}f(u) \\ u_t^- - \Gamma u_x^- &= \tfrac{1}{2}M(u^+ - u^-) + \tfrac{1}{2}f(u) \end{aligned} \tag{13.28}$$

on $[0, l]$ with zero Neumann conditions

$$u^+(t,0) = u^-(t,0), \quad u^-(t,l) = u^+(t,l). \tag{13.29}$$

As before we assume that there is a stationary solution of the underlying ordinary differential equation, $f(\bar{u}) = 0$. Let $f'(\bar{u}) = A$. Then the linearized system is

$$\begin{aligned} u_t^+ + \Gamma u_x^+ &= \tfrac{1}{2}M(u^- - u^+) + \tfrac{1}{2}Au \\ u_t^- - \Gamma u_x^- &= \tfrac{1}{2}M(u^+ - u^-) + \tfrac{1}{2}Au \end{aligned} \tag{13.30}$$

with the same boundary condition (13.29). The spectral problem is

$$\begin{aligned} \lambda u^+ + \Gamma u_x^+ &= M(u^- - u^+) + \tfrac{1}{2}Au \\ \lambda u^- - \Gamma u_x^- &= M(u^+ - u^-) + \tfrac{1}{2}Au \end{aligned} \tag{13.31}$$

again with the same boundary condition. In another formulation (13.31) reads

$$\begin{aligned} \lambda u + v_x &= Au \\ \lambda v + \Gamma^2 u_x &= -Mv \\ v(0) &= v(l) = 0. \end{aligned} \tag{13.32}$$

Similar to the scalar case (6.5), the differential equation can be carried into

$$\Gamma^2 u_{xx} = (M + \lambda I)(\lambda I - A)u. \tag{13.33}$$

Again assume that $n = 2$, and the matrix A is given as in (13.9) satisfying (13.22). Define the fictitious diffusion coefficients

$$d_i = \frac{\gamma_i^2}{\mu_i}, \quad i = 1, 2, \tag{13.34}$$

(these are the diffusion coefficients of the diffusion approximation). Since, away from the diffusion limit, the hyperbolic system behaves rather differently from the parabolic system, the following result of Hillen [47] is somewhat astonishing: *Let* $a_{11} < \mu_1$. *Then the stability conditions for the hyperbolic system* (13.28) *are exactly the same as for the parabolic system* (13.1).

The proof is much more involved than in the parabolic case and requires tedious discussion of Hurwitz determinants of order four depending on the parameters.

Nothing is said for the case $a_{11} > \mu_1$.

14 Other applications and extensions

Here we collect some extensions and applications of hyperbolic systems to biological problems.

Density-dependent diffusion: In some biological models in the form of reaction-diffusion systems it is assumed that the diffusion coefficient $D = D(u)$ depends on the density u [36]. Then the diffusion term must be written in divergence form

$$u_t = \operatorname{div}(D(u)\operatorname{grad} u) + f(u). \tag{14.1}$$

Of course one can study more general problems where D depends also on the gradient. Typically one will have situations where $D(u)$ is a decreasing or an increasing function of u. In vector valued-problems such as predator-prey models the diffusion rate of one species may depend on the densities of others. This situation is sometimes called cross-diffusion [71].

Similar effects can be introduced into the hyperbolic models. In the random walk problems the particle speed $\gamma = \gamma(u)$ and the turning rate $\mu = \mu(u)$ may depend on u [36],

$$\begin{aligned} u_t + (\gamma(u)u^+)_x &= \tfrac{1}{2}\mu(u)u + \tfrac{1}{2}f(u) \\ u_t - (\gamma(u)u^-)_x &= \tfrac{1}{2}\mu(u)u + \tfrac{1}{2}f(u). \end{aligned} \tag{14.2}$$

Travelling front solutions of this system have been investigated in [35]. *Nerve axon equations*: The classical model for the excitation of nerve cells and for conduction of pulses along the nerve axon is the Hodgkin-Huxley model. A simplified though mathematically more transparent version is the Fitzhugh-Nagumo system

$$\begin{aligned} u_t &= f(u) - \delta w + u_{xx} \\ v_t &= u - \nu w \end{aligned} \tag{14.3}$$

for the voltage u and the "membrane activation" w. It has been remarked repeatedly (see [68]) that the original Hodgkin-Huxley model has been derived as a hyperbolic problem with a small parameter (self-inductancy) and that systems of the form (14.3) arise if this parameter is put to zero. A small parameter can be introduced in several ways. In [22] hyperbolic versions of (14.3) of the form

$$\begin{aligned} \epsilon u_{tt} + u_t &= f(u) - \delta w + u_{xx} \\ v_t &= u - \nu w \end{aligned} \tag{14.4}$$

are studied, and in [26] a general system which includes

$$\begin{aligned} & u_t + v_x = f(u) - \delta w \\ & \tau v_t + u_x + v = 0 \\ & w_t = u - \nu w. \end{aligned} \tag{14.5}$$

Age structure: It is a standard problem in ecology to study populations structured by age and by location in space. Usually spread in space is described by diffusion. A standard form of model for a function $u = u(t, a, x)$ of age a and space x

$$\begin{aligned} & u_t + u_a + \mu(a, x)u = Du_{xx} \\ & u(t, 0, x) = \textstyle\int_0^\infty b(a, x)u(t, a, x)da \end{aligned} \tag{14.6}$$

together with conditions on the boundary of the spatial domain. A similar system in the correlated random walk setting is

$$\begin{aligned} u_t + u_a + (\gamma(a, x)u)_x + \mu(a)u &= \tfrac{1}{2}\nu(u^- - u^+) \\ u_t + u_a - (\gamma(a, x)u)_x + \mu(a)u &= \tfrac{1}{2}\nu(u^+ - u^-) \\ u(t, 0, x) = \textstyle\int_0^\infty b(a, x)u(t, a, x)da, &\quad s = \pm. \end{aligned} \tag{14.7}$$

Viscoeleastic damping: In the Cattaneo system (1.26) one can consider the term v as a damping term because this term gives rise the u_t term in (1.28). Thus one can try to add another "positive" damping $-\Delta v$. Then we arrive at a system

$$\begin{aligned} u_t + \operatorname{div} v &= 0 \\ \tau v_t + D\operatorname{grad} u + v - \alpha\Delta v &= 0. \end{aligned} \tag{14.8}$$

The first equation is a conservation law. The second equation is a parabolic system. The Laplacian is applied to the variable v componentwise, α is a nonnegative constant. Altogether (14.7) is a parabolic system. The vector v can be seen as a flow driven by the gradient of u with some additional fluctuations which become stronger for large α. Both variables v and u show the effect of infinitely fast propagation. If we eliminate v as in (1.28) then we arrive at

$$\tau u_{tt} + u_t - \alpha\Delta u_t = D\Delta u \tag{14.9}$$

which is a so-called strongly damped wave equation describing viscoelastic damping of vibrating strings and membranes. If we keep D and τ fixed and move α from 0 to large values, then the hyperbolic character of the system is less and less apparent. This becomes obvious from the following two observations. If in (14.8) we introduce a new variable $w = u + (\alpha/D)u_t$ then we get again a first order system

$$\begin{aligned} u_t &= \tfrac{D}{\alpha}(w-u) \\ \tau w_t &= \alpha\Delta w + (1 - \tau\tfrac{D}{\alpha})(u-w). \end{aligned} \tag{14.10}$$

This system preserves positivity with respect to the cone $u \geq 0, w \geq 0$ if and only if

$$\alpha \geq \tau D \tag{14.11}$$

saying that the system has a maximum principle if in the parameter space it is sufficiently far away from the hyperbolic case $\alpha = 0$. Mechanical systems like (14.9) can show overdamping, i.e., the damping may become so strong that all oscillating modes disappear. J.Hale has suggested to the author that the conditions for overdamping and for the appearance of a maximum principle may be the same. This is indeed the case. If we apply a formal spectral analysis to (14.8), i.e., if we replace $u_t = \lambda u$, $v_t = \lambda v$ then we get

$$-\Delta u = \kappa u, \quad \kappa = \frac{\lambda + \lambda^2\tau}{D + \alpha\lambda}u. \tag{14.12}$$

Then (14.11) is the exact condition such that for nonnegative κ the corresponding values of λ are real and nonnegative. The following is a reaction transport system with strongly damped transport operator,

$$\begin{aligned} u_t + \operatorname{div} v &= f(u) \\ \tau v_t + D\operatorname{grad} u + v - \alpha\Delta v &= 0. \end{aligned} \tag{14.13}$$

For qualitative properties of the corresponding strongly damped nonlinear wave equations and their relation to coupled oscillators, see [43] and [23].

Free boundary value problems: The classical Stefan problem of the diffusion equation describes, in biological terms, some species diffusing in a one-dimensional domain which moves the boundary of that domain by interacting with the environment. For the diffusion equation $u_t = Du_{xx}$ there is an initial datum $u(0,x) = u_0(x)$ on an interval $[0, s_0)$, boundary data on the fixed boundary $u(t,0) = \varphi(t)$, and two boundary conditions on the free boundary $x = s(t)$, e.g. a Dirichlet condition $u(t,s(t)) = 0$

and a Stefan condition $\dot{s}(t) = -au_x(t, s(t))$, where $a > 0$ is a constant. One looks for two functions, the free boundary $s(t)$ with $s(0) = s_0$ and the function $u(t, x)$ in the domain described by $t > 0$, $0 < x < s(t)$.

There are several ways in which Stefan problems can be formulated for the telegraph equation or for first order hyperbolic systems. In [30] a problem is studied which is motivated by a detailed physical analysis. In [24] existence of solutions is proved for a problem that can be seen (in our notation) as a Stefan problem for a one-dimensional Cattaneo system

$$\begin{aligned} u_t + v_x &= 0 \\ \tau v_t + u_x + v &= 0 \end{aligned} \tag{14.14}$$

with conditions at the free boundary

$$\dot{s}(1 + u) = v, \quad \tau \dot{s} v = u. \tag{14.15}$$

In [64] a related problem is studied in terms correlated random walks. The differential system is given by (1.2) or (1.3). At the free boundary there is one condition

$$u^-(t, s(t)) = \kappa u^+(t, s(t)) \tag{14.16}$$

saying that a fraction of the u^+ particles arriving at the free boundary are changing direction and the remaining fraction is absorbed. The second condition describes how the boundary is moved by the arriving particles, e.g.,

$$\dot{s}(t) = \frac{(1 - \kappa)u^+(t, s(t))}{1 + (1 - \kappa)u^+(t, s(t))}\gamma. \tag{14.17}$$

15 Models in discrete space

The solutions of reaction diffusion equations can be computed by numerical methods, e.g. by finite element methods or difference schemes. These are closely connected to discrete models for reaction and diffusion. In simple cases of low order approximation the discretized system can be seen as a discrete model. In particular, in one space dimension, such systems are so-called stepping stone models, where particles can move on a one-dimensional grid from one grid point to its neighbors. Thus there is a close connection between one-dimensional reaction diffusion equations and Sturm-Liouville problems on the one side and tridiagonal matrices on the other side. Indeed, consider the linear diffusion equation

$$u_t = a(x)u_{xx} + b(x)u_x + c(x)u \tag{15.1}$$

or an equation in divergence form

$$u_t = (\tilde{a}(x)u_x)_x + \tilde{c}(x)u \tag{15.2}$$

with, say, zero Dirichlet conditions on some bounded interval $u(t, 0) = u(t, l) = 0$. The eigenvalue problem can always be cast into the form of a selfadjoint generalized problem

$$-(p(x)u_x)_x + q(x)u = \lambda r(x)u, \quad u(0) = u(l) = 0 \tag{15.3}$$

using a multiplicator function. An explicit difference scheme with constant step sizes h and k applied to (15.1) yields difference equations

$$u_i^{(\nu+1)} = \frac{k}{h^2}(a_i - \frac{h}{2}b_i)u_{i-1}^{(\nu)} + (1 - \frac{2k}{h^2}a_i + kc_i)u_i^{(\nu)} + \frac{k}{h^2}(a_i + \frac{h}{2}b_i)u_{i+1}^{(\nu)} \tag{15.4}$$

for $i = 1, \ldots, n$, with $u_0 = u_{n+1} = 0$. In matrix notation, with $u = (u_1, \ldots, u_n)^T$, the system assumes the form

$$u^{(\nu+1)} = u^{(\nu)} - \frac{k}{h^2}Au^{(\nu)} + kDu^{(\nu)}, \tag{15.5}$$

where A is a tridiagonal matrix $(-\gamma_i, \alpha_i, -\beta_i)$ (in the usual notation) with $\gamma_i = (a_i - hb_i/2)$, $\alpha_i = 2a_i$, $\beta_i = (a_i + hb_i/2)$. In view of $a > 0$, for h small, one has $\alpha_i > 0$, $\beta_i > 0$, $\gamma_i > 0$, $\alpha_i - \beta_i - \gamma_i \geq 0$.

The most obvious properties of tridiagonal matrices are the following. If all β_i and γ_i have the same sign, then one can achieve symmetry by a similarity transform. In that case all eigenvalues are real. The characteristic polynomials of a descending sequence of principal submatrices form a three terms recursion. The eigenvalues of the matrix of order n and those of a principal submatrix of order $n-1$ separate each other strongly. These properties relate tridiagonal matrices to the theory of orthogonal polynomials. Indeed, the problem above with constant coefficients $a = 1$, $b = c = 0$ leads to the tridiagonal matrix $(-1, 2, 1)$ and to the Tchebysheff polynomials of the first kind.

A hyperbolic analogue of (15.2) must be written in divergence form

$$\begin{aligned} u_t^+ + (\gamma(x)u^+)_x &= \mu(x)(u^- - u^+) \\ u_t^- - (\gamma(x)u^-)_x &= \mu(x)(u^+ - u^-). \end{aligned} \tag{15.6}$$

In order to incorporate this model and other related models for spatial spread, we consider a system of a general form

$$\begin{aligned} u_t^+ + a^+(x)u_x^+ &= b^+(x)u^+ + c^+(x)u^- \\ u_t^- - a^-(x)u_x^- &= b^-(x)u^- + c^-(x)u^+ \end{aligned} \tag{15.7}$$

on an interval $[0, l]$ with zero Dirichlet boundary conditions (3.3). As in the parabolic case we apply an explicit difference scheme with step sizes h and k,

$$\begin{aligned} u_i^{+(\nu+1)} &= u_i^{+(\nu)} - \tfrac{k}{h}a_i^+(u_i^{+(\nu)} - u_{i-1}^{+(\nu)}) + kb_i^+u_i^{+(\nu)} + kc_i^+u_i^{-(\nu)} \\ u_i^{-(\nu+1)} &= u_i^{-(\nu)} + \tfrac{k}{h}a_i^-(u_{i+1}^{-(\nu)} - u_i^{-(\nu)}) + kb_i^-u_i^{-(\nu)} + kc_i^-u_i^{+(\nu)} \end{aligned} \tag{15.8}$$

for $i = 1, \ldots, n$. Again the boundary condition is incorporated by setting $u_0^+ = 0$, $u_{n+1}^- = 0$. It is not necessary to introduce u_{n+1}^+ or u_0^- because the hyperbolic system takes up initial data only along inbound characteristics. We introduce matrix notation $u = (u_1^+, \ldots, u_n^+, u_1^-, \ldots, u_n^-)^T$,

$$u^{(\nu+1)} = u^{(\nu)} + \frac{k}{h}Au^{(\nu)}. \tag{15.9}$$

In the matrix

$$A = \begin{pmatrix} A^+ & D^+ \\ D^- & A^- \end{pmatrix} \tag{15.10}$$

the off-diagonal blocks are diagonal matrices $D^\pm = (c_i^\pm \delta_{ij})$ and the diagonal blocks $A^\pm$ are bidiagonal matrices with diagonal elements $-a_i^\pm + hb_i^\pm$. In A^+ the subdiagonal elements are a_i^+, in A^- the superdiagonal elements are a_i^-.

The structure appearing in (15.9) is a most natural "hyperbolic" analogue of a tridiagonal matrix. Because (15.9) is an approximation to hyperbolic systems of the form (15.7) for which the generator has complex eigenvalues, the matrix A will have, in general, complex eigenvalues. Thus some properties of tridiagonal matrices will be lost whereas others can be recovered, as we shall show now.

We study in general matrices of the form (15.10) where $D^\pm = (\gamma_i^\pm \delta_{ij})$ and

$$A^+ = \begin{pmatrix} \alpha_1^+ & & & & \\ \beta_2^+ & \alpha_2^+ & & & \\ & \beta_3^+ & \ddots & & \\ & & \ddots & & \\ & & & \beta_n^+ & \alpha_n^+ \end{pmatrix}, \quad A^- = \begin{pmatrix} \alpha_1^- & \beta_2^- & & & \\ & \alpha_2^- & \beta_3^- & & \\ & & \ddots & \ddots & \\ & & & & \beta_n^- \\ & & & & \alpha_n^- \end{pmatrix}. \tag{15.11}$$

The structure of the matrix A reflects the division of the population into particles going to the right and particles going to the left. Now we reorder and collect components with the same number, i.e., we apply a permutation of the variables. We order them as $u_1^+, u_1^-, u_2^+, u_2^-, \ldots, u_n^+, u_n^-$. Of course, this permutation correponds to a permutation of the rows and columns of the matrix. We get a five-band matrix where the side bands contain zeros in some regular way. We cannot use this fact to further simplify the matrix but we can use it in determining the characteristic polynomial. Instead of $\tilde{A}$ consider $\det(\tilde{A} - \lambda I)$. In the determinant exchange the first and second column, the third and fourth column, a.s.o. (but not the rows). This will change only a sign. Then we get a "continuant" or "tridiagonal determinant" with, however, the parameter λ appearing off the diagonal,

$$\begin{vmatrix} \gamma_1^+ & \alpha_1^+ - \lambda & & & & & & & \\ \alpha_1^- - \lambda & \gamma_1^- & \beta_2^- & & & & & & \\ & \beta_2^+ & \gamma_2 & \alpha_2^+ - \lambda & & & & & \\ & & \alpha_2^- - \lambda & \gamma_2^- & & & & & \\ & & & \beta_3^+ & \ddots & & & & \\ & & & & \gamma_{n-1}^+ & \alpha_{n-1}^+ - \lambda & & & \\ & & & & \alpha_{n-1}^- - \lambda & \gamma_{n-1}^- & \beta_n^- & & \\ & & & & & \beta_n^+ & \gamma_n^+ & \alpha_n^+ - \lambda \\ & & & & & & \alpha_n^- - \lambda & \gamma_n^- \end{vmatrix}. \tag{15.12}$$

This determinant we call $P_n(\lambda)$, and the determinant with the last row and column deleted we call $Q_n(\lambda)$.

Thus the characteristic polynomial $P_n(\lambda)$ of the matrix A can be generated by the three terms vector recursion. With $Q_0(\lambda) = 0$, $P_0(\lambda) = 1$,

$$\begin{aligned} Q_n(\lambda) &= \gamma_n^+ P_{n-1}(\lambda) - \beta_n^+ \beta_n^- Q_{n-1}(\lambda) \\ P_n(\lambda) &= \gamma_n^- Q_n(\lambda) - (\lambda - \alpha_n^+)(\lambda - \alpha_n^-) P_{n-1}(\lambda). \end{aligned} \tag{15.13}$$

Always P_n has degree $2n$ and Q_n has degree $2n-2$.

In the special case of a discretization of the system (15.8) with constant coefficients γ and μ we get $\alpha_i^{\pm} = -\gamma - \mu$, $\beta_i^{\pm} = \gamma$, $\gamma_i^{\pm} = \mu$. The formulae (15.13) assume the form $Q_0(\lambda) = 0$, $P_0(\lambda) = 1$,

$$\begin{aligned} Q_n(\lambda) &= \mu P_{n-1}(\lambda) - \gamma^2 Q_{n-1}(\lambda) \\ P_n(\lambda) &= \mu Q_n(\lambda) - (\lambda+\gamma+\mu)^2 P_{n-1}(\lambda). \end{aligned} \tag{15.14}$$

In this case the P_n are actually polynomials of the variable $z = (\lambda+\gamma+\mu)^2$. Define polynomials $S_n(z)$, $R_n(z)$ of degree n by $R_0(z) = 0$, $S_0(z) = 1$,

$$\begin{aligned} R_n(z) &= \mu S_{n-1}(z) - \gamma^2 R_{n-1}(z) \\ S_n(z) &= \mu R_n(z) - z S_{n-1}(z) \end{aligned} \tag{15.15}$$

Then

$$P_n(\lambda) = S_n((\lambda+\gamma+\mu)^2). \tag{15.16}$$

Thus we have a similar situation as in the parabolic or tridiagonal case. There we know that each fixed eigenvalue of the Stürm-Liouville problem is approximated by the discrete eigenvalues, but the eigenvalues of the matrix problem have a totally different asymptotics. In particular, the matrix spectrum is symmetric (zeros of Tchebycheff polynomial) with respect to some parallel to the imaginary axis (in the tridiagonal case the spectrum is real). Also in the hyperbolic case the spectrum is symmetric with respect to the real axis and also with respect to the line $\operatorname{Re}\lambda = -(\mu+\gamma)$. The zeros of these polynomials play a role in the stability analysis of "stepping stone" models of the form (15.9).

References

[1] Alt, W., Biased Random walk models for chemotaxis and related diffusion approximations. *J. Math. Biol.* **9** (1980) 147-177

[2] Aronson, D.G., The asymptotic speed of propagation of a simple epidemic. In: W. E. Fitzgibbon, H. F. Walker (eds), Nonlinear Diffusion. Pitman Research Notes in Mathematics **14** (1977) 1-23

[3] Aronson, D.G., Weinberger, H.F., Nonlinear diffusion in population genetics, combustion, and nerve propagation. Lect. Notes in Math. **446**, p.5-49, Springer Verlag 1975

[4] Bartlett, M.S., A note on random walks at constant speed. *Adv. Appl. Prob.* **10** (1978) 704-707

[5] Beals, R., Protopopescu, V., Abstract time-dependent transport equations. *J. Math. Anal. Appl.* **121**, (1987) 370-405

[6] Berg, H.C., Brown, D.A., Chemotaxis in Escherichia coli analyzed by three-dimensional tracking. In: Antibiotics and Chemotherapy. Vol. 19, Basel, Karger (1974) 55-78

[7] Brayton, R., Miranker, W., A stability theory for nonlinear mixed initial boundary value problems. *Arch. Rat. Mech. Anal.* **17** (1964) 358-376

[8] Britton, N., Reaction-diffusion equations and their application to biology. Academic Press 1986

[9] Broadwell, J.E., Shock structure in a simple velocity gas. *Phys. Fluids* **7** (1964) 1243-1247

[10] Cattaneo, C., Sulla conduzione del calore. *Atti del Semin. Mat. e Fis. Univ. Modena* **3** (1948) 83-101

[11] Cercignani, C., The Boltzmann Equation and its Applications. Springer Verlag 1988

[12] Cercignani, C., Illner, R., Pulvirenti, M., The Mathematical Theory of Dilute Gases. Springer Verlag 1994

[13] Crank, J., The Mathematics of Diffusion. 2nd ed., Clarendon Press, Oxford 1975

[14] Diekmann, O., Thresholds and travelling waves for the geographical spread of infection. *J. Math. Biol.* **6** (1978), 109-130

[15] Dunbar, S., A branching random evolution and a nonlinear hyperbolic equation. *SIAM J. Appl. Math.* **48** (1988) 1510-1526

[16] Dunbar, S., Othmer, H., On a nonlinear hyperbolic equation describing transmission lines, cell movement, and branching random walks. In: H.G.Othmer (ed.) Nonlinear Oscillations in Biology and Chemistry. *Lect. Notes in Biomath.* **66**, Springer Verlag 1986

[17] Einstein, A., Zur Theorie der Brownschen Bewegung. *Annalen der Physik* **19** (1906) 371-381

[18] Feireisl, E., Attractors for semilinear damped wave equation on $\mathbb{R}^3$. *Nonlinear Analysis TMA* **23** (1994) 187-195

[19] Fife, P.C., Mathematical Aspects of Reacting and Diffusing Systems. Lect. Notes in Biomath. **28**, Springer Verlag 1979

[20] Fife, P.C., McLeod, J.B., The approach of solutions of nonlinear diffusion equations to travelling front solutions. *Arch. Rat. Mech. Anal.* **65** (1977) 335-361

[21] Fisher, R.A., The advance of advantageous genes. *Ann. Eugenics* **7** (1937) 355-361

[22] Fitzgibbon, W.E., Parrot, M.E., Convergence of singularly perturbed Hodgkin--Huxley systems. *J. Nonlin. Anal. TMA* **22** (1994) 363-379

[23] Fitzgibbon, W.E., Parrot, M.E., Convergence of singular perturbations of strongly damped nonlinear wave equations. *J. Nonl. Anal. TMA* **28** (1997) 165-174

[24] Friedman, A., Bei Hu, The Stefan problem for a hyperbolic heat equation. *J. Math. Anal. Appl.* **138** (1989) 249-279

[25] Fürth, R., Die Brownsche Bewegung bei Berücksichtigung einer Persistenz der Bewegungsrichtung. *Zeitschr. für Physik* **2** (1920) 244-256

[26] Fusco, D., Manganaro, N., A method for finding exact solutions to hyperbolic systems of first-order PDEs. *IMA J. Appl. Math.* **57** (1996), 223-242

[27] Gierer, A., Meinhardt, H., A theory of biological pattern formation. *Kybernetik* **12** (1972) 30-39

[28] Glansdorff, P., Prigogine, I., Thermodynamic Theory of Structure, Stability, and Fluctuations. Wiley, London 1971

[29] Goldstein, S., On diffusion by discontinuous movements and the telegraph equation. *Quart. J. Mech. Appl. Math.* **4** (1951) 129-156

[30] Greenberg, J.M., A hyperbolic heat transfer problem with phase change. *IMA J. Appl. Math.* **38** (1988) 1-21

[31] Greiner, G., Spectral properties and asymptotic behavior of the linear transport equation. *Math. Zeitschr.* **185** (1984) 167-177

[32] Gurtin, M.E., Pipkin, A.C., A general theory of heat conduction with finite wave speeds. *Arch. Rat. Mech. Anal.* **31** (1968) 113-126

[33] Hadeler, K.P., Hyperbolic travelling fronts. *Proc. Edinburgh Math. Soc.* **31** (1988) 89-97

[34] Hadeler, K.P., Travelling fronts for correlated random walks. *Canad. Appl. Math. Quart.* **2** (1994) 27-43

[35] Hadeler, K.P., Travelling epidemic waves and correlated random walks. In: M. Martelli et al. (eds.) Differential Equations and Applications to Biology and Industry. Proc. Conf. Claremont 1994 World Scientific 1995

[36] Hadeler, K.P., Reaction-telegraph equations with density-dependent coefficients. In: G. Lumer, S. Nicaise, B.-W. Schulze (eds) Partial Differential equations, Models in Physics and Biology, Mathematical Research **82**, Akademie-Verlag, Berlin 1994, p. 152-158

[37] Hadeler, K.P., Travelling fronts in random walk systems. *Forma* (Tokyo) **10** (1995) 223-233

[38] Hadeler, K.P., Reaction telegraph equations and random walk systems. In: S. van Strien, S. Verduyn Lunel (eds), Stochastic and spatial structures of dynamical systems. Roy. Acad. of the Netherlands. North Holland, Amsterdam (1996), 133-161

[39] Hadeler, K.P., Spatial epidemic spread by correlated random walk, the case of slow infectives. In: R. A. Jarvis et al., Ordinary and partial differential equations. Proc. Conf. Dundee 1996 (1998)

[40] Hadeler, K.P., Nonlinear propagation in reaction transport systems. In: S. Ruan, G. Wolkowicz (eds) Differential equations with Applications to Biology. Fields Institute Communications, Amer. Math. Soc. 1998

[41] Hadeler, K.P., Illner, R., van den Driessche, P., A disease transport model. In preparation.

[42] Hadeler, K.P., Rothe, F., Travelling fronts in nonlinear diffusion equations. *J. Math. Biol.* **2** (1975) 251-263

[43] Hale, J.K., Asymptotic Behavior of Dissipative Systems. Amer. Math. Soc., Providence R.I. 1988

[44] Hale, J.K., Diffusive coupling, dissipation, and synchronization. *J. Dynamics Diff. Equ.* **9** (1997) 1-52

[45] Henry, D., Geometric Theory of Semilinear Parabolic Equations. *Lect. Notes in Math.* **840** Springer Verlag 1981

[46] Herrero, M.A., Velázquez, J.J.L., Singularity patterns in a chemotaxis model. *Math. Ann.* **306**, (1996) 583-623

[47] Hillen, T., A Turing model with correlated random walk. *J. Math. Biol.* **35** (1996) 49-72

[48] Hillen, T., Nonlinear hyperbolic systems describing random motion and their application to the Turing model. *Dissertation Summaries in Math.* **1** (1996) 121-128

[49] Hillen, T., Qualitative Analysis of hyperbolic random walk systems. Preprint SFB 382, No. 43 (1996)

[50] Hillen, T., Invariance principles for hyperbolic random walk system. *J. Math. Anal. Appl.* **210** (1997) 360-374

[51] Hillen, T., Qualitative Analysis of semilinear Cattaneo systems. *Math. Models Methods Appl. Sci.* **3** (1998)

[52] Hillen, T., Stevens, A., A random walk model with coefficients depending on an external signal as a model for chemotaxis. In preparation.

[53] Holmes, E.E., Are diffusion models too simple? A comparison with telegraph models of invasion. *Amer. Naturalist* **142** (1993) 779-795

[54] Jörgens, K., An asymptotic expansion in the theory of neutron transport. *Comm. Pure Appl. Math.* **XI** (1958) 219-242

[55] Joseph, D.D., Preziosi, L., Heat waves. *Reviews of Modern Physics* **61** (1988) 41-73

[56] Kac, M., A stochastic model related to the telegrapher's equation. (1956)

[57] reprinted in *Rocky Mtn. Math. J.* **4** (1974) 497-509

[58] Källén, A., Thresholds and travelling waves in an epidemic model for rabies. *Nonlinear Analysis TMA* **8** (1984) 851-856

[59] Kaper, H.G., Lekkerkerker, C.G., Hejtmanek, J., Spectral Methods in Linear Transport Theory. Birkhäuser Verlag 1982

[60] Keller, E.F., Segel, L.A., Initiation of slime mold aggregation viewed as an instability. *J. Theor. Biol.* **26** (1970), 399-415

[61] Keller, E.F., Segel, L.A., Traveling bands of chemotactic bacteria, a theoretical analysis. J. Theor. Biol. **30** (1971) 235-248

[62] Kendall, D.G., Mathematical models of the spread of infection. *Mathematics and Computer Science in Biology and Medicine,* p. 213-225. Medical Research Council H.M.S.O., London 1965

[63] Kermack, W.O., McKendrick, A.G., A contribution to the mathematical theory of epidemics I. *Proc. Roy. Soc. London* **A115** (1927) 700-721

[64] Kolmogorov, A., Petrovskij, I., Piskunov, N., Etude de l'équation de la diffusion avec croissance de la quantité de la matière et son application a une problème biologique. *Bull. Univ. Moscou*, Ser. Int., Sec A **1,6** (1937) 1-25

[65] Kuttler, C., A free boundary value problem for correlated random walk. Preprint University of Tübingen, in preparation.

[66] Larsen, E.W., Zweifel, P.F., On the spectrum of the linear transport operator. *J. Math. Physics* **15** (1974) 1987-1997

[67] Lauffenburger, D.A., Chemotaxis and cell aggregation. In: W. Jäger, J. D. Murray (eds) Modelling of Patterns in Space and Time. Lect. Notes in Biomath. **55**, Springer Verlag 1984

[68] Levine, H.A., Sleeman, B.D., A system of reaction diffusion equations arising in the theory of reinforced random walks. *SIAM J. Appl. Math.* **57** (1997), 683-730

[69] Lieberstein, H.M., Mathematical Physiology. Blood flow and electrically active cells. Amer. Elsevier Co., New York, Amsterdam, London 1973

[70] MacNab, R., Koshland D.E. jr., The gradient-sensing system in bacterial chemotaxis. *Proc. Nat. Acad. Sci. USA* **69** (1972) 2509-2512

[71] McKean, H.P., Application of Brownian motion to the equation of Kolmogorov-Petrovskij-Piskunov. *Comm. Pure Appl. Math.* **28** (1975) 323-331, **29** (1976) 553-554

[72] Mimura, M., Kawasaki, K., Spatial segregation in competitive interaction-diffusion equations. *J. Math. Biol.* **9** (1980) 49-64

[73] Mollison, D., Spatial contact models for ecological and epidemic spread. *J. Roy. Statist. Soc.* Ser. **B 39** (1977) 283-326

[74] Müller, I., Hyperbolic equations for diffusion. Classical Mechanics and Relativity: Relationship and Consistency. Monographs and Textbooks in Physical Science, Bibliopolis, Napoli 1991, p. 121-133

[75] Murray, J.D., Mathematical Biology. Springer Verlag 1989.

[76] Nagai, T., Blow up of radially symmetric solutions to a chemotaxis system. *Adv. Math. Sci. Appl.* **5** (1995), 581-601

[77] Neves, A.F., Ribeiro, H., Lopes, O., On the spectrum of evolution operators generated by hyperbolic systems. *J. Funct. Anal.* **67** (1986) 320-344

[78] Okubo, A., Diffusion and Ecological Problems: Mathematical Models. Biomathematics 10, Springer Verlag 1980

[79] Orsingher, E., A planar random motion governed by the two-dimensional telegraph equation. *J. Appl. Prob.* **23** (1986) 385-397

[80] Othmer, H.G., Dunbar, S.R., Alt, W., Models of dispersal in biological systems. *J. Math. Biol.* **26** (1988) 263-298

[81] Othmer, H.G., Stevens, A., Aggregation, blowup and collapse: The ABCs of taxis in reinforced random walks. *SIAM J. Appl. Math.* **57** (1997) 1044-1081

[82] Papanicolaou, G.C., Asymptotic analysis of transport processes. *Bull. AMS* **81** (1975) 330-392

[83] Patlak, C., Random walk with persistence and external bias. *Bull. Math. Biophysics* **15** (1953) 311-318

[84] Pearson, K., Nature **72** (1905) 294

[85] Poincaré, H., Sur la propagation de l'électricité. *Compt. Rend. Ac. Sci.* **107**, 1027-1032. Œuvres IX, 278-283

[86] Rascle, M., Ziti, C., Finite time blow up in some model of chemotaxis. *J. Math. Biol.* **33** (1995), 388-414

[87] Rivero, M.A., Tranquillo, R.T., Buettner, H.M., Lauffenburger, D.A.,

[88] Transport models for chemotactic cell populations based on individual cell behavior. *Chemical Engin. Sci.* **44** (1989)

[89] Rothe, F., Global Solutions of Reaction-Diffusion Systems. Lecture Notes in Mathematics 1072, Springer Verlag 1984

[90] Ruggieri, T., Cattaneo equation and relativistic extended thermodynamics. Classical Mechanics and Relativity: Relationship and Consistency. Monographs and Textbooks in Physical Science, Bibliopolis, Napoli 1991, p. 135-150

[91] Schaaf, R., Global behaviour of solution branches for some Neumann problems depending on one or several parameters. *J. Reine Angew. Math.* **346** (1984) 1-31

[92] Schwetlick, H., On the minimal speed of travelling waves in reaction transport equations. Preprint University of Tübingen. SFB 382, No. (1997)

[93] Senba, T., Blow-up of radially symmetric solutions to some systems of partial differential equations modelling chemotaxis. *Adv. Math. Sciences Appl.* (Tokyo) **7** (1997), 79-92

[94] Sharov, O.I., Random walks in the euclidean space R^n associated with the telegraph equation. *Theor. Prob. Math. Statist.* **49** (1994) 165-171

[95] Skellam, J.G., The formulation and interpretation of mathematical models of diffusional processes in population biology. In: M. S. Bartlett, R. W. Hiorns (eds) The Mathematical Theory of the Dynamics of Biological Populations. Academic Press 1973, p. 63-85

[96] Smoller, J., Shock Waves and Reaction-Diffusion Equations. Springer Verlag 1982

[97] Stadje, W., Exact probability distributions for noncorrelated random walk models. *J. Stat. Physics* **56** (1989), 415-435

[98] Stevens, A., Trail following and aggregation of myxobacteria. *J. Biol. Systems* **3** (1995), 1059-1068

[99] Tang, Y., Othmer, H.G., Excitation, oscillations and wave propagation in a G-protein based model of signal transduction in *Dictyostelium discoideum*. *Phil. Trans. R. Soc. London B* **349** (1995), 179-195

[100] Taylor, G.I., Diffusion by continuous movements. *Proc. London Math. Soc.* **20** (1920) 196-212

[101] Temam, R., Infinite-dimensional systems in Mechanics and Physics.

[102] Springer Verlag 1988

[103] Turing, A.M., The chemical basis of morphogenesis. *Phil. Trans. Roy. Soc. London B* **237** (1952) 37-72

[104] Voigt, J. Spectral properties of the neutron transport equation. *J. Math. Anal. Appl.* **106** (1985) 140-153

[105] Webb, G., Existence and asymptotic behavior for a strongly damped nonlinear wave equation. *Canad. J. Math.* **32** (1980) 631-643

[106] Weiss, G.H., Aspects and applications of the random walk. North Holland Publ., Amsterdam 1994

[107] Witt, I., Existence and continuity of the attractor for a singularly perturbed hyperbolic equation. *J. Dynamics Diff. Equ.* **7** (1995), 591-639

[108] Zauderer, E., Partial Differential Equations of Applied Mathematics. Wiley, New York 1983

Mathematical Models in Morphogenesis

Philip K. Maini
Centre for Mathematical Biology
Mathematical Institute
24-29 St Giles'
Oxford OX1 3LB
England

Contents

Introduction

Spatial and spatio-temporal patterns occur widely in chemistry and biology. In many cases, these patterns seem to be generated spontaneously. The best known oscillatory reaction is the Belousov-Zhabotinsky reaction, in which bromate ions oxidise malonic acid in a reaction catalysed by cerium, which has the states Ce^{3+} and Ce^{4+}. Sustained periodic oscillations are observed in the cerium ions. If, instead, one uses the catalyst Fe^{2+} and Fe^{3+} and phenanthroline, the periodic oscillations are visualised as colour changes between reddish-orange and blue (see, for example, Murray, 1993, Johnson and Scott, 1996 for review). This system can also exhibit a number of different types of wave structures such as propagating fronts, spiral waves, target patterns and toroidal scrolls (Zaikin and Zhabotinskii, 1970, Winfree, 1972, 1974, Müller *et al.*, 1985, Welsh *et al.*, 1983, Zykov, 1987). Such oscillatory and wave-like patterns also arise in physiology and one of the most widely-studied and important areas of wave propagation concerns the electrical activity in the heart (Panfilov and Holden, 1997). These phenomena have motivated a great deal of mathematical modelling and the analysis of the resultant systems (coupled ordinary differential equations and/or partial differential equations) has led to a greater understanding of the underlying mechanisms involved and suggested control strategies in the case of medical applications (see, for example, Goldbeter, 1996).

The development of spatial pattern and form is one of the central issues in embryology. Although genes control pattern formation, genetics does not give us an understanding of the actual mechanisms involved in patterning. Many models of how different processes can conspire to produce pattern have been proposed and analysed. They range from gradient-type models involving a simple source-sink mechanism (Wolpert, 1969); to cellular automata models in which the tissue is discretised and rules are introduced as to how different elements interact with each other (see, for example, Bard, 1981); to more complicated models which incorporate more sophisticated chemistry and biology. In this chapter we shall focus on some models from the latter category which are based on two very different principles of pattern formation. Broadly speaking there are two ways pattern may form: (1) A spatial pattern in some chemical (morphogen) may be set up which in turn determines cell differentiation. Thus the pattern or structure that we observe is due to the underlying pre-pattern in morphogen. (2) A spatial pattern in cell density is set up and cells in high density aggregations then differentiate. Thus the pattern we observe is due directly to the underlying pattern in cell density.

In the next section we consider the most widely studied pre-pattern model, namely, reaction-diffusion (RD). The model is formulated and analysed using standard linear and nonlinear techniques. As these techniques carry over to the other models in this chapter, they are considered only for the reaction-diffusion model. The properties of the spatial patterns exhibited by the model are presented and some recent results for modified RD models are discussed. Sections 2 and 3 discuss respectively, cell-chemotactic and mechanical models, which are based on (2) above, namely, they hypothesize that patterning occurs due to cell motion. The models are formulated and in Section 2.2 propagating patterns are considered.

Section 4 considers some examples of the application of these models to biology. Specifically, we focus on skeletal patterning in the vertebrate limb, pigmentation patterns in reptiles and avian skin organ formation. The coupling of patterning models is also discussed. Conclusions are presented in Section 5.

1 Reaction-Diffusion Models

1.1 Model Formulation and Linear Analysis

Let $c(\mathbf{x}, t)$ be the concentration of a chemical at position $\mathbf{x} \in \mathbf{R}^3$ and time $t \in [0, \infty)$. Consider an arbitrary volume $V \subset \mathbf{R}^3$. Then
rate of change of chemical in $V = -$ flux + net production i.e.

$$\frac{d}{dt}\int_V c dv = -\int_{\partial V} \mathbf{F}.d\mathbf{S} + \int_V f(c) dv \tag{1.1}$$

where $\mathbf{F}$ is the flux of chemical per unit area and $f(c)$ is net chemical production per unit volume. Using the divergence theorem, (1.1) becomes

$$\int_V \left\{ \frac{\partial c}{\partial t} + \nabla.\mathbf{F} - f(c) \right\} dv = 0. \tag{1.2}$$

As this is true for all arbitrary volumes V, it follows that

$$\frac{\partial c}{\partial t} = -\nabla.\mathbf{F} + f(c). \tag{1.3}$$

We now need an expression for the flux of chemical in terms of chemical concentration. We use Fick's Law, which states that chemical flux is proportional to the concentration gradient, i.e.

$$\mathbf{F} = -D\nabla c \tag{1.4}$$

where D, the diffusion coefficient, is assumed constant (positive). This models flux from high concentrations to low concentrations. Substituting (1.4) into (1.3) we obtain the reaction-diffusion equation

$$\frac{\partial c}{\partial t} = D\nabla^2 c + f(c). \tag{1.5}$$

There are many ways to derive (1.5). For example, Turing (1952) considered a one-dimensional row of discrete cells with chemical flow between cells, moving from cells with high chemical concentrations to cells with low chemical concentrations. His model was a system of coupled discrete-differential equations and, when averaged over continuous space, gives (1.5). Alternatively, his system could be viewed as a finite difference discretization of (1.5).

Note that in obtaining (1.5) we have assumed that D is constant and that the net production term, f, depends only on c. More generally, D may also be a function of c (density-dependent diffusion) and both D and f may also have spatio-temporal variation.

To complete the model formulation we need to specify initial conditions, $c(\mathbf{x}, 0) = c_0(\mathbf{x})$ and boundary conditions. The latter may typically be written in the form

$$\theta_1(\mathbf{n}.\nabla)c + (1 - \theta_1)c = \theta_2 \text{ on } \partial V \tag{1.6}$$

where $\mathbf{n}$ is the outward normal to the surface ∂V of the volume V, and θ_1 and θ_2 are constants. For example, if $\theta_1 = 1$, $\theta_2 = 0$ then we have zero flux (Neumann) conditions. If, on the other hand, we set $\theta_1 = 0$, then we have fixed (Dirichlet) boundary conditions. If the model was to be solved on a ring, then the appropriate boundary condition would be periodic, $c(0) = c(L)$, where L is the circumference of the ring.

For a system of interacting chemicals (1.5) generalises to

$$\frac{\partial \mathbf{u}}{\partial t} = \mathbf{D}\nabla^2\mathbf{u} + \mathbf{f}(\mathbf{u}), \tag{1.7}$$

where $\mathbf{u}$ is a vector of chemical concentrations, $\mathbf{u} = (u_1, u_2, ..., u_n)^T$; $\mathbf{f} = (f_1(\mathbf{u}), f_2(\mathbf{u}), ..., f_n(\mathbf{u}))^T$ and models chemical interaction; and $\mathbf{D}$ is an $n \times n$ diffusion matrix. In the simplest examples, $\mathbf{D}$ is a diagonal matrix. More generally, $\mathbf{D}$ can have off-diagonal terms to model cross-diffusion.

The classical reaction-diffusion (RD) system which will be analysed in this section is a system of two chemicals, u and v, reacting and diffusing as follows:

$$\frac{\partial u}{\partial t} = D_1\nabla^2 u + f(u, v) \tag{1.8a}$$

$$\frac{\partial v}{\partial t} = D_2\nabla^2 v + g(u, v). \tag{1.8b}$$

We will assume zero flux boundary conditions. The functions f and g are rational functions of u and v (see examples later).

Definition. A uniform steady state of (1.8) is a state $(u, v) = (u_0, v_0)$ where u_0 and v_0 are constants in time and space, satisfying (1.8) and the boundary conditions.

Zero flux boundary conditions are trivially satisfied by any (u_0, v_0), and equations (1.8) are satisfied by

$$f(u_0, v_0) = g(u_0, v_0) = 0.$$

As u and v represent chemical concentrations, we consider only non-negative solutions to these equations.

Definition. Diffusion-driven instability (or Turing instability) occurs when a steady state, stable in the absence of diffusion, goes unstable when diffusion is present.

We now carry out a linear stability analysis to derive the conditions under which a Turing instability can arise.

Let $u = u_0 + \tilde{u}$, $v = v_0 + \tilde{v}$ where $\tilde{u}$ and $\tilde{v}$ are small perturbations from the steady state values u_0, v_0 of u and v respectively.

Substituting into (1.8) and using a Taylor expansion (ignoring quadratic and higher order terms), the equations for $\tilde{u}$ and $\tilde{v}$ become

$$\frac{\partial \tilde{u}}{\partial t} = D_1 \nabla^2 \tilde{u} + f_u \tilde{u} + f_v \tilde{v} \tag{1.9a}$$

$$\frac{\partial \tilde{v}}{\partial t} = D_2 \nabla^2 \tilde{v} + g_u \tilde{u} + g_v \tilde{v} \tag{1.9b}$$

with zero flux boundary conditions, where the partial derivatives f_u, f_v, g_u, g_v are evaluated at (u_0, v_0). We look for a separable solution to (1.9) of the form

$$\tilde{u} = a e^{\lambda t} \phi(\mathbf{x}), \quad \tilde{v} = b e^{\lambda t} \phi(\mathbf{x}) \tag{1.10}$$

where

$$\nabla^2 \phi + k^2 \phi = 0 \tag{1.11}$$

and ϕ satisfies the boundary conditions. The property (1.11) reduces the partial differential equation system (1.9) to an ordinary differential equation system and is equivalent to looking for a Fourier series solution.

Substituting (1.10) into (1.9) leads to the pair of simultaneous equations

$$\begin{pmatrix} \lambda + D_1 k^2 - f_u & -f_v \\ -g_u & \lambda + D_2 k^2 - g_v \end{pmatrix} \begin{pmatrix} a \\ b \end{pmatrix} = \begin{pmatrix} 0 \\ 0 \end{pmatrix} \tag{1.12}$$

which will have non-trivial solutions $(a, b)^T$ if and only if λ satisfies the dispersion relation

$$\lambda^2 - \{f_u + g_v - k^2(D_1 + D_2)\}\lambda + h(k^2) = 0 \tag{1.13}$$

where

$$h(k^2) = D_1 D_2 k^4 - (D_1 g_v + D_2 f_u)k^2 + f_u g_v - f_v g_u. \tag{1.14}$$

The uniform steady state will be linearly stable if Rl $\lambda(k^2) < 0 \ \forall \ k^2 > 0$ and linearly unstable if $\exists \ k^2 > 0$, such that Rl $\lambda(k^2) > 0$. For diffusion-driven instability we require, firstly, that Rl $\lambda(k^2) < 0$ for $k^2 = 0$, that is, the uniform steady state is stable in the absence of diffusion. Hence we require the roots of the quadratic equation

$$\lambda^2(0) - (f_u + g_v)\lambda(0) + f_u g_v - f_v g_u = 0 \tag{1.15}$$

to have negative real part. This occurs iff

$$f_u + g_v < 0 \text{ and } f_u g_v - f_v g_u > 0. \tag{1.16}$$

Secondly, we require there to exist a positive k^2 for which $\lambda(k^2)$ has positive real part. Given (1.16), this can only occur if $h(k^2) < 0$ for some $k^2 > 0$. Hence we require

$$D_1 g_v + D_2 f_u > 0, \tag{1.17}$$

so that $h(k^2)$ has a minimum at a positive value of k^2 $\left(k^2_{\text{min}} = \dfrac{D_1 g_v + D_2 f_u}{2D_1 D_2}\right)$, and

$$D_1 g_v + D_2 f_u > 2\sqrt{D_1 D_2 (f_u g_v - f_v g_u)} \tag{1.18}$$

so that this minimum is negative.

Therefore, the conditions for diffusion-driven instability are:

(C.1) $f_u + g_v < 0$

(C.2) $f_u g_v - f_v g_u > 0$

(C.3) $D_1 g_v + D_2 f_u > 0$

(C.4) $D_1 g_v + D_2 f_u > 2\sqrt{D_1 D_2 (f_u g_v - f_v g_u)}$

Under these conditions, at the onset of instability, λ is purely real so that the instability is stationary. [If (1.13) has complex conjugate roots with a positive real part, then the instability is oscillatory.]

Remark 1: (C.1) and (C.3) $\Rightarrow D_1 \neq D_2$.

Remark 2: (C.1) and (C.3) imply that f_u and g_v have opposite signs. This observation, together with (C.2), implies that, to an arbitrary relabelling of species, any two-component kinetic mechanism that can lead to diffusion-driven instability must give rise to a Jacobian in kinetic terms at (u_0, v_0) with the following sign structure:

$$K_p \equiv \begin{bmatrix} - & + \\ - & + \end{bmatrix}, \quad K_c \equiv \begin{bmatrix} - & - \\ + & + \end{bmatrix} \tag{1.19}$$

Definition. A kinetic mechanism for which the Jacobian is of type K_p (type K_c) is said to be a pure (cross) activator-inhibitor mechanism at (u_0, v_0). Note that the type of a mechanism may vary with (u_0, v_0).

Remark 3: (C.1)-(C.4) ensure that there exist wavenumbers $k^2 > 0$ such that $\lambda(k^2) > 0$. For the uniform steady state to be unstable, at least one of these wavenumbers must lead to an *admissible* solution. That is, the corresponding function ϕ must satisfy the boundary conditions. Hence (C.1)-(C.4) are necessary but not sufficient conditions for diffusion-driven instability.

Example. Suppose we are on the one-dimensional domain $[0, L]$. Then, for zero flux boundary conditions, the functions ϕ are $\cos \frac{n\pi x}{L}$, that is, the admissible wavenumbers, k, are $k_n = \frac{n\pi}{L}$, $n = 1, 2, 3, \ldots$.

Remark 4: The onset of instability is termed a *bifurcation* point. If instability is to an oscillatory solution it is termed a Hopf bifurcation. Note that a Hopf bifurcation can only occur in (1.8) at $k = 0$, that is, the bifurcation can only yield temporally oscillating solutions when the stationary point of the associated space-independent equations loses its stability (A full statement of the Hopf bifurcation theorem can be found in any standard bifurcation textbook).

Consider now the mechanism K_p. Here $g_u < 0$ so that u *inhibits* v, while $f_v > 0$, so that v *activates* u. Furthermore, condition (C.1) $\Rightarrow |f_u| > |g_v|$, so from (C.3) $D_1 > D_2$. That is, the activator diffuses more slowly than the inhibitor. This is an example of the classic property of many self-organising systems, namely *short-range activation, long-range inhibition.*

In Turing's original model, f and g were linear so that, if the uniform steady state became unstable, then the chemical concentrations would grow exponentially. This, of course, is biologically unrealistic. Since Turing's paper, a number of models have been proposed wherein f and g are nonlinear so that when the uniform steady state becomes unstable, it may or may not evolve to a bounded, stationary, spatially non-uniform, steady state (a *spatial pattern*) depending on the nonlinear terms.

These models may be classified into four types:

(i) Phenomenological Models: The functions f and g are chosen so that one of the chemicals is an activator, the other an inhibitor. An example is the Gierer-Meinhardt model (1972):

$$\begin{aligned}\frac{du}{dt} &= \underbrace{\alpha}_{\text{source}} - \underbrace{\beta u}_{\substack{\text{linear}\\ \text{degradation}}} + \underbrace{\frac{\gamma u^2}{v}}_{\substack{\text{autocatalysis in } u/\\ \text{inhibition from } v}} \\ \frac{dv}{dt} &= \underbrace{\delta u^2}_{\substack{\text{activation}\\ \text{by } v}} - \underbrace{\eta v}_{\substack{\text{linear}\\ \text{degradation}}}\end{aligned} \tag{1.20}$$

where $\alpha, \beta, \gamma, \delta$ and η are positive constants.

(ii) Hypothetical Models: Derived from a hypothetically proposed series of chemical reactions. For example, Schnakenberg (1979) proposed a series of trimolecular autocatalytic reactions involving two chemicals as follows

$$X \underset{k_2}{\overset{k_1}{\rightleftharpoons}} A, \quad B \xrightarrow{k_3} Y, \quad 2X + Y \xrightarrow{k_4} 3X.$$

Using the Law of Mass Action, which states that the rate of reaction is directly proportional to the product of the active concentrations of the reactants, and denoting the concentrations of X, Y, A and B by u, v, a and b, respectively, we have

$$f(u,v) = k_2 a - k_1 u + k_4 u^2 v, \quad g(u,v) = k_3 b - k_4 u^2 v \tag{1.21}$$

where $k_1, ..., k_4$ are (positive) rate constants. Assuming that there is an abundance of A and B, a and b can be considered to be approximately constant.

(iii) Empirical Models: The kinetics are fitted to experimental data. For example, the Thomas (1975) immobilized-enzyme substrate-inhibition mechanism involves the reaction of uric acid (concentration u) with oxygen (concentration v). Both reactants diffuse from a reservoir maintained at constant concentration u_0 and v_0, respectively, onto a membrane containing the immobilized enzyme uricase. They react in the presence of the enzyme with empirical rate $\frac{V_m uv}{K_m+u+u^2/K_s}$, so that

$$f(u,v) = \alpha(u_0 - u) - \frac{V_m uv}{K_m + u + u^2/K_s}, \quad g(u,v) = \beta(v_0 - v) - \frac{V_m uv}{K_m + u + u^2/K_s} \tag{1.22}$$

where α, β, V_m, K_m and K_s are positive constants.

(iv) Actual Chemical Reactions: Although Turing predicted, in 1952, the spatial patterning potential of chemical reactions, this phenomenon has only recently been realised in actual chemical reactions. Therefore, it is now possible, in certain cases, to write down detailed reaction schemes and derive, using the Law of Mass Action, the kinetic terms.

The first Turing patterns were observed in the chlorite-iodide-malonic acid starch reaction (CIMA reaction) (Castets *et al.*, 1990, De Kepper *et al.*, 1991). The model proposed by Lengyel and Epstein (1991) stresses three processes: the reaction between malonic acid (MA) and iodine to create iodide, and the reactions between chlorite and iodide and chloride and iodide. These reactions take the form

$$MA + I_2 \rightarrow IMA + I^- + H^+$$
$$ClO_2 + I^- \rightarrow ClO_2^- + \frac{1}{2}I_2$$
$$ClO_2^- + 4I^- + 4H^+ \rightarrow Cl^- + 2I_2 + 2H_2O.$$

The rates of these reactions can be determined experimentally. By making the experimentally realistic assumption that the concentration of malonic acid, chlorine dioxide and iodine are constant, Lengyel and Epstein derived the following model:

$$\begin{aligned}\frac{\partial u}{\partial t} &= k_1 - u - \frac{4uv}{1+u^2} + \nabla^2 u \\ \frac{\partial v}{\partial t} &= k_2\left[k_3\left(u - \frac{uv}{1+u^2}\right) + c\nabla^2 v\right]\end{aligned}$$

where u, v are the concentrations of iodide and chlorite, respectively and k_1, k_2, k_3 and c are positive constants.

Murray (1982) calculates and compares the parameter space determined by (C.1)-(C.4) for instability in the Gierer-Meinhardt, Schnakenberg and Thomas models.

Remark 5: A key problem in the verification of Turing structures is the required variation of diffusion coefficients. For a general reaction-diffusion system, the ratio may be changed as follows: consider a standard two-species reaction-diffusion system of the form

$$\begin{aligned}\frac{\partial u}{\partial t} &= f(u,v) + D_1\nabla^2 u,\\ \frac{\partial v}{\partial t} &= g(u,v) + D_2\nabla^2 v,\end{aligned}$$

where u is the activator and v the inhibitor. We make the additional assumption that the activator is involved in a reaction of the form:

$$U + S \underset{r_2}{\overset{r_1}{\rightleftharpoons}} C.$$

Assuming that both S and C are immobile, the RD system is now modified to:

$$\begin{aligned}\frac{\partial u}{\partial t} &= f(u,v) - r_1 us + r_2 c + D_1\nabla^2 u\\ \frac{\partial v}{\partial t} &= g(u,v) + D_2\nabla^2 v\\ \frac{\partial c}{\partial t} &= r_1 us - r_2 c\end{aligned}$$

where s and c are the concentrations of S and C, respectively, and r_1, r_2 are rate constants. If r_1 and r_2 are large, then using singular perturbation theory, c can be approximated in terms of u by $c \equiv ru$, where $r = s_0 r_1/r_2$ and we have assumed that the concentration of S remains close to its initial value, s_0.

On addition of the first and third equations above, we obtain the following equation for the activator:

$$(1+r)\frac{\partial u}{\partial t} = f(u,v) + D_1\nabla^2 u$$

thus when $r \gg 1$ the diffusion of the activator is greatly reduced.

This demonstrates how the formation of an immobile complex can reduce the effective diffusion rate of the activator species. It was this type of approach that was first used by Lengyel and Epstein (1991) to explain how Turing structures develop in the CIMA reaction. In this case, starch forms a stable complex with triiodide ions via the reaction

$$S + I^- + I_2 \rightleftharpoons SI_3^-$$

and the high molecular weight of the complex reduces the rate of diffusion.

Remark 6: A characteristic of Turing patterns is the intrinsic relation between the average diffusion coefficient of the reactants and the wavelength of the pattern. This characteristic differentiates Turing patterns from other patterning phenomena. Turing demonstrated that near the bifurcation from a uniform steady state to Turing patterns, the wavelength of the pattern is predicted to be $\sqrt{2\pi TD}$, where $D = \sqrt{D_1 D_2}$, and D_1, D_2 are the diffusion coefficients. T is the period of the limit cycle when the system is at the onset of Hopf bifurcations (temporally-varying pattern).

Using two types of gel, and varying the concentrations of the gels, it is possible to experimentally test if Turing's rule is obeyed for the CIMA reaction (Ouyang *et al.*, 1995). With the above variations, pattern wavelengths can be measured when D is varied over a factor of three. The corresponding plot of average diffusion coefficient against the experimental wavelength confirms Turing's prediction. Experimental measurement of the period of limit cycles at the onset of Hopf bifurcation is also in good agreement with the theoretical predictions.

1.2 Nondimensionalised system

To reduce the number of parameters in a model, appropriate non-dimensionalisation may be used. For example, in the Schnakenberg model (1.21), set $\mathbf{x}^* = \frac{\mathbf{x}}{L}$, $\tau = \frac{t}{T}$, $u^* = \frac{u}{U}$, $v^* = \frac{v}{V}$. Then (1.21) becomes

$$\frac{\partial u^*}{\partial \tau} = \frac{k_2 T a}{U} - k_1 T u^* + k_4 T U V u^{*2} v^* + \frac{D_1 T}{L^2} \nabla^{*2} u^* \tag{1.23a}$$

$$\frac{\partial v^*}{\partial \tau} = \frac{k_3 T}{V} b - k_4 T U^2 u^{*2} v^* + \frac{D_2 T}{L^2} \nabla^{*2} v^* \tag{1.23b}$$

Choosing $T = \frac{1}{k_1}$, $U = V = \sqrt{\frac{k_1}{k_4}}$, $L = \sqrt{\frac{D_1}{k_1}}$, reduces this system to (dropping * for notational convenience):

$$\frac{\partial u}{\partial \tau} = \alpha - u + u^2 v + \nabla^2 u \tag{1.24a}$$

$$\frac{\partial v}{\partial \tau} = \beta - u^2 v + \delta \nabla^2 v \tag{1.24b}$$

where $\alpha = a\frac{k_2}{k_1}\sqrt{\frac{k_4}{k_1}}$, $\beta = b\frac{k_3}{k_1}\sqrt{\frac{k_4}{k_1}}$ and $\delta = \frac{D_2}{D_1}$. The number of parameters has been reduced from eight to three. Of course, the nondimensionalisation is not unique.

1.3 Properties of Spatial Patterns

Conditions (C.1)-(C.4) determine domains in parameter space wherein diffusion-driven instability is possible. From linear analysis, a number of predictions can be made on the properties exhibited by spatially patterned solutions to the reaction-diffusion system (1.8):

P.1 Pattern complexity depends on domain size:

Under the conditions (C.1)-(C.4), there exists a range of wavenumbers, k, of possible spatial patterns satisfying

$$k_-^2 < k^2 < k_+^2 \tag{1.25}$$

where k_-^2 and k_+^2 are solutions to Rl $\lambda(k^2) = 0$, and Rl $\lambda(k^2) > 0$ for $k^2 \in (k_-^2, k_+^2)$. The values of $k_\pm^2$ depend on the parameters. On one-dimensional domains with zero flux boundary conditions, or Dirichlet conditions fixed at the spatially uniform steady state, admissible wave numbers are of the form $k = \frac{n\pi}{L}$ where L = domain length. Therefore as L increases, n must increase in order for $\frac{n^2\pi^2}{L^2} \in [k_-^2, k_+^2]$. That is, pattern complexity increases with L. On the two-dimensional domain $[0, L_x] \times [0, L_y]$ with zero flux boundary conditions, the spatial component of the linear solution takes the form

$$\phi_{nm} = \cos\frac{n\pi x}{L_x}\cos\frac{m\pi y}{L_y} \tag{1.26}$$

where n and m are integers (at least one of which is non-zero).

Hence admissible k are now of the form $k_{nm}^2 = \left(\frac{n^2}{L_x^2} + \frac{m^2}{L_y^2}\right)\pi^2$. Therefore, if the domain is long and narrow, i.e. $L_x \gg L_y$, then (1.25) will only be satisfied if $m = 0$, i.e., the pattern depends only on the x coordinate.

P.2 Phase relationship between solutions:

At a primary bifurcation point, from equation (1.12),

$$a = \frac{f_v b}{D_1 k^2 - f_u}. \tag{1.27}$$

Therefore, for a pure activator-inhibitor mechanism sgn a = sgn b, while for a cross activator-inhibitor mechanism sgn $a = -$ sgn b. Hence solutions for the pure (cross) activator-inhibitor mechanism are in (out of) phase, at least in the vicinity of the primary bifurcation point.

1.4 Nonlinear Analysis

All the above results are based on linear theory. As the solution begins to grow, however, nonlinear terms become important and linear theory is no longer valid, i.e. linear theory holds on a short time scale but breaks down on a long time scale. It is possible to calculate the solution in the vicinity of a primary bifurcation point as follows (for full details see, for example, Fife, 1979, Britton 1986, Grindrod, 1996):

Consider the two species RD system in the one space dimension $[0,\pi]$

$$\frac{\partial \mathbf{u}}{\partial t} = \mathbf{f}(\mathbf{u}) + \mathbf{D}\frac{\partial^2 \mathbf{u}}{\partial x^2} \tag{1.28}$$

with zero flux boundary conditions, where $\mathbf{D} = \begin{bmatrix} 1 & 0 \\ 0 & \delta \end{bmatrix}$. Suppose that δ is the bifurcation parameter with critical value δ_c such that the uniform steady state $\mathbf{u}_0$ loses linear stability for $\delta < \delta_c$. Assume that at this point, $\phi_m = \cos mx$ is the first mode to have positive growth rate in time for some integer m.

Set $\delta = \delta_c - \delta_1 \varepsilon$, where $\varepsilon \ll 1$ and δ_1 is a constant, and expand any equilibrium solutions via:

$$\mathbf{u}(x) = \mathbf{u}_0 + \sum_{n=1}^{\infty} \varepsilon^{\xi n}\mathbf{u}_n(x) \tag{1.29}$$

where ξ is a positive constant to be determined, and $\mathbf{u}_n = (u_n, v_n)^t$.

Substituting (1.29) into (1.28) and expanding in Taylor series, we have

$$\begin{aligned}
\mathbf{f}(\mathbf{u}) &= \mathbf{f}(\mathbf{u}_0) + \varepsilon^{\xi} d\mathbf{f}(\mathbf{u}_0).\mathbf{u}_1 + \varepsilon^{2\xi}(d\mathbf{f}(\mathbf{u}_0).\mathbf{u}_2 + \frac{1}{2}u_1^2\mathbf{f}_{uu} + u_1v_1\mathbf{f}_{uv} + \frac{1}{2}v_1^2\mathbf{f}_{vv}) \\
&\quad + \varepsilon^{3\xi}(d\mathbf{f}(\mathbf{u}_0).\mathbf{u}_3 + \frac{1}{2}u_1u_2\mathbf{f}_{uu} + (u_1v_2 + u_2v_1)\mathbf{f}_{uv} + \frac{1}{2}v_1v_2\mathbf{f}_{uv} \\
&\quad + \frac{1}{6}u_1^3\mathbf{f}_{uuu} + \frac{1}{2}u_1^2v_1\mathbf{f}_{uuv} + \frac{1}{2}u_1v_1^2\mathbf{f}_{uvv} + \frac{1}{6}v_1^3\mathbf{f}_{vvv}) \\
&\quad + O(\varepsilon^{4\xi}),
\end{aligned}$$

where $d\mathbf{f}(\mathbf{u}_0)$ is the Jacobian of $\mathbf{f}$ evaluated at $\mathbf{u}_0$, and $\mathbf{f}_{uu}, \mathbf{f}_{uv}$ etc. denote the vectors obtained by partially differentiating $\mathbf{f}$ componentwise, evaluated at $\mathbf{u}_0$.

Equating powers of ε we find that $\xi = \frac{1}{2}$. Denoting by L the linear opeartor

$$L\mathbf{u} = \left\{\mathbf{D}\frac{\partial^2}{\partial x^2} + d\mathbf{f}(\mathbf{u}_0)\right\}\mathbf{u}$$

we find that at $O(\varepsilon^{1/2})$,

$$L\mathbf{u}_1 = 0.$$

Thus $\mathbf{u}_1 = A\mathbf{a}\cos mx$ where $\mathbf{a}$ is an eigenvector of the matrix

$$L_m = d\mathbf{f}(\mathbf{u}_0) - m^2\mathbf{D}$$

and A is a real constant to be determined. At $O(\varepsilon)$, $L\mathbf{u}_2 = \mathbf{R}_2$ which can be solved to give $\mathbf{u}_2 = B\mathbf{a}\cos mx + \mathbf{b}\cos 2mx$, where B is some constant and $\mathbf{b}$ can be found in terms of the components of $\mathbf{a}$ and the constant A.

At $O(\varepsilon^{3/2})$ we have

$$L\mathbf{u}_3 = \mathbf{R}_3$$

where $\mathbf{R}_3$ contains secular terms. Using the Fredholm Alternative, the solvability condition is that $\mathbf{R}_3$ must be orthogonal to $\mathbf{a}^* \cos mx$ as functions in $L_2((0,\pi),\mathbf{R}^2)$, where $\mathbf{a}^*$ is an eigenvector of the adjoint of L_m. That is,

$$\int_0^\pi \mathbf{R}_3^t.\mathbf{a} \cos mx dx = 0.$$

This leads to an equation of the form

$$0 = A(\delta_1 + lA^2)$$

where the Landau constant l can be found in terms of the parameters of the system (but is independent of the bifurcation parameter δ_1). Hence we have the solution

$$\mathbf{u} = \mathbf{u}_0 \pm \sqrt{\frac{\delta_c - \delta}{\delta_1}} |A_0| \bar{\mathbf{a}} \cos mx + O(\delta_c - \delta), \tag{1.30}$$

where $A_0^2 = -\delta_1/l$ (if the latter is positive). Thus, $l < 0$ yields the existence of a stable spatially period pattern for $\delta_1 > 0$ (a supercritical bifurcation), while $l > 0$ yields the existence of an unstable spatially periodic pattern for $\delta_1 < 0$ (a subcritical bifurcation). See, for example, Sattinger (1972).

Remark 1. Note that this analysis holds only in the vicinity of a bifurcation point, that is, for ε small, and is termed a *weakly nonlinear analysis*. To study solution behaviour far away from the steady state one must use other techniques, for example, numerical continuation and bifurcation techniques. The software package AUTO (Doedel, 1986), for example, discretizes the steady state equations using finite differences and solves the resulting nonlinear algebraic system.

Remark 2. This type of analysis can be carried out on domains of more than one space dimension. In this case, the problem of degeneracy can arise. For example, consider the domain $[0, 2\pi] \times [0, 2\pi]$ with zero flux boundary conditions and assume without loss of generality that the first unstable mode occurs at the wavenumber $k = 1$. Then

$$\mathbf{u}_1 = \mathbf{a}(A_1 \cos x + A_2 \cos y)$$

where A_1 and A_2 are arbitrary constants to be determined. Note that $A_1 = 0$, $A_2 \neq 0$ corresponds to a 'stripe' parallel to the x axis, $A_1 \neq 0$, $A_2 = 0$ corresponds to a 'stripe' parallel to the y axis, while if both A_1 and A_2 are non-zero and in particular, equal, then we have a 'spot'. Extending the above weakly nonlinear analysis to such a case in two dimensions Ermentrout, 1991, showed that both types of solution could exist but that they were mutually exclusive as stable patterns. Specifically, in a symmetric system with no quadratic terms and only cubic terms, stripes are always selected over spots. Spots can only stably exist if quadratic terms are present in the nonlinearities. However, Benson *et al.*, (1997), have shown that when a spatially varying diffusion is included it is possible to force an RD system that would exhibit stable spots to exhibit stable stripes.

1.5 Inhomogeneous Domains and the Role of boundary conditions

The linear analysis presented in Section 1.1 holds for the case of spatially uniform parameters and zero flux boundary conditions (or Dirichlet conditions, fixed at the spatially uniform steady state). Here we consider two cases where neither of these conditions hold.

(a) Inhomogeneous domain. Let us consider a one-dimensional domain, $x \in [0,1]$, where the diffusion coefficients of one of the chemicals is spatially non-uniform. For simplicity, let us assume that in the general two chemical RD system we have nondimensionalised the equations such that $D_1 = 1$, and that $D_2 = D(x)$ is the step function

$$D(x) = \begin{cases} D^-, & 0 \leq x < \xi \\ D^+, & \xi < x \leq 1 \end{cases}$$

where $\xi \in (0,1)$ and $D^- < D^+$.

The requirements for (u_0, v_0) to be stable to spatially homogeneous perturbations remain as before. To derive the analogues of (C.3) and (C.4), we linearise the model about the steady state (u_0, v_0), and look for separable solutions of this linearized system, in the form $u - u_0 = e^{\lambda t} X_u(x)$, $v - v_0 = e^{\lambda t} X_v(x)$. Substituting into the linearized model gives coupled ordinary differential equations for X_u and X_v:

$$X_u'' + (a - \lambda)X_u + bX_v = 0 \tag{1.31a}$$

$$[D(x)X_v']' + cX_u + (d - \lambda)X_v = 0; \tag{1.31b}$$

here prime denotes d/dx and, for notational simplicity, we denote by a, b, c, d, the values of f_u, f_v, g_u, g_v, respectively, at (u_0, v_0). We consider these equations separately on $[0,\xi)$ and $(\xi,1]$. In the former case, adding (1.31a) to s^-/D^- times (1.31b) gives:

$$(X_u+s^-X_v)''+\left[a-\lambda+\frac{cs^-}{D^-}\right]\left[X_u+\frac{[b+(d-\lambda)s^-/D^-]}{[a-\lambda+cs^-/D^-]}X_v\right]=0. \tag{1.32}$$

We choose s^- such that:

$$\frac{b+(d-\lambda)s^-/D^-}{a-\lambda+cs^-/D^-}=s^-. \tag{1.33}$$

which is a quadratic equation for s^-, with roots s_1^- and s_2^- say. Equation (1.32) then becomes a single equation in $X_u+s_j^-X_v$, for $j=1,2$, with general solution $C_j\cos(\alpha_j^-x)+D_j\sin(\alpha_j^-x)$. Here C_j and D_j are constants of integration, and $\alpha_j^-=[a-\lambda+cs_j^-/D^-]^{1/2}$, $j=1,2$. We therefore have two simultaneous equations for $X_u(x)$ and $X_v(x)$ in $[0,\xi)$. Solving these and applying zero flux boundary conditions at $x=0$ gives:

$$X_u(x)=\frac{1}{(s_2^--s_1^-)}\left[\frac{(\Gamma_u+s_1^-\Gamma_v)s_2^-}{\cos(\xi\alpha_1^-)}\cos(\alpha_1^-x)-\frac{(\Gamma_u+s_2^-\Gamma_v)s_1^-}{\cos(\xi\alpha_2^-)}\cos(\alpha_2^-x)\right] \tag{1.34a}$$

$$X_v(x)=\frac{1}{(s_2^--s_1^-)}\left[\frac{(\Gamma_u+s_2^-\Gamma_v)}{\cos(\xi\alpha_2^-)}\cos(\alpha_2^-x)-\frac{(\Gamma_u+s_1^-\Gamma_v)}{\cos(\xi\alpha_1^-)}\cos(\alpha_1^-x)\right] \tag{1.34b}$$

on $[0,\xi)$, where $\Gamma_u=X_u(\xi)$, $\Gamma_v=X_v(\xi)$. In (1.34), we are assuming that $s_1^-\neq s_2^-$ and $\cos(\xi\alpha_j^-)\neq 0$ for 1,2; these special cases are discussed in Benson *et al.*, (1993). Similarly, on $(\xi,1]$:

$$X_u(x)=\frac{1}{(s_2^+-s_1^+)}\left[\frac{(\Gamma_u+s_1^+\Gamma_v)s_2^+}{\cos((1-\xi)\alpha_1^+)}\cos(\alpha_1^+(1-x))-\frac{(\Gamma_u+s_2^+\Gamma_v)s_1^+}{\cos((1-\xi)\alpha_2^+)}\cos(\alpha_2^+(1-x))\right] \tag{1.34c}$$

$$X_v(x)=\frac{1}{(s_2^+-s_1^+)}\left[\frac{(\Gamma_u+s_2^+\Gamma_v)}{\cos((1-\xi)\alpha_2^+)}\cos(\alpha_2^+(1-x))-\frac{(\Gamma_u+s_1^+\Gamma_v)}{\cos((1-\xi)\alpha_1^+)}\cos(\alpha_1^+(1-x))\right] \tag{1.34d}$$

By design, this solution is continuous at $x=\xi$, but we also require it to satisfy continuity of flux, that is:

$$\lim_{x\to\xi^-}X_u'(x)=\lim_{x\to\xi^+}X_u'(x)\quad \lim_{x\to\xi^-}D^-X_v'(x)=\lim_{x\to\xi^+}D^+X_v'(x). \tag{1.35}$$

Substituting the solutions (1.34) into (1.35) gives:

$$P(\lambda)\Gamma_u+Q(\lambda)\Gamma_v=0$$

$$R(\lambda)\Gamma_u+S(\lambda)\Gamma_v=0,$$

where

$$\begin{aligned}
P(\lambda) &= (s_1^- T_2^- - s_2^- T_1^-)/(s_2^- - s_1^-) + (s_1^+ T_2^+ - s_2^+ T_1^+)/(s_2^+ - s_1^+) \\
Q(\lambda) &= s_1^- s_2^- (T_2^- - T_1^-)/(s_2^- - s_1^-) + s_1^+ s_2^+ (T_2^+ - T_1^+)/(s_2^+ - s_1^+) \\
R(\lambda) &= D^-(T_1^- - T_2^-)/(s_2^- - s_1^-) + D^+(T_1^+ - T_2^+)/(s_2^+ - s_1^+) \\
S(\lambda) &= D^-(s_1^- T_1^- - s_2^- T_2^-)/(s_2^- - s_1^-) + D^+(s_1^+ T_1^+ - s_2^+ T_2^+)/(s_2^+ - s_1^+)
\end{aligned}$$

and $T_j^- = \alpha_j^- \tan(\xi\alpha_j^-)$, $T_j^+ = \alpha_j^+ \tan((1-\xi)\alpha_j^+)$, for $j = 1, 2$. Now from (1.34), $\Gamma_u = \Gamma_v = 0$ implies that $X_u(x) \equiv X_v(x) \equiv 0$. Thus for non-trivial X_u and X_v, we require:

$$F(\lambda) \equiv P(\lambda)S(\lambda) - Q(\lambda)R(\lambda) = 0. \tag{1.36}$$

This is the dispersion relation, relating growth rates of instabilities to the model parameter values. The model system will exhibit diffusion-driven instability provided (C.1) and (C.2) are satisfied, and provided this dispersion relation has a solution with positive real part. Our derivation of (1.34) assumes that $\cos(\xi\alpha_1^-)$, $\cos(\xi\alpha_2^-)$, $\cos((1-\xi)\alpha_1^+)$, $\cos((1-\xi)\alpha_2^+)$ and $(s_1^+ - s_2^+)$ are all non-zero. Similar analysis can be done in the cases when one or more of these is zero, but the solutions for u and v cannot in general satisfy continuity of flux at $x = \xi$. One notable exception to this, however, is the homogeneous case $D^- = D^+$. The solutions of the dispersion relation given by the standard analysis (see above) satisfy $\alpha_j^\pm = n\pi$ for some $n \in [1, 2, 3, ...\}$ and either $j = 1$ or $j = 2$. Thus with $\xi = 1/2$, half of the eigenvalues λ are not roots of (1.33), since $\cos(n\pi/2) = 0$ when n is even. These roots can be retrieved, however, either by investigating the above special cases, or by taking more general values of ξ: the value of ξ is irrelevant when $D^- = D^+$.

A typical functional form of $F(\lambda)$ is illustrated in Fig. 1.1 (see Maini *et al.*, 1992 and Benson *et al.*, 1993 for full details).

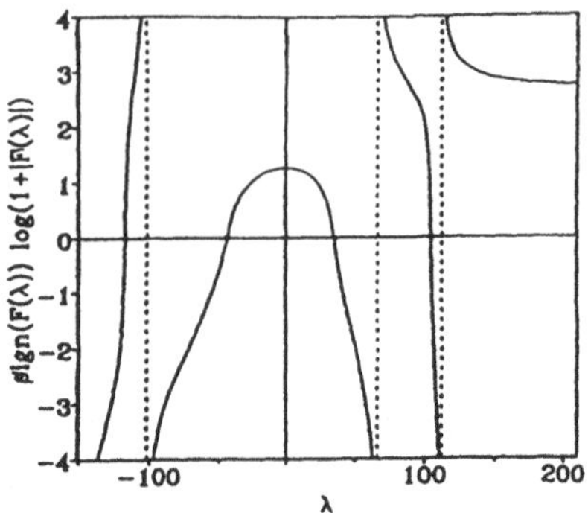

Figure 1.1: A typical functional form of the dispersion relation $F(\lambda)$ defined in (1.36) for Schnakenberg kinetics. (See Maini *et al.*, 1992, Figure 2, for full details). Reproduced from Maini *et al.*, 1992, by permission of Oxford University Press.

(b) Mixed boundary conditions. For boundary conditions of the form (1.6), a uniform steady state will not, in general, exist and therefore the linear analysis of Section 1.1 cannot be carried out. Here we consider a special case where a uniform steady state does exist but each chemical satisfies a different boundary condition. Consider the one-dimensional domain [0,1] and suppose that u is fixed at u_0 on the boundary, while v satisfies zero flux boundary conditions. The linearised system (1.9) now has solutions of the form

$$\begin{pmatrix} \tilde{u} \\ \tilde{v} \end{pmatrix} = e^{\lambda t}\Phi(x)$$

where

$$\Phi = \begin{pmatrix} \sum_{m=1}^{\infty} A_m \sin(m\pi x) \\ \sum_{n=0}^{\infty} B_n \cos(n\pi x) \end{pmatrix}. \tag{1.37}$$

By contrast, in the classical linear problem with scalar homogeneous Neumann conditions, the nth eigenfunction is of the form

$$\Phi = \begin{pmatrix} A_n \\ B_n \end{pmatrix} \cos(n\pi x). \tag{1.38}$$

Substituting (1.37) into the linearised system we obtain

$$\sum_{m=1}^{\infty} [\lambda + v(m\pi)^2 - a] A_m \sin(m\pi x) = \sum_{n=0}^{\infty} bB_n \cos(n\pi x)$$

$$\sum_{n=0}^{\infty}[\lambda + v\delta(n\pi)^2 - d]B_n\cos(n\pi x) = \sum_{m=1}^{\infty} cA_m\sin(m\pi x). \tag{1.39}$$

Multiplying through by $\sin(m\pi x)$ and integrating over [0,1] leads to

$$\begin{aligned} [\lambda + v(m\pi)^2 - a]A_m &= \sum_{n=0}^{\infty} bB_n\alpha_{nm} \\ cA_m &= \sum_{n=0}^{\infty}[\lambda + v\delta(n\pi)^2 - d]B_n\alpha_{nm} \end{aligned} \tag{1.40}$$

where

$$\alpha_{nm} = \frac{\langle\cos(n\pi x), \sin(m\pi x)\rangle}{\langle\sin(m\pi x), \sin(m\pi x)\rangle} \text{ and } \langle f, g\rangle = \int_0^1 f(x)g(x)dx.$$

System (1.40) is an infinite system of linear equations for the infinite number of unknowns A_m, B_n, $m = 1, 2, 3, \ldots, n = 0, 1, 2, 3, \ldots$. To solve this system we make a finite dimensional approximation (FDA) by considering only values of m up to M and truncating the sums at $n = M - 1$. This leads to a system of $2M$ equations for the $2M$ unknowns $(A_1, A_2, \ldots, A_M, B_0, B_1, \ldots, B_{M-1})$. We may rewrite the truncated system as the generalized eigenvalue problem

$$\mathbf{Px} = \lambda\mathbf{Qx} \tag{1.41}$$

where $\mathbf{x} = (A_1, A_2, \ldots, A_M, B_1, \ldots, B_{M-1})$ and $\mathbf{P}$ and $\mathbf{Q}$ are $2M \times 2M$ matrices which have the block structure

$$\mathbf{P} \equiv \begin{bmatrix} \mathbf{P}_1 & \mathbf{P}_2 \\ \mathbf{P}_3 & \mathbf{P}_4 \end{bmatrix}, \quad \mathbf{Q} \equiv \begin{bmatrix} -\mathbf{I} & 0 \\ 0 & \mathbf{Q}_4 \end{bmatrix}.$$

Here $\mathbf{P}_i$ and $\mathbf{Q}_i$ are $M \times M$ matrices given by

$$\begin{aligned} (\mathbf{P}_1)_{ij} &= \begin{cases} 0 & \text{if } i \neq j \\ v(m\pi)^2 - a & \text{if } i = j \end{cases} \\ (\mathbf{P}_2)_{ij} &= -b\alpha_{j-1,i} \quad \mathbf{P}_3 = c\mathbf{I} \\ (\mathbf{P}_4)_{ij} &= (v\delta(j-1)^2\pi^2 - d)\alpha_{j-1,i} \\ (\mathbf{Q}_4)_{ij} &= \alpha_{j-1,i} \end{aligned}$$

where $i, j = 1, 2, \ldots, M$, and $\mathbf{I}$ is the $M \times M$ unit matrix.

The solution of (1.41) leads to $2M$ eigenvalues λ_i with corresponding eigenvectors $\mathbf{x}_i$. Thus the M-dimensional approximation to the solution is

$$\begin{pmatrix} \tilde{u}_1 \\ \tilde{v}_2 \end{pmatrix} = \sum_{i=1}^{2M} \begin{pmatrix} \sum_{m=1}^{M} A_{mi}\sin(m\pi x) \\ \sum_{n=0}^{M-1} B_n\cos(n\pi x) \end{pmatrix} e^{\lambda_i\tau} \tag{1.42}$$

where $(A_{1i}, A_{2i}, ..., A_{Mi}, B_{0i}, B_{1i}, ..., B_{(M-1)i})$ is the eigenvector with eigenvalue λ_i.

As the dimension of the FDA is increased, the values of the previously calculated eigenvalues and eigenvectors will change and more eigenvalues and corresponding eigenvectors will be generated. We use the following criteria as stopping tests.

(i) The approximation for a chosen λ_i and its corresponding eigenvector must converge as the dimension of the FDA is increased. In the case of the eigenvectors this also means that higher order terms are insignificant. For any given λ_i there is an M_c such that the computed eigenvalue and eigenvector are sufficiently accurate for $M = M_c$.

(ii) The eigenvalues introduced for $M > M_c$ have real part negative, and thus correspond to temporally decaying solutions of the linearised system.

If both these criteria are satisfied, then we are assured that the FDA only ignores exponentially decaying terms in time and insignificantly small terms in the trigonometric expansion of the spatial component of the solution to the linearised problem. Furthermore, if some of the eigenvalues obtained by the FDA have positive real part, then the uniform steady state is unstable and we postulate that the solution will evolve to a spatially varying solution of the form (1.42) with temporal growth rates given by the real part of the positive eigenvalues (assuming, of course, that a bounded solution exists). This procedure is illustrated in Dillon *et al.*, (1994).

We can gain some insight into the connection with the Neumann problem as follows. Note that the system (1.40) may be reduced by eliminating A_m to

$$\sum_{n=0}^{\infty}[(\lambda + v(m\pi)^2 - a)(\lambda + v\delta(n\pi)^2 - d) - bc]\alpha_{nm}B_n = 0 \qquad (1.43)$$

for $m = 1, 2, 3, \ldots$.

This may be written as

$$\sum_{n=0}^{\infty}\{\lambda^2 + [v(m\pi)^2 + \delta v(n\pi)^2 - \text{trace } K]\lambda + \delta v^2(m\pi)^2(n\pi)^2 - [av(m\pi)^2 + dv(n\pi)^2] + \det K\}B_n\alpha_{nm} = 0,$$

where K is the Jacobian matrix $\begin{bmatrix} a & b \\ c & d \end{bmatrix}$.

As $\alpha_{nm} = 0$ for $n + m = 2p$, the infinite system represented by (1.43) can be written as

$$\Omega(\lambda)\mathbf{B} = \begin{bmatrix} \Omega_1(\lambda) & 0 \\ 0 & \Omega_2(\lambda) \end{bmatrix}\begin{bmatrix} \mathbf{B}_e \\ \mathbf{B}_o \end{bmatrix}.$$

Here $\mathbf{B}_e = (B_0, B_2 \ldots)^T$ and $\mathbf{B}_o = (B_1, B_3 \ldots)^T$. Thus det $\Omega(\lambda) =$ det $\Omega_1(\lambda)$ det $\Omega_2(\lambda)$ and the eigenvectors decompose into those with only $B_j \neq 0$ for j even and those with only $B_j \neq 0$ for j odd.

The above analysis can be used to locate bifurcation points as a certain parameter p is varied using the method of bisection as follows: we choose a low dimensional FDA, a value of the parameter p, say p_1, at which all eigenvalues have real part negative and another value p_2 at which at least one eigenvalue has positive real part. Assume, without loss of generality, that $p_2 > p_1$. Clearly, a certain number of eigenvalues must cross the axis in (p_1, p_2). By examining the signs of the eigenvalues at the midpoint $(p_1 + p_2)/2$ of the interval we can easily determine in which half of the interval the bifurcation point lies. We can continue this procedure to find the bifurcation point to the required degree of accuracy. By going to a higher dimensional FDA we may obtain a more accurate value of the parameter at the bifurcation point.

2 Cell-Chemotactic Models

2.1 Model Formulation

The model involves two dependent variables, cell density, $n(\mathbf{x}, t)$, and chemoattractant concentration, $c(\mathbf{x}, t)$, where $\mathbf{x}$ and t are the spatial coordinate and time, respectively. Following Section 1.1, these variables satisfy the equations

$$\frac{\partial n}{\partial t} = -\nabla.\mathbf{J}_n + f(n, c) \tag{2.1}$$

$$\frac{\partial c}{\partial t} = -\nabla.\mathbf{J}_c + g(n, c) \tag{2.2}$$

where $\mathbf{J}_n, \mathbf{J}_c$ and $f(n, c), g(n, c)$ are cell, chemical flux and net production, respectively. It is assumed that two processes contribute to cell flux: random motion, and motion in response to gradients in chemical concentration. Hence

$$\mathbf{J}_n = \mathbf{J}_n^d + \mathbf{J}_n^c \tag{2.3}$$

where $\mathbf{J}_n^d = -D_n \nabla n$, that is, Fickian type motion with diffusion coefficient $D_n (> 0)$, and $\mathbf{J}_n^c = \chi(c) n \nabla c$, modelling chemotactic motion. The function $\chi(c)$ measures chemotactic sensitivity and can take a number of forms. For example, the simplest form for $\chi(c)$ is a constant. This assumes that the sensitivity of cells to attractant is independent of attractant concentration. If cell sensitivity to c is known to decrease with c, then one possible form for $\chi(c)$ is α/c where α is a constant (Keller and Segel,

1971). The action of cell surface receptors has been modelled by setting $\chi(c) = \frac{\alpha}{(\alpha_1+c)^2}$ where α and α_1 are constants (Lapidus and Schiller, 1976; Ford and Lauffenberger, 1991). This modelling approach is phenomenological. Recently, Othmer and Stevens (1997) have derived macroscopic chemotaxis equations based on microscopic rules using a random walk approach. They show how specific assumptions at a microscopic level can give rise to the types of phenomenological macroscopic models mentioned above.

Typically, net cell production is assumed to follow logistic growth, that is, $f(n, c) = rn(N - n)$ where r and N are non-negative constants. Hence, at low cell densities, growth is exponential, while at high cell densities (approaching N), cell growth tends to zero.

The chemical is assumed to diffuse according to Fick's Law, hence $\mathbf{J}_c = -D_c \nabla c$, where D_c is the (non-negative) diffusion coefficient. Net chemical production is composed of two parts: production by the cells minus degradation. There are a number of ways to model these terms depending on the biological situation. As an example, chemical production may be a saturating function of cell density, modelled by the term $\frac{Sn}{\beta+n}$, where S and β are positive constants. Here, chemical production rate per unit cell $\frac{S}{\beta+n}$, decreases with cell density, to account for the process of contact inhibition whereby at high cell densities various metabolic pathways are turned off. Degradation may be simply of the form $-\gamma c$, where γ is the rate of linear degradation.

The system (2.1)-(2.2) may be analysed in an analogous fashion to that of (1.8) and shown to have the ability to produce spatial pattern. In this case, activation is due to cells producing chemical which attracts more cells by chemoattraction, while inhibition is due to cell depletion in the neighbourhood of a cell aggregation. However, the nonlinear flux term means that the model is not as well behaved as the standard RD system and cell-chemotactic models can exhibit blow-up (see, for example, Childress and Percus, 1981 and Othmer and Stevens, 1997). A detailed numerical bifurcation analysis of a version of (2.1)-(2.2) in two spatial dimensions was carried out by Maini *et al.*, (1991). Their results are considered in Section 4.3. The model can also exhibit propagating patterns and these will be the focus of study in the next section.

2.2 Propagating Patterns

We consider here the non-dimensionalised version of (2.1)-(2.2) studied, in one spatial dimension, by Myerscough and Murray, 1992:

$$\frac{\partial n}{\partial t} = D_n \frac{\partial^2 n}{\partial x^2} - \alpha \frac{\partial}{\partial x}\left(n \frac{\partial c}{\partial x}\right) \tag{2.4a}$$

$$\frac{\partial c}{\partial t} = \frac{\partial^2 c}{\partial x^2} + \frac{n}{1+n} - c \tag{2.4b}$$

Here, it has been assumed that the chemotactic sensitivity is constant (α) and that, on the timescale of interest, there is no net change in cell density.

Equations (2.4) have a one parameter family of homogeneous steady states

$$n = n_0, \quad c = c_0 = \frac{n_0}{1+n_0} \tag{2.5}$$

where n_0, the average cell density, is a constant parameter. If we assume zero flux boundary conditions, then the initial cell density determines n_0, which can then be considered as a fixed parameter. Linearising (2.4) about (2.5), and solving the resulting equations gives the dispersion relation

$$\lambda^2 + [k^2(D_n+1)+1]\lambda + k^2\left[D_n(k^2+1) - \frac{\alpha n_0}{(1+n_0)^2}\right] = 0. \tag{2.6}$$

Note that in this case $\lambda = 0$ at $k = 0$, so the standard weakly nonlinear theory presented in the previous section does not automatically apply (see Grindrod *et al.*, 1989).

If the system is set initially to a spatially uniform steady state and then a perturbation in cell density is imposed at one end, a regular pattern of standing peaks and troughs is generated progressively. The wavelength of the pattern and its speed of spread appear constant. Myerscough and Murray analysed this propagating pattern with a method based on that developed by Dee and Langer (1983). At the leading edge of the pattern, the amplitude of the disturbance is small and linear theory applies. Therefore, the solution to (2.4) at the leading edge may be written as the integral of Fourier modes

$$n(x,t) = \int_{-\infty}^{\infty} A(k)\exp[ikx + \lambda(k^2)t]dk \tag{2.7}$$

where $\lambda(k^2)$ is given by (2.6) and $A(k)$ is determined by the initial conditions.

This solution is valid at the leading edge and, assuming that the speed of propagation is v, a constant, we have

$$n(vt,t) = \int_{-\infty}^{\infty} A(k)\exp[tg(k)]dk \tag{2.8}$$

where $g(k) = ikv + \lambda(k^2)$. Using the method of steepest descents (see, for example, Murray, 1984), this integral can be evaluated asymptotically for large t and x to yield

$$n \sim \frac{F(k^*)}{\sqrt{t}} \exp[t(ik^*v + \lambda(k^{*2})] \tag{2.9}$$

where F is a function of k which is not important for this particular analysis and k^* is a saddle point of $g(k)$ in the complex k-plane (i.e. a solution to $\frac{dg}{dk} = 0$), chosen so that $\mathrm{Re}(ik^*v + \lambda(k^{*2})) > \mathrm{Re}(ikv + \lambda(k^2))|_{k\in\alpha}$ where α is the set of all saddle points other than k^*.

As the envelope of the pattern is constant in shape far from the initial perturbation, it is assumed that $\mathrm{Re}(g(k^*)) = 0$. This is called the marginal stability hypothesis. Hence, v and k^* satisfy the equations:

$$iv + \frac{d\lambda}{dk} = 0 \tag{2.10}$$

and

$$\mathrm{Re}(ik^*v + \lambda(k^{*2})) = 0. \tag{2.11}$$

Now consider the pattern in the moving frame of the envelope. In this frame, we see an oscillating pattern at the leading edge with frequency of oscillation

$$\Omega = \mathrm{Im}(g(k)). \tag{2.12}$$

This is the frequency at which nodes are created at the front of the envelope. Assuming that peaks do not coalesce and are conserved, this is also the frequency of oscillation far from the leading edge. If k' is the wave number of the pattern far behind the leading edge, then $\Omega = k'v$, so

$$k' = Re(k^*) + Im\frac{\lambda(k^{*2})}{v}. \tag{2.13}$$

This method assumes that the equations are weakly nonlinear in the vicinity of the leading edge and that the behaviour of the solution is determined by the leading edge.

Myerscough and Murray carried out a comparison of the analytical results with numerical simulations and found that for small amplitude solutions there was good quantitative agreement but for large amplitude solutions the agreement was only qualitative. In the latter, nonlinear effects play a crucial role. For example, peaks produced near the leading edge tend to grow and coalesce, hence invalidating the peak counting argument.

An application of the type of propagating patterns discussed in this section is presented in Section 4.3.

3 Mechanical Models

3.1 Model Formulation

The mechanical model proposed by Oster *et al.*, (1983) has three dependent variables: $n(\mathbf{x},t)$, $\rho(\mathbf{x},t)$ and $\mathbf{u}(\mathbf{x},t)$ which represent, respectively, cell density, matrix density and matrix displacement at position $\mathbf{x}$ and time t. The cell and matrix equations take the general form (1.3) and we consider them first.

The cell equation is

$$\frac{\partial n}{\partial t} = -\nabla.\mathbf{J} + rn(N-n) \tag{3.1}$$

where J is cell flux and cell growth is assumed to be of logistic form. In this case, cell flux is due to three processes.

(i) Random motion, modelled as usual by $\mathbf{J}_d = -D\nabla n$

(ii) Haptotaxis: Cells move by attaching cell processes to adhesive sites within the matrix and crawling along. As cells exert forces on the extracellular matrix (ECM) they generate adhesive gradients which serve as guidance cues to motion. The movement up such gradients is termed haptotaxis. Assuming that the number of adhesive sites is proportional to ECM density, we have that $\mathbf{J}_h = \alpha n\nabla\rho$, where α is the haptotactic coefficient, assumed to be a non-negative constant.

(iii) Advection: As the matrix is deformed, cells may be carried or dragged passively along. This is termed advection and contributes a flux $\mathbf{J}_a = n\frac{\partial \mathbf{u}}{\partial t}$, where $\mathbf{u}$ is the displacement of a material point of ECM.

Hence, the total cell flux, J, is given by $\mathbf{J} = \mathbf{J}_d + \mathbf{J}_h + \mathbf{J}_a$, so that the equation for cell motion is

$$\frac{\partial n}{\partial t} = D\nabla^2 n - \alpha\nabla.(n\nabla\rho) - \nabla.\left(n\frac{\partial \mathbf{u}}{\partial t}\right) + rn(N-n) \tag{3.2}$$

The matrix equation is much simpler as the only contribution to matrix flux is advection, and matrix secretion is assumed negligible. Hence, ρ satisfies

$$\frac{\partial \rho}{\partial t} = -\nabla.\left(\rho\frac{\partial \mathbf{u}}{\partial t}\right) \tag{3.3}$$

To derive the equation for the matrix displacement, $\mathbf{u}(\mathbf{x},t)$, we first note that for cellular and embryonic processes, inertial terms are negligible in comparison to viscous and elastic forces, that is, motion ceases instantly when the applied forces are turned off. Hence the traction forces generated by the cells are balanced by the viscoelastic forces within the ECM. Therefore the equilibrium equations are

$$\nabla.\sigma + \rho\mathbf{F} = \mathbf{0} \tag{3.4}$$

where σ is the composite stress tensor of the cell-ECM milieu and $\rho\mathbf{F}$ accounts for body forces.

Oster *et al.*, (1983) model the cell-matrix composite as a viscoelastic material with stress tensor

$$\sigma = \sigma_p + \sigma_n. \tag{3.5}$$

Here σ_p is the usual viscoelastic stress tensor (see, for example, Landau and Lifshitz, 1970),

$$\sigma_p = \underbrace{\mu_1\frac{\partial\varepsilon}{\partial t} + \mu_2\frac{\partial\theta}{\partial t}\mathbf{I}}_{\text{viscous}} + \underbrace{\frac{E}{1+\nu}\left(\varepsilon + \frac{\nu}{1-2\nu}\theta\mathbf{I}\right)}_{\text{elastic}} \tag{3.6}$$

where: $\theta = \nabla.\mathbf{u}$ is the dilatation, $\varepsilon = \frac{1}{2}[\nabla\mathbf{u} + \nabla\mathbf{u}^T]$ is the stress tensor, $\mathbf{I}$ is the unit tensor, μ_1, μ_2 are the shear and bulk viscosities, respectively, and E, ν are the Young's modulus and the Poisson ratio, respectively.

The stress due to cell traction is modelled by

$$\sigma_n = \frac{\tau n \rho}{1 + \lambda n}\mathbf{I} \tag{3.7}$$

where τ and λ are positive constants. This satisfies the conditions that there is no traction without matrix and that traction per cell decreases with increasing cell density (contact inhibition).

If the cell-matrix composite is attached to an external substatum, for example a subdermal basement layer, then the body force will be

$$\mathbf{F} = s\mathbf{u} \tag{3.8}$$

where s is the modulus of elasticity of the substrate to which the composite is attached.

With appropriate boundary conditions [for example zero flux on n and p, with $\mathbf{u}$ fixed], (3.1)-(3.8) define a simple version of the more complicated mechanical model presented by Oster *et al.*, (1983). The model equations can be analysed in much the same way as the RD system discussed in Section 1 but due to its complexity it is less amenable to analysis and further simplifying assumptions need to be made.

Linear and nonlinear analyses, plus numerical simulation, show that models within this general mechanical framework can exhibit steady-state spatial patterns (Perelson *et al.*, 1986) and spatio-temporal patterns (Ngwa and Maini, 1995).

4 Biological Applications

4.1 Developmental Constraints

Although the models discussed in Sections 1-3 are based on very different biological hypotheses, they share many common mathematical features. In particular, in the vicinity of primary bifurcation points from the uniform steady states, linear analysis predicts spatial patterns that are eigenfunctions of the Laplacian operator with the appropriate boundary conditions. Moreover, the mechanism of "short-range activation, long-range inhibition" that leads to patterning in RD systems may be generalised to the cell-chemotaxis (CC) and mechanical models discussed in Sections 2 and 3. In the CC model, short-range activation is due to chemoattractant secretion by cells increasing with cell density resulting in chemical gradients attracting cells to centres of high cell density. In turn, this creates zones of recruitment separated by regions of virtually zero cell density which inhibit these zones from growing even further. In the mechanical models, short-range activation is due to cell traction dragging cells to areas of high cell density while the long-range inhibition is due to the elastic restoring forces of the external tethering and the paucity of cells in surrounding regions.

Therefore, taking into account these mechanistic and mathematical similarities, it is not surprising that these models exhibit many similarities in their patterning properties. In other words, many of these properties are mechanism independent and are therefore developmental constraints that patterns must satisfy, regardless of their origin (Oster *et al.*, 1988).

We now illustrate the application of the above models to three very different types of biological pattern formation. In light of the above discussion it should be noted that more than one model could account for each of these patterning processes.

4.2 Skeletal patterning in the vertebrate limb

The formation of skeletal patterning in the vertebrate limb has been the focus of a great deal of experimental and theoretical research (for a review, see Maini and Solursh, 1991). Recently, it has been shown that a number of Hox genes are switched on in a precise spatio-temporal manner in the developing chick limb. Although these are exciting advances, they still beg the question of how this patterning of activity is initiated. RD theory, as a model for the generation of such robust processes as digit formation, has been heavily criticised. For example, Bard and Lauder (1974) showed that the qualitative form of the model solutions could be greatly influenced by minor perturbations in the system such as a small change in length. In such an application, an essential requirement for a model is that it must be able to produce a limited number of patterns in a very robust way. In this respect, the models discussed in this chapter are almost too sophisticated because they exhibit a vast variety of patterns, many of which are never observed. Hence, one is forced to turn the question of pattern formation on its head and ask, how can one *not* generate so much pattern? One way to address this issue is to investigate the role of boundary conditions. The key point here is that certain types of boundary conditions preclude many patterns from forming while extending the domains of stability of the remaining patterns. This has been shown for an RD system in one dimension (Dillon *et al.*, 1994). Figure 4.1 shows a comparison of the patterns formed under zero flux boundary conditions with conditions in which the boundary is a sink for one of the morphogens while the other still satisfies zero flux. The model now selects only patterns that are internal to the domain, and exhibits a patterning sequence that is consistent with that observed in the limb. The insight gained here then, is that the boundary plays an active role in the patterning mechanism, rather than simply being a passive impermeable membrane.

In 1990, Wolpert and Hornbruch performed an experiment in an attempt to prove that limb development in chick could not possibly arise as a consequence of a RD or mechanical mechanism of the sort discussed in this chapter. They removed the posterior half of a host limb bud and replaced it by the anterior half of a donor limb bud so that the resultant double-anterior recombinant limb bud was the same size as a normal limb bud. This experiment was performed at a sufficiently early stage in development that no pattern was visible. The limbs developed two humeri instead of one. This contradicted both models, due to the fact that the model solutions are size-dependent, that is, if the domain size is unaltered, the patterns produced are unaltered.

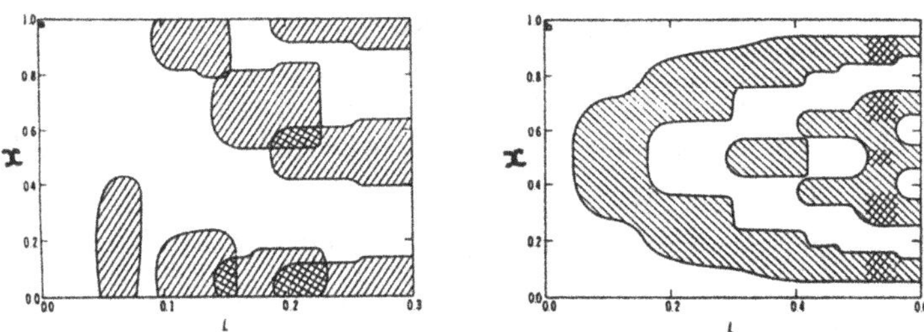

Figure 4.1: Subintervals of $(0, 1)$ in which the v component of a typical RD system exceeds a fixed threshold as a function of length scale, L, for the cases (a) Neumann

boundary conditions on both species, (b) Neumann boundary conditions for u, homogeneous Dirichlet boundary conditions for v. (See Dillon *et al.*, 1994, for full details). Reproduced from Dillon *et al.*, 1994, by permission of Springer-Verlag.

However, this result is consistent with the theory if the model is modified. The nondimensionalisation in Section 1.2 shows that diffusion and length scale are intimately linked. Therefore, if we assume that the diffusion coefficient of the morphogen varies appropriately across the domain, this essentially sets up an internal scaling, where the length scale in one part of the domain can be made sufficiently small that it cannot support Turing structures. Therefore, although combining two anterior halves results in a limb bud of normal size, it might actually consist of doubling the patterning sub-domain and hence result in more complicated patterns. The prediction of the modelling was that there must be a variation of diffusion across the anterior-posterior axis of the chick limb (see Fig. 4.2 and Maini *et al.*, 1992). This actually agrees with experimental results which show that gap junction permeability varies across the anterior-posterior axis of the chick limb (see, for example, Brümmer *et al.*, 1991).

The above results are all for a one-dimensional domain. Recently, we have shown that the crucial patterning selection properties of different types of boundary conditions and internal scaling by spatially varying diffusion coefficients can be carried over to two-dimensional domains, producing patterns consistent with those observed in the limb (Myerscough *et al.*, 1997).

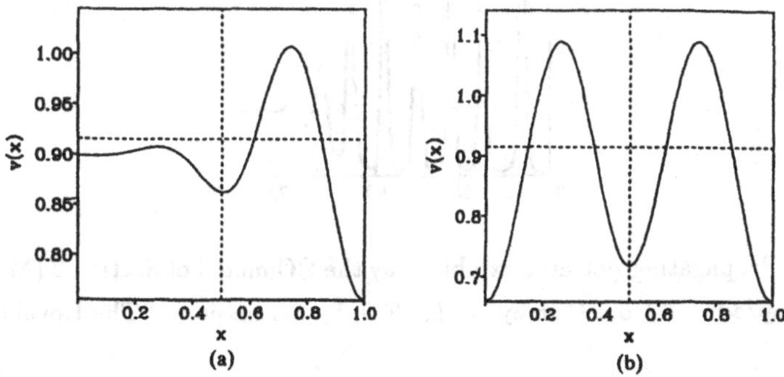

Figure 4.2: (a) Steady state solution for $v(x)$ to the Schnakenberg RD model with a spatially-varying diffusion coefficient of the form $D(x) = c_0 cosh\delta x/cosh\delta$. (b) Solution for the case where the $D(x)$ is symmetrical about $x = 1/2$ but identical to the case (a) on $(1/2, 1]$. For a suitable choice of threshold concentration (- - -) the prepattern in (a) specifies a single structure whereas that in (b) specifies two

structures even though the domain size has remained unchanged. (See Maini *et al.*, 1992, for full details). Reproduced from Maini *et al.*, 1992, by permission of Oxford University Press.

4.3 Pigmentation patterns in reptiles

The steady-state patterns exhibited by the cell-chemotactic (CC) model presented in Section 2 have been extensively studied in two dimensions and shown to be consistent with many characteristic skin pigmentation patterns on snakes (Maini *et al.*, 1991, Murray and Myerscough, 1991). However, here we focus on a different type of patterning, namely propagating patterns, with particular application to the pigmentation patterns of brown/black and white stripes on hatchling alligators. The white stripes are due to an absence of melanocytes (Murray *et al.*, 1990) and therefore it appears that a cell movement model in which high cell density results in pigmented regions while low cell densities lead to an absence of pigmentation, is a more realistic model than one based on reaction-diffusion. The pattern is initiated from the head and propagates down the head-tail axis. Murray *et al.*, (1990) showed that a CC model of the form illustrated in Section 2 can produce such propagating pattern (see Fig. 4.3). Note that the sharpness of the peaks is characteristic of CC models.

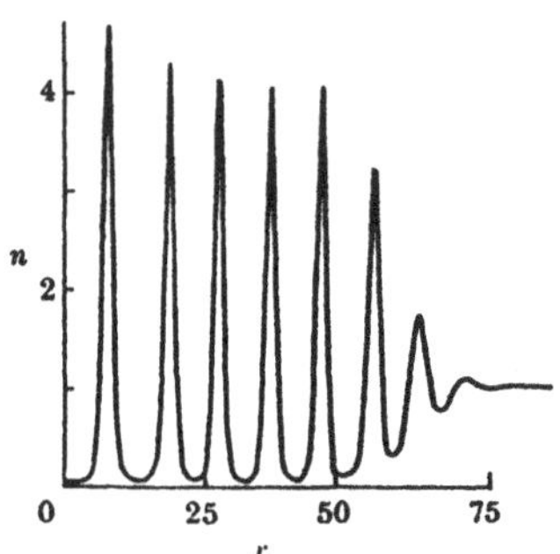

Figure 4.3: Propagating patterns exhibited by the CC model of Section 2 (Murray *et al.*, 1990). Reproduced from Murray *et al.*, 1990, by permission of The Royal Society, London.

4.4 Skin Organ Formation

During morphogenesis regular patterns often develop behind a frontier of pattern formation which travels across the prospective tissue. A simple example of such patterning was considered in Section 4.3. More complicated patterns can arise in, for example, the structure of feather primordia. A feather primordium consists of a thickening (placode) of the epidermis overlying an aggregation of cells (papilla) in the dermis. The primordium initiates the formation of a feather. On the dorsal surface of the chick, feather primordia form in a specific sequence. Firstly, a row of primordia form along the dorsal midline. Subsequent rows form lateral to this initial row with the primordia in each row roughly 180° out of phase with those in the previous row, so that a rhombic pattern appears. This patterning process requires the interaction of both epidermis and dermis.

Perelson *et al.*, (1986) considered a version of the mechanical model of Section 3 and showed that it could give rise to sequential patterning in the dermis consistent with that observed experimentally. Once the initial row of pattern is set up due, for example, to the uniform steady state going unstable, it, in turn, sets up a strain field that causes cell aggregations in the adjacent rows to be situated 180° out of phase with those of the dorsal midline.

A more biologically realistic model was proposed by Cruywagen and Murray (1992) which includes tissue-tissue interaction. This model encorporates many of the key features of the models discussed in Sections 2 and 3 by coupling chemical interaction with mechanical effects in a mechanochemical model. Briefly, dermal cells secrete a chemical which diffuses into the epidermis where it stimulates cell traction. Epidermal cells secrete a chemical which diffuses into the dermis where it acts as a chemoattractant. Cruywagen *et al.*, 1992, show that tissue-tissue interaction is essential for this model to produce pattern and that the form of the full two-dimensional pattern is determined by the pattern along the dorsal midline. In other words, the specification of a simple quasi-one-dimensional pattern is all that is required to determine a complex two-dimensional pattern (see Fig. 4.4).

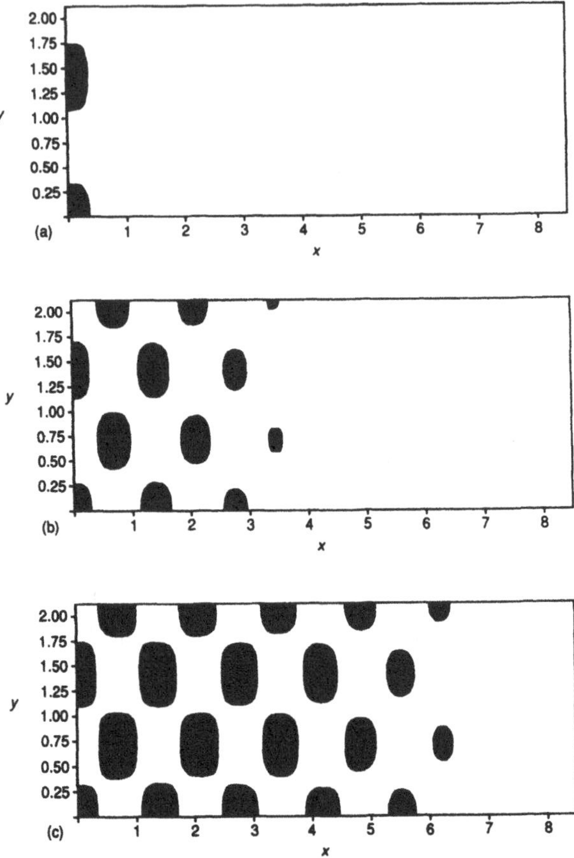

Figure 4.4: Sequential pattern formation in a tissue-tissue interaction mechanochemical model. Regions of high cell density are shaded. (a) Initially, a single row of spots of high cell density is specified at one end of a rectangular domain. As the system evolves, more rows are added sequentially as the pattern propagates across the domain; (b) and (c) show density profiles at subsequent times. (For full details, see Cruywagen *et al.*, 1992). Reproduced from Cruywagen *et al.*, 1992, by permission of Oxford University Press.

4.5 Coupled Pattern Generators

In Section 4.4 we considered a coupled mechanical-chemotaxis system. This model coupled the mechanical properties of the epidermis, with the chemotactic properties of dermal cells. Each tissue on its own was unable to produce patterns. The coupling was crucial to pattern formation: the dermal-epidermal interaction triggered "activation" in the epidermis by initiating cell traction; the epidermal-dermal interaction triggered "activation" in the dermis through chemotaxis. Here we briefly consider a coupled RD-CC model in which each model on its own can produce pattern. In this case the coupling enhances the complexity of the patterns formed (Painter *et al.*, in prep.).

Suppose that a cell population responds to two chemicals which themselves form a RD system. A possible model for this situation is the following:

$$\frac{\partial n}{\partial t} = D_n \nabla^2 n - \nabla . \{ n\chi_u(u,v)\nabla u + n\chi_v(u,v)\nabla v \} \tag{4.1a}$$

$$\frac{\partial u}{\partial t} = D_1 \nabla^2 u + f(u,v,n) \tag{4.1b}$$

$$\frac{\partial v}{\partial t} = D_2 \nabla^2 v + g(u,v,n) \tag{4.1c}$$

where n, u and v are cell density and chemical concentrations, respectively, D_n, D_1, D_2 are diffusion coefficients and χ_u, χ_v are cell chemotactic responses to gradients in u and v, respectively.

For simplicity, let us assume that f and g do not depend on n, and that the time scale for the chemical dynamics is much faster than that of the cell dynamics. Then, in the appropriate parameter space, u and v will evolve to spatially-varying steady state patterns. If we restrict our attention for the moment to the one-dimensional domain $x \in [0,1]$, the steady state equation for n is

$$0 = D_n \frac{d^2 n}{dx^2} - \frac{d}{dx}\left(n\chi_u(u,v)\frac{du}{dx} + n\chi_v(u,v)\frac{dv}{dx} \right) \tag{4.2}$$

where u and v are at their steady states. Imposing zero flux boundary conditions, this equation can be integrated twice to give

$$n(x) = k \exp\left(\frac{1}{D_n} \int_0^1 \left(\chi_u \frac{du}{dx} + \chi_v \frac{dv}{dx} \right) dx \right) \tag{4.3}$$

where k can be evaluated using conservation of cell density.

One way to measure complexity of pattern is to calculate the number of turning points for a particular pattern. Critical points for $n(x)$ are given by

$$\chi_u \frac{du}{dx} + \chi_v \frac{dv}{dx} = 0. \tag{4.4}$$

For the parameter regimes wherein the subsystem (4.1b)-(4.1c) is close to a primary bifurcation point, we can use nonlinear analysis (Section 1.5) to derive analytic approximations for u and v. We can then substitute these expressions into (4.4) and, for the different chemotactic sensitivities described in Section 2, find if (4.4) has more solutions (corresponding to more complex pattern) than the number of critical points for the chemical concentrations. In this way, we can precisely measure how the complexity of pattern is enhanced by different modes of coupling.

The above analysis applies only to a one-dimensional domain and in the vicinity of a primary bifurcation point. A more extensive investigation of the solution properties of this model can be carried out using numerical simulations. Fig. 4.5 illustrates some typical patterns. Fig. 4.5a shows a one-dimensional pattern corresponding to thick and thin stripes. This kind of pattern can be generated by a pure CC model (Maini *et al.*, 1991). However, the pattern in Figure 4.5b cannot, to our knowledge, be exhibited by a simple pattern generator. Note that this pattern captures the intricate features of some animal coat markings, such as those on the jaguar.

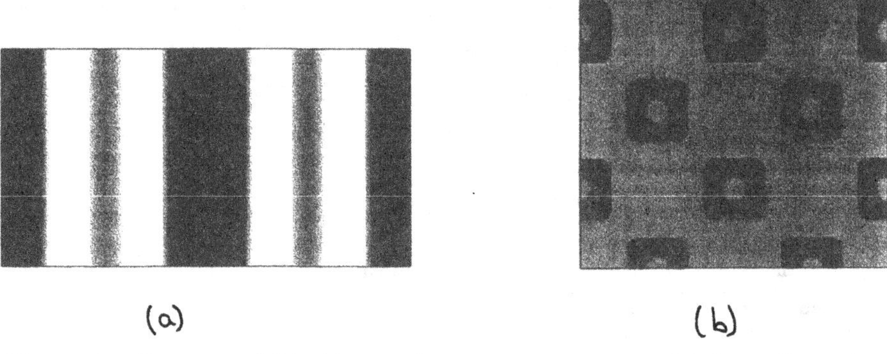

(a) (b)

Figure 4.5: Patterns exhibited by coupling pattern generators. (a) A one-dimensional pattern in cell density corresponding to thick and thin stripes. (b) A two-dimensional pattern in which the intensity of shading represents cell density.

5 Conclusions

The study of coupled systems of reaction-diffusion equations was inspired by a mathematician wishing to understand the mechanisms underlying biological pattern formation. This led to the prediction that interacting chemicals could, under special conditions, lead to patterns in chemical concentrations. It has only recently been discovered that this is indeed the case and many patterns predicted by Turing type models have now been found experimentally (see Maini *et al.*, 1997, for a review of recent experimental results on spatial pattern formation in chemistry).

Since Turing's seminal paper, a number of other mechanisms have been proposed as pattern generators, most notable being the chemotactic and mechanical models described in this chapter. As mentioned in Section 4.1, many features of the patterns exhibited by these models are generic. This has the advantage of allowing one to make some general preditions irrespective of the detailed biology, but has the disadvantage of not allowing us to distinguish between biological mechanisms. There are some exceptions to this. For example, as discussed in Section 4.3, cell movement seems to play a key role in reptile pigmentation patterns suggesting that RD models are probably not appropriate. However, in such circumstances, a study of the simplest and most mathematically tractable model can still yield valuable insight due to the mechanism-independent mathematical properties of many of the patterns.

Acknowledgement: I would like to thank Brenda Willoughby for typing this manuscript.

References

J. Bard, A model for generating aspects of zebra and other mammalian coat patterns, *J. theor. Biol.*, **93**, 363-385 (1981)

J. Bard and I. Lauder, How well does Turing's theory of morphogenesis work? *J. theor. Biol.*, **45**, 501–531 (1974)

D.L. Benson, J.A. Sherratt and P.K. Maini, Diffusion driven instability in an inhomogeneous domain, *Bull. Math. Biol.*, **55**, 365-384 (1993)

D.L. Benson, P.K. Maini and J.A. Sherratt, Unravelling the Turing bifurcation using spatially varying diffusion coefficients, *J. Math. Biol.*, (1998) (to appear)

N. F. Britton, *Reaction-Diffusion Equations and Their Applications to Biology*, Academic Press, London, 1986

F. Brümmer, G. Zempel, P. Buhle, J.-C. Stein and D. F. Hulser, Retinoic acid modulates gap junction permeability: A comparative study of dye spreading and ionic coupling in cultured cells, *Exp. Cell. Res.*, **196**, 158–163 (1991)

V. Castets, E. Dulos, J. Boissonade and P. De Kepper, Experimental evidence of a sustained Turing-type equilibrium chemical pattern, *Phys. Rev. Lett.*, **64**(3), 2953–2956 (1990)

S. Childress and J.K. Percus, Nonlinear aspects of chemotaxis, *Math. Biosciences,* **56**, 217-237 (1981)

G.C. Cruywagen, P.K. Maini and J.D. Murray, Sequential pattern formation in a model for skin morphogenesis, *IMA J.Math.Appl.Med. & Biol.*, **9**, 227-248 (1992)

G.C. Cruywagen and J.D. Murray, On a tissue interaction model for skin pattern formation, *J. Nonlinear Sci.*, **2**, 217-240 (1992)

G. Dee and J.S. Langer, Propagating pattern selection, *Phys. Rev. Letts.*, **50(6)**, 383-386 (1983)

P. De Kepper, V. Castets, E. Dulos and J. Boissonade, Turing-type chemical patterns in the chlorite-iodide-malonic acid reaction, *Physica D,* **49**, 161–169 (1991)

R. Dillon, P.K. Maini and H.G. Othmer, Pattern formation in generalised Turing systems: I. Steady-state patterns in systems with mixed boundary conditions, *J. Math. Biol.,* **32**, 345-393 (1994)

E. Doedel, AUTO: Software for continuation and bifurcation problems in ordinary differential equations. *Technical Report, Cal. Tech.*, 1986

B. Ermentrout, Stripes or spots? Nonlinear effects in bifurcation of reaction-diffusion equations on the square, *Proc. Roy. Soc. Lond.,* **A434**, 413-417 (1991)

P. Fife, *Mathematical Aspects of Reacting and Diffusing Systems, Lect. Notes in Biomath.*, **28**, Springer-Verlag, Berlin, Heidelberg, New York, 1979.

R.M. Ford, and D.A. Lauffenburger, Analysis of chemotactic bacterial distributions in population migration assays using a mathematical model applicable to steep or shallow attractant gradients, *Bull. Math. Biol.*, **53**, 721–749 (1991)

A. Gierer and H. Meinhardt, A theory of biological pattern formation, *Kybernetik,* **12**, 30-39 (1972)

A. Goldbeter, *Biochemical Oscillations and Cellular Rhythms: The molecular bases of periodic and chaotic behaviour,* Cambridge University Press (1996)

P. Grindrod, *The Theory of Applications of Reaction-Diffusion Equations: Pattern and Waves*, Oxford University Press, 1996

P. Grindrod, J.D. Murray and S. Sinha, Steady-state spatial patterns in a cell-chemotaxis model, *IMA J. Appl. Math. Med. & Biol.*, **6**, 69-79 (1989)

B.R. Johnson and S.K. Scott, New approaches to chemical patterns, *Chem. Soc. Rev.*, 265-273 (1996)

E.F. Keller and L.A. Segel, Travelling bands of bacteria: a theoretical analysis, *J. theor. Biol.*, **30**, 235-248 (1971)

L. Landau and E. Lipshitz, *Theory of Elasticity*, 2nd edn., New York, Pergamon, 1970

I.R. Lapidus, and R. Schiller, A model for the chemotactic response of a bacterial population, *Biophys. J.*, **16**, 779–789 (1976)

I. Lengyel and I.R. Epstein, Modeling of Turing structures in the chlorite-iodide-malonic acid-starch reaction system, *Science*, **251**, 650-652 (1991)

P.K. Maini, D.L. Benson and J.A. Sherratt, Pattern formation in reaction diffusion models with spatially inhomogeneous diffusion coefficients, *IMA J.Math.Appl.Med. & Biol.* 9, 197-213 (1992)

P.K. Maini, M.R. Myerscough, K.H. Winters and J.D.Murray, Bifurcating spatially heterogeneous solutions in a chemotaxis model for biological pattern formation, *Bull. Math. Biol.*, **53**, 701–719 (1991)

P.K. Maini, K.J. Painter and H. Chau, Spatial pattern formation in chemical and biological systems, *Faraday Transactions*, **93**, 3601–3610 (1997)

P.K. Maini and M. Solursh, Cellular mechanisms of pattern formation in the developing limb, *Int. Rev. Cytology*, 129, 91-133 (1991)

S.C. Müller, T. Plesser and B. Hess, The structure of the core of the spiral wave in the Belousov-Zhabotinskii reaction, *Science*, **230**, 661-663 (1985)

J.D. Murray, Parameter space for Turing instability in reaction diffusion mechanisms: a comparison of models, *J. theor. Biol.*, **98**, 143-163 (1982)

J.D. Murray, *Asymptotic Analysis*, 2nd edn., Berlin Heidelberg New York Tokyo, Springer (1984)

J.D. Murray, *Mathematical Biology*, 2nd edn., Berlin, Heidelberg, New York, London. Paris, Tokyo, Springer-Verlag (1993)

J.D. Murray, D.C. Deeming and M.W.J. Ferguson, Size dependent pigmentation pattern formation in embryos of Alligator mississippiensis: time of initiation of pattern generation mechanism, *Proc. Roy. Soc. Lond.*, **B 239**, 279-293 (1990)

J.D. Murray and M.R. Myerscough, Pigmentation pattern formation on snakes, *J. theor. Biol.*, **149**, 339-360 (1991)

M.R. Myerscough, P.K. Maini and K.J. Painter, Pattern formation in a generalised chemotactic model, *Bull. Math., Biol.*, (1997) (to appear)

M.R. Myerscough and J.D. Murray, Analysis of propagating pattern in a chemotaxis system, *Bull. Math. Biol.*, **54**, 77-94 (1992)

G.A. Ngwa and P.K. Maini, Spatio-temporal patterns in a mechanical model for mesenchymal morphogenesis, *J. Math. Biol.*, 33, 489-520 (1995)

H.G. Othmer and A. Stevens, Aggregation, Blow-up and Collapse: The ABCs of taxis inreinforced random walks, *SIAM J.Appl.Math* (to appear)

Q. Ouyang, R. Li, G. Li and H.L. Swinney, Dependence of Turing pattern wavelength on diffusion rate, *J. Chem. Phys.*, **102**(6), 2551-2555 (1995)

G.F. Oster, J.D. Murray and A.K. Harris, Mechanical aspects of mesenchymal morphogenesis, *J. Embryol.exp. Morph.*, **78**, 83-125 (1983)

G.F. Oster, N. Shubin, J.D. Murray and P. Alberch, Evolution and morphogenetic rules. The shape of the vertebrate limb in ontogeny and phylogeny, *Evolution,* **45**, 862-884 (1988)

K.J. Painter, P.K. Maini and H.G. Othmer, Spatial pattern formation in coupled cell-chemotactic-reaction-diffusion systems (in prep.)

A.V. Panfilov and A.V. Holden (eds), *Computational Biology of the Heart,* John Wiley & Sons (1997)

A.S. Perelson, P.K. Maini, J.D. Murray, J.M. Hyman and G.F. Oster, Nonlinear pattern selection in a mechanical model for morphogenesis, *J. Math. Biol.*, 24, 525-541 (1986)

D.H. Sattinger, Six lectures on the transition to instability, in *Lecture Notes in Mathematics*, **322**, Berlin, Springer-Verlag (1972)

J. Schnakenberg, Simple chemical reaction systems with limit cycle behaviour, *J. theor. Biol.*, **81**, 389-400 (1979)

D. Thomas, Artifical enzyme membranes, transport, memory and oscillatory phenomena, in ***Analysis and Control of Immobilized Enzyme Systems***, ed. D. Thomas and J.-P. Kernevez, Springer, Berlin, Heidelberg, New York, 115-150 (1975)

A. M. Turing, The chemical basis of morphogenesis, *Phil. Trans. Roy. Soc. Lond,* **B327**, 37-72 (1952)

B.J. Welsh, J. Gomatam and A.E. Burgess, Three-dimensional chemical waves in the Belousov-Zhabotinskii reaction, *Nature,* **340**, 611-614 (1983)

A.T. Winfree, Spiral waves of chemical activity, *Science,* **175**, 634-636 (1972)

A.T. Winfree, Rotating chemical reactions, *Sci. Amer.*, **230**(6), 82-95 (1974)

L. Wolpert, Positional information and the spatial pattern of cellular differentiation, *J. theor. Biol.*, **25**, 1-47 (1969)

L. Wolpert and A. Hornbruch, Double anterior chick limb buds and models for cartilage rudiment specification, *Development,* **109**, 961-966 (1990)

A.N. Zaikin and A.M. Zhabotinskii, Concentration wave propagation in two-dimensional liquid-phase self-organising system, *Nature,* **225**, 535-537 (1970)

V.S. Zykov, *Simulation of Wave Processes in Excitable Media,* Manchester University Press (1987)

Dynamics of Competition

H.L. Smith
Department of Mathematics
Arizona State University
Tempe, AZ 85287–1804

Contents

1 Introduction

Comparison principles and monotonicity methods applied to differential equations have a long history but it is safe to say that a real resurgence of interest in these methods has occurred over the past several decades due to their frequent applications to mathematical models in the biological sciences, particularly to population biology. What is now called monotone dynamical systems theory was largely inspired by studies in mathematical biology. The well-known construction of Smale [40] showed that, contrary to popular belief in the early 70's, mathematical models of competition between species could lead to differential equations with complicated dynamics. This in turn led to a famous series of papers by M.W. Hirsch [17], [18], [19], [20], [21], [22], [23], [24] who showed, roughly, that the dynamics of competitive (and cooperative) systems can be no worse than that of completely general systems of one less dimension. The most exciting fallout of this result was to the study of three-dimensional competitive systems to which much of the classical Poincaré-Bendixson theory has been extended [22], [23], [24], [42], [43], [53], [58]. The work of deMottoni and Schiaffino [10] must also be mentioned here for while it dealt with a specific system, the periodic Lotka-Volterra model of two-species competition, its results were more general than originally thought and quite influential. From these and other works, a real geometric theory of dynamics arose out of what had been merely a useful collection of comparison methods. In summary, this is mathematics inspired by biology. The purpose of these lectures is to provide an introduction to this mathematics.

There has been an even more impressive body of results obtained for so-called cooperative systems or more generally, monotone systems of differential equations and monotone mappings. However, it is not our purpose to treat these results here since our focus is primarily on competitive systems which are generally not monotone systems. The interested reader may consult [3], [17]–[19], [43] for work on cooperative systems.

A glance at the table of contents will cause the reader to doubt that the subject of these lectures deals with competitive systems for some of the applications deal with the growth of phytoplankton in a chemostat, predator-prey models, models of the signaling system of the slime mold, and of the cell cycle. These models are described by differential equation systems which, after various change of variables and reductions, have the mathematical form of more classical systems of competitive type. This is one of the central messages of these lectures: systems of differential equations and difference equations which can be transformed to competitive systems are pervasive in the biological sciences. As one can prove strong results on the long-time behavior of solutions of competitive systems, it pays to be on the lookout for these systems and to know how to recognize them.

We begin by describing the basic order relations and comparison results which are fundamental to our subject. Then we consider, in the order of increasing dimension, results and examples of competitive systems. A beautiful geometric theory of the dynamics of competitive maps in the plane was initiated by deMottoni and Schiaffino [10] which we describe in Section 3. An example is treated in Section 4. The theory of autonomous ordinary differential equations of competitive type, treated in the papers of Hirsch noted above, is the subject of Section 5, with applications discussed in Section 6. A general, model-independent, theory of competition between two populations, following the paper of Hsu et al. [27] which was inspired by a similar work of Lazer and Hess [16], is dealt with in Section 7. This theory is applied to a reaction-diffusion Lotka-Volterra model of competition in Section 8.

We have often stated special cases or less general results than are known in order to avoid technicalities in these lectures. The reader interested in more on the mathematics of competition and cooperation can consult the monograph [43] and the references in it for continuous-time dynamics and the recent paper of Tereščák [55] and references therein, for discrete-time dynamics.

The author thanks Mary Ballyk who carefully read the manuscript and caught many errors.

2 Some Notation and Fundamental Results

2.1 Cones and Partial Orders

Let X be a Banach space. A *cone* in X is a non-empty, closed subset K of X satisfying:

(i) Closure under addition: $K + K \subset K$

(ii) Closure under positive scalar multiplication: $(\mathbb{R}_+)K \subset K$

(iii) $K \cap (-K) = \{0\}$

If K is a cone in X and $x, y \in X$, we write $x \leq_K y$ if $y - x \in K$, $x <_K y$ if $x \leq_K y$ and $x \neq y$, and $x \ll_K y$ if $y - x \in K^0$, the interior of K. The reader should check that the relation $\leq_K$ is a partial order on X. The usual ordering on $X = \mathbb{R}^n$ is generated by the cone $K = \mathbb{R}^n_+$ of vectors with non-negative components. Then, $x \leq y$ if and only if $x_i \leq y_i$ for all i. We will drop the subscript "K" on the order relation for this natural ordering. We use similar notation for matrices, writing $A \geq 0$ ($A \gg 0$) if every entry of the matrix A is non-negative (positive). The square matrix A is called *quasi-positive* if its off-diagonal entries are non-negative, or equivalently, if $A + sI \geq 0$ for all large real s.

Any other orthant of $\mathbb{R}^n$ will serve as a cone. Given $m \in \mathbb{R}^n$ with entries $m_i \in \{0, 1\}$, let $K_m = \{x : (-1)^{m_i} x_i \geq 0\}$. Then K_m is a cone in $\mathbb{R}^n$. If Q_m denotes the $n \times n$ diagonal matrix with diagonal entries $(-1)^{m_i}$, then $Q_m^2 = I$, $Q_m \mathbb{R}^n_+ = K_m$, and $Q_m K_m = \mathbb{R}^n_+$. As a mapping, Q_m is an order isomorphism, taking the standard ordering $\leq$ to the ordering $\leq_{K_m}$ in the sense that $x \leq y$ if and only if $Q_m x \leq_{K_m} Q_m y$ and $x \leq_{K_m} y$ if and only if $Q_m x \leq Q_m y$. Hereafter, we drop the subscript m on $\leq_{K_m}$ in the notation $x \leq_K y$.

A particularly important case for us is the cone $K = K_m = \{(x_1, x_2) : x_1 \geq 0, x_2 \leq 0\} \subset \mathbb{R}^2$ $(m = (0, 1))$. It generates the so-called "south-east" ordering: $x \leq_K y$ if and only if y is south east of x in the plane (the vertical axis points north, the horizontal axis points east). The associated matrix $Q = Q_m = \text{diag}(1, -1)$.

An important example of a cone in the infinite dimensional space of continuous functions $X = C(\bar{\Omega})$ on the closure of the open set $\Omega \subset \mathbb{R}^n$ is given by $K = C_+(\bar{\Omega})$, the cone of non-negative functions. Here, $f \leq_K g$ if $f(x) \leq g(x)$ for $x \in \bar{\Omega}$.

Some notation will be useful. If K is a cone in X and $x, y \in X$ satisfy $x <_K y$ then $[x, y] = \{u \in X : x \leq_K u \leq_K y\}$ is called an order interval.

An important notion for square matrices is the idea of irreducibility. An $n \times n$ matrix A is *irreducible* if for every non-empty proper subset I of $N = \{1, 2, \ldots, n\}$, there is an $i \in I$ and $j \in J = N \setminus I$ such that $a_{ij} \neq 0$. A matrix is called reducible if there is such a partition I, J such that $a_{ij} = 0$ for $i \in I, j \in J$. The easiest way to test for irreducibility uses the incidence graph of A. It has vertices P_i, $1 \leq i \leq n$, with a directed edge $P_i P_j$ joining P_i to P_j if $a_{ij} \neq 0$. It can be shown ([2]), that A is irreducible if for each pair (P_i, P_j) of vertices there is a directed path $P_i P_{k_1}, P_{k_1} P_{k_2}, \ldots, P_{k_{r-1}} P_{k_r}, P_{k_r} P_j$ joining P_i to P_j.

The matrices:

$$A_1 = \begin{pmatrix} 0 & 1 \\ 1 & 0 \end{pmatrix}$$

and

$$A_2 = \begin{pmatrix} 0 & 0 & 1 \\ 1 & 0 & 0 \\ 0 & 1 & 0 \end{pmatrix}$$

are irreducible while

$$A_3 = \begin{pmatrix} 1 & 1 \\ 0 & 1 \end{pmatrix}$$

is reducible. The reader should sketch the incidence graphs of the A_i.

2.2 Basic Comparison Principle for ODEs

Let $f : \mathbb{R} \times D \to \mathbb{R}^n$, where D is a subset of $\mathbb{R}^n$, be a vector-valued function and consider the system of differential equations

$$x' = f(t, x) \ . \tag{1}$$

Our aim is to state the fundamental comparison theorem for ordinary differential equations. f is said to be of *type-K* in D if for each i and all t, $f_i(t, a) \le f_i(t, b)$ for any two points a and b in D satisfying $a \le b$ and $a_i = b_i$. The type-K condition is no restriction at all for scalar maps ($n = 1$) but is a rather severe restriction for $n \ge 2$.

We can now state the fundamental comparison theorem.

Theorem 1 (Kamke-Müller Theorem) *Let f be continuous on $\mathbb{R} \times D$ and of type-K. Assume that solutions of initial value problems for (1) are unique. Let $x(t)$ and $y(t)$ be solutions of (1) defined on $[a, b]$. If $x(a) \le y(a)$, then $x(t) \le y(t)$ for all t in $[a, b]$.*

See [7], [50], [43] for a proof. It is easy to see that the type-K condition is the correct one for order preservation. Assume for simplicity that $x(t) \le y(t)$ for $t \in [a, t_0)$ but that $x_i(t_0) = y_i(t_0)$ for some i. Of course, $x_j(t_0) \le y_j(t_0)$ for all j. Now observe that the type-K condition implies $x_i'(t_0) \le y_i'(t_0)$ suggesting that $x_i(t) \le y_i(t)$ for $t \ge t_0$.

It is easier to recognize type-K systems by their Jacobians. If D is open and convex, $D_x f(t,x)$ is continuous and quasi-positive:

$$\frac{\partial f_i}{\partial x_j}(t,x) \geq 0, \qquad i \neq j, \quad (t,x) \in D$$

then f is called *cooperative* in D. If f is cooperative, then it is type-K in D. In fact, if $a \leq b$ and $a_i = b_i$, then

$$f_i(t,b) - f_i(t,a) = \int_0^1 \sum_{j \neq i} \frac{\partial f_i}{\partial x_j}(t, a + r(b-a))(b_j - a_j)dr \geq 0.$$

System (1) is cooperative if f is cooperative.

Denote by $x(t,s,\xi)$ the solution map for (1), that is, the solution $x(t)$ of (1) satisfying the initial value problem $x(s) = \xi$. Recalling that the map $\xi \to x(t,s,\xi)$ is a homeomorphism for each $t > s$, the following is an immediate corollary of Thm. 1.

Corollary 1 *Let (1) be a cooperative system on D. If $s < t$, then:*

(1) $\xi_1 \leq \xi_2$ implies $x(t,s,\xi_1) \leq x(t,s,\xi_2)$.

(2) $\xi_1 < \xi_2$ implies $x(t,s,\xi_1) < x(t,s,\xi_2)$.

(3) $\xi_1 \ll \xi_2$ implies $x(t,s,\xi_1) \ll x(t,s,\xi_2)$.

Equation (1) is said to be *cooperative and irreducible* in D if it is a cooperative system and $D_x f(t,x)$ is an irreducible matrix for each $(t,x) \in \mathbb{R} \times D$. The irreducibility condition ensures that the solution map carries the weak ordering to the strong ordering because a strict inequality (positivity) in any one component is immediately communicated to all others.

Theorem 2 *Let (1) be a cooperative and irreducible system in $\mathbb{R} \times D$. Then*

$$\frac{\partial x}{\partial \xi}(t,s,\xi) \gg 0$$

for $t > s$. Furthermore, if $\xi_1, \xi_2 \in D$ satisfy $\xi_1 < \xi_2$ then

$$x(t,s,\xi_1) \ll x(t,s,\xi_2)$$

for $t > s$.

Observe that the second assertion of Thm. 2 follows immediately from the first by the fundamental theorem of calculus:

$$x(t,s,\xi_2) - x(t,s,\xi_1) = \int_0^1 \frac{\partial x}{\partial \xi}(t,s,r\xi_2 + (1-r)\xi_1)(\xi_2 - \xi_1)dr. \quad (2)$$

Thus, Thm. 2 really concerns the positivity properties of solutions of linear systems because the Jacobian matrix $X(t,s) \equiv \frac{\partial x}{\partial \xi}(t,s,\xi)$ is the matrix solution of:

$$X'(t) = A(t)X(t), \quad X(s) = I$$

where $A(t) = D_x f(x(t,s,\xi))$. More generally, if the continuous matrix function $A(t)$ is quasi-positive and irreducible for each t, then its fundamental matrix satisfies $X(t,s) \gg 0$ for $t > s$. For the matrices A_i displayed at the end of section 2.1, the reader should check that

$$X_i(t,s) = \exp(A_i(t-s)) \gg 0, \quad t > s$$

when $i = 1,2$ but not for $i = 3$ (although $X_3(t,s) \geq 0$ for $t \geq s$). The power series representation of the exponential function is useful for this.

The three comparison results above involve the standard ordering $\leq$ generated by the standard cone $\mathbb{R}^n_+$. It is a straightforward exercise to adopt these results to the partial order $\leq_K$ generated by the cone K_m. We describe this very briefly in the case that D is convex and f is continuously differentiable with respect to x. After permuting coordinates, if necessary, we can assume that there is a k satisfying $1 \leq k < n$ such that $K_m = \{x \in \mathbb{R}^n : x_i \geq 0, 1 \leq i \leq k, x_j \leq 0, k+1 \leq j \leq n\}$. Let $Q = \text{diag}(1,1,\dots,1,-1,-1,\dots,-1)$ (k ones followed by $n-k$ minus ones) be the order isomorphism associated with K_m. Suppose that at each $(t,x) \in \mathbb{R} \times D$, the Jacobian matrix of f has the structure

$$D_x f(t,x) = \begin{pmatrix} A & B \\ C & D \end{pmatrix}$$

where A is a $k \times k$ quasi-positive matrix, D is an $n-k \times n-k$ quasi-positive matrix, and $C, B \leq 0$. Of course, each matrix A,B,C,D depends on (t,x). Then Thm. 1 and its Cor. 1 hold with $\leq_K, <_K, \ll_K$ in place of $\leq, <, \ll$. The proof follows immediately upon making the transformation of variables $y = Q_m x$ in (1).

See [43] for a simple algorithm to check whether a given autonomous system can be transformed to a competitive or cooperative system by a change of variables as above.

2.3 Comparison Principles for Parabolic PDEs

Our goal in this section is to state some fundamental comparison results for parabolic systems in just enough generality to handle the simple example treated in Section 7. For this reason, we stick to the Laplace operator and Neumann boundary conditions for this brief treatment. Much more general results than those quoted here are known. See [38], [36], [41], [29], [43].

Let $\Omega \subset \mathbb{R}^m$ be a bounded, connected, open set with the property that $\partial\Omega$ is a C^2-smooth hypersurface and let $n(x)$, $x \in \partial\Omega$ be the outward pointing, unit normal vector field. As usual, the Laplacian, denoted by Δ, is defined by $\Delta = \sum_{i=1}^m \frac{\partial^2}{\partial x_i^2}$. The following positivity lemma follows from the strong maximum principle.

Lemma 1 *Let $T, d > 0$, the function $u(x,t)$ be continuous on $\bar{\Omega} \times [0,T]$ and let the derivatives $\frac{\partial u}{\partial x_i}, \frac{\partial^2 u}{\partial x_i \partial x_j}, \frac{\partial u}{\partial t}$ exist and be continuous on $\Omega \times (0,T]$. Finally, assume that u satisfies*

$$
\begin{aligned}
\frac{\partial u}{\partial t} - d\Delta u + c(x,t)u &\geq 0, \quad (x,t) \in \Omega \times (0,T] \\
\frac{\partial u}{\partial n} &\geq 0, \quad (x,t) \in \partial\Omega \times (0,T] \qquad (3) \\
u(x,0) &\geq 0 \quad x \in \Omega
\end{aligned}
$$

where c is bounded on $\Omega \times (0,T]$. Then $u(x,t) \geq 0$ in $\bar{\Omega} \times [0,T]$. Moreover, $u(x,t) > 0$ on $\bar{\Omega} \times (0,T]$ if it is not identically zero.

We are really interested in systems of parabolic equations. Let $d_i > 0$ for $1 \leq i \leq n$ and let Λ be the diagonal matrix $\Lambda = \mathrm{diag}(d_1, \ldots, d_n)$. If $f : D \to \mathbb{R}^n$, then we consider the reaction-diffusion system

$$
\begin{aligned}
\frac{\partial u}{\partial t} &= \Lambda\Delta u + f(u), \quad (x,t) \in \Omega \times (0,T] \\
\frac{\partial u}{\partial n} &= 0, \quad (x,t) \in \partial\Omega \times (0,T] \qquad (4) \\
u(x,0) &= \phi(x), \quad x \in \Omega
\end{aligned}
$$

where ϕ takes values in D. Assume that $f, f^-, f^+ : D \to \mathbb{R}^n$, where D is a convex subset of $\mathbb{R}^n$ and suppose that f, f^-, f^+ are each cooperative in D. Let $v^+(x,t)$ and $v^-(x,t)$ be continuous on $\bar{\Omega} \times [0,\tau)$, continuously differentiable on $\bar{\Omega} \times (0, \tau)$ and twice continuously differentiable in $x \in \Omega$ for $t > 0$. Furthermore, assume that $v^-(x,t), v^+(x,t) \in D$ and satisfy

$$v^-(x,t) \leq v^+(x,t) \quad (x,t) \in \bar{\Omega} \times [0, \tau) .$$

Finally, assume that $v^\pm$ satisfy the differential inequalities

$$\begin{aligned} \frac{\partial v^+}{\partial t} &\geq \Lambda \Delta v^+ + f^+(v^+), \ (x,t) \in \Omega \times (0,T] \\ \frac{\partial v^+}{\partial \nu} &\geq 0, \ (x,t) \in \partial\Omega \times (0,T] \end{aligned} \tag{5}$$

and

$$\begin{aligned} \frac{\partial v^-}{\partial t} &\leq \Lambda \Delta v^- + f^-(v^-), \ (x,t) \in \Omega \times (0,T] \\ \frac{\partial v^-}{\partial \nu} &\leq 0, \ (x,t) \in \partial\Omega \times (0,T] \end{aligned} \tag{6}$$

The function v^+ is called a *super-solution* and v^- is called a *sub-solution* in the literature of partial differential equations. The next result is a special case of a fundamental comparison technique.

Theorem 3 *In addition to the above, suppose that*

$$f^-(u) \leq f(u) \leq f^+(u), \quad u \in D$$

and ϕ satisfies

$$v^-(x,0) \leq \phi(x) \leq v^+(x,0), \quad x \in \bar{\Omega} .$$

Then the unique solution of (4) exists on $[0, \sigma)$ where $\sigma \geq \tau$ and

$$v^-(x,t) \leq u(x,t) \leq v^+(x,t), \quad (x,t) \in \bar{\Omega} \times [0, \tau) .$$

If $v^-(\bullet, 0) \neq \phi$, then $v^-(\bullet, t) \neq u(\bullet, t)$ for $t \geq 0$. An analogous result holds for v^+. Finally, if in addition $f = f^-$ is cooperative and irreducible and if $v^-(\bullet, 0) \neq \phi$, then $v^-(x,t) \ll u(x,t)$ for $x \in \bar{\Omega}$ and $t > 0$ where both solutions are defined. An analogous result holds for v^+ if $f = f^+$ is cooperative and irreducible.

If, in Thm. 3, we assume only the existence of v^+ (v^-) satisfying (5) ((6)) and only the inequalities involving v^+ (v^-) are retained in the statement of the theorem, then the inequality $u(x,t) \leq v^+(x,t)$ ($v^-(x,t) \leq u(x,t)$) holds but only on the common interval of existence of the two functions.

See [43], [29] for a proof which is based on the maximum principle for weakly coupled parabolic systems [38].

3 Planar Competitive Systems

3.1 Periodically Forced Competitive Systems

Equation (1) is *competitive* in D if for all $(t,x) \in \mathbb{R} \times D$

$$\frac{\partial f_i}{\partial x_j}(t,x) \leq 0, \quad i \neq j \ .$$

A solution $x(t)$ of a competitive system gives rise to a solution $y(t) = x(t_0 - t)$ of the cooperative system

$$y' = -f(t_0 - t, y)$$

and conversely. In particular, Thm. 1 implies that if $x(t,s,\xi)$ is a solution of a competitive system and if $x(t,s,\xi_1) \leq x(t,s,\xi_2)$, then $\xi_1 \leq \xi_2$ for $t > s$. That is, if two states of the system are ordered in the future, then they were ordered in the past.

In this section, we are interested in planar systems ($n = 2$) which are periodic:

$$f(t,x) = f(t+\omega, x), \quad (t,x) \in \mathbb{R} \times D$$

where $D \subset \mathbb{R}^2$ and ω is some positive number. A famous example is the periodic Lotka-Volterra system studied by deMottoni and Schiaffino in [10]:

$$\begin{aligned} x' &= x[r(t) - a(t)x - b(t)y] \\ y' &= y[s(t) - c(t)x - d(t)y] \end{aligned} \tag{7}$$

where r, s, a, b, c, d are periodic functions (of the same period ω) and $a, b, c, d \geq 0$.

The *period map* associated to (1) is defined by:

$$Px_0 = x(\omega, 0, x_0) \ .$$

The period map captures the dynamics of (1). Its fixed points are the initial data of the ω-periodic solutions. Hereafter, we drop the modifier "ω", in front of periodic; a periodic solution will always mean an ω-periodic solution.

Lemma 2 *Suppose that (1) is a planar, periodic, competitive system. Let K denote the South-East Cone in $\mathbb{R}^2$. If $x_0 \leq_K y_0$, then $Px_0 \leq_K Py_0$. If $Px_0 \leq Py_0$, then $x_0 \leq y_0$.*

Proof: We use Thm. 1 and the order isomorphism $Q = \text{diag}(1, -1)$. The change of variables $z = Qx$ in (1) yields $z' = g(t, z)$ where $g(t, z) \equiv Qf(t, Qz)$. This system is cooperative since $D_z g(t, z) = QD_x f(t, Qz)Q$ is easily seen to be quasi-positive. Thus, if $x_0 \leq_K y_0$, then $u_0 \equiv Qx_0 \leq Qy_0 \equiv v_0$ so $z(t, s, u_0) \leq z(t, s, v_0)$ for $t > s$ and $Qz(t, s, u_0) \leq_K Qz(t, s, v_0)$. But $x(t, s, x_0) = Qz(t, s, u_0)$ and $x(t, s, y_0) = Qz(t, s, v_0)$ so we have $x(t, s, x_0) \leq_K x(t, s, y_0)$ as claimed. Now suppose that $Px_0 \leq Py_0$, or equivalently, $x(\omega, 0, x_0) \leq x(\omega, 0, y_0)$. Set $u(t) = x(\omega - t, 0, x_0)$ and $v(t) = x(\omega - t, 0, y_0)$. They satisfy the cooperative system $z' = -f(\omega - t, z)$ and $u(0) \leq v(0)$. Consequently, $u(\omega) \leq v(\omega)$, or equivalently, $x_0 \leq y_0$. □

Remark 1 *It is useful to note explicitly a corollary of the proof of the lemma, which does not require the assumption that the system is periodic. If $x(t, s, \xi)$ denotes the solution map of a planar, competitive system, then the relation $\xi_1 \leq_K \xi_2$ implies that $x(t, s, \xi_1) \leq_K x(t, s, \xi_2)$ for $t > s$ and similarly for the stronger relations $<_K, \ll_K$.*

A similar result (interchange $\leq$ and $\leq_K$ in Lemma 2) holds for planar, periodic, cooperative systems.

3.2 Competitive Maps

It is useful to take a more general point of view and study planar *competitive maps*. Consider a continuous map $P : A \to A$ where $A \subset \mathbb{R}^2$. We say that P is competitive if $Px \leq_K Py$ whenever $x \leq_K y$ where K is the south-east ordering. We have just seen from Lemma 2 that the period map of a planar, periodic, competitive system is a competitive map. Hereafter, we assume only that P is a continuous competitive map on A (it need not be a period map).

Given distinct points $x, y \in \mathbb{R}^2$, exactly one of the relations $x \ll y$, $y \ll x$, $x \leq_K y$, or $y \leq_K x$ hold. This simple fact will be exploited repeatedly.

Lemma 3 *If $x, y \in A$ satisfy $Px \ll Py$, then either $x \ll y$ or $y \ll x$.*

Proof: If neither $x \ll y$ nor $y \ll x$ hold, then $x \leq_K y$ or $y \leq_K x$ holds. But $x \leq_K y$ implies $Px \leq_K Py$, so if $(a,b) = Px$ and $(c,d) = Py$, then $b \geq d$ and $b < d$ must hold, a contradiction. A similar contradiction is obtained from $y \leq_K x$. □

In the special case that P is a period map of a planar, periodic, competitive system, Lemma 2 implies that the former inequality holds. Lemma 3 suggests placing one of the following additional assumptions on P in the general case.

$(O+)$ If $x, y \in A$ and $Px \ll Py$, then $x \leq y$.

$(O-)$ If $x, y \in A$ and $Px \ll Py$, then $y \leq x$.

Roughly, if P is orientation preserving, then $(O+)$ holds and if it is orientation reversing, then $(O-)$ holds. A sequence $\{x_n = (u_n, v_n)\} \subset \mathbb{R}^2$ is *eventually componentwise monotone* if there exists a positive integer N such that either $u_n \leq u_{n+1}$ for all $n \geq N$ or $u_{n+1} \leq u_n$ for all $n \geq N$ and similarly for v_n.

The following result was first proved by DeMottoni and Schiaffino [10] for the period map of a periodic competitive Lotka-Volterra system of differential equations. The idea of the proof was later generalized to competitive and cooperative maps by Hale and Somolinas [13] and by the author [46], [47], [49].

Theorem 4 *If P is a competitive map for which $(O+)$ holds then for all $x \in A$, $\{P^n x\}_{n\geq 0}$ is eventually componentwise monotone. If the orbit of x has compact closure in A, then it converges to a fixed point of P. If, instead, $(O-)$ holds then for all $x \in A$, $\{P^{2n} x\}_{n\geq 0}$ is eventually componentwise monotone. If the orbit of x has compact closure in A, then its omega limit set is either a period-two orbit or a fixed point.*

Proof: We first note that if P is competitive and $(O-)$ holds then P^2 is competitive and $(O+)$ holds (use Lemma 3) so the second conclusion of the theorem follows from the first.

Suppose that $(O+)$ holds. If $P^n x \leq_K P^{n+1} x$ or $P^{n+1} x \leq_K P^n x$ holds for some $n \geq 1$, then it holds for all larger n so the conclusion is obvious. Therefore, we assume that this is not the case. It follows that for each $n \geq 1$ either (a) $P^n x \ll P^{n+1} x$ or (b) $P^{n+1} x \ll P^n x$. We claim that either (a) holds for all n or (b) holds for all n. Assume $x \ll Px$ (the argument is similar in the other case). If the claim is false, then there is an $n \geq 1$ such that

$$x \ll Px \ll \ldots \ll P^{n-1}x \ll P^n x$$

but $P^{n+1}x \ll P^n x$. But $(O+)$ implies $P^n x \leq P^{n-1}x$ contradicting the displayed inequality. □

Of course, an entirely analogous result holds for planar cooperative maps, defined in the obvious manner. One simply swaps the ordering $\leq$ for $\leq_K$ everywhere. Without one of the assumptions $(O+)$ or $(O-)$, one cannot obtain such a strong result. In [48], [49] we show how to imbed the standard scalar quadratic map into a competitive planar map, implying that chaos is possible for competitive maps. Finally, we remark that by Lemma 3, the period map of a planar, periodic, competitive or cooperative system satisfies $(O+)$ so Theorem 4 applies. In the special case that the system (1) is autonomous, one can show that every solution defined on $[0, \infty)$ has the property that each of its components is eventually monotone.

Discrete-time models of two-population competition generally do not satisfy the hypothesis $(O+)$ except under restrictive conditions (see [49] for several examples). Consider for example the well-known Lotka-Volterra model:

$$\begin{aligned} u_{n+1} &= u_n \exp[r(1 - u_n - bv_n)] \\ v_{n+1} &= v_n \exp[s(1 - cu_n - v_n)] \end{aligned} \tag{8}$$

where $r, s, c, b > 0$. Denote by $P : \mathbb{R}^2_+ \to \mathbb{R}^2_+$ the map defined by the right side of (8). It is easily seen that

$$P(\mathbb{R}^2_+) \subset [0, r^{-1}\exp(r-1)] \times [0, s^{-1}\exp(s-1)].$$

The calculation $\frac{\partial P_1}{\partial u} = (1 - ru)\exp[r(1 - u - bv)]$ and others like it show that the Jacobian matrix has the sign structure

$$DP(u, v) = \begin{pmatrix} + & - \\ - & + \end{pmatrix}$$

on $D \equiv [0, r^{-1}] \times [0, s^{-1}]$. The reader should be able to mimic the argument in equation (2) showing that a cooperative system of differential equations is type-K, to show that P is strictly order preserving with respect to $<_K$ on D. Therefore, if $r, s \leq 1$, then $P : D \to D$ is a competitive map and every orbit of (8) starting in $\mathbb{R}^2_+$ enters and remains in D. It is not difficult to see that (O+) need not hold in the case of strong inter-specific competition $b, c > 1$.

4 Phytoplankton Growth in Periodic Chemostat

In this chapter, we apply the results of the previous one to the Droop model of phytoplankton growth in a chemostat environment, periodically forced by nutrient feed. The chemostat (see [50], also called a continuously stirred tank reactor in chemical engineering) is a well-stirred tank into which fresh nutrient is supplied at a constant rate from an external source and from which the well-mixed contents are removed at an equal rate so that the volume of the contents of the chemostat are kept constant. It provides the simplest "open" environment. The key feature of the Droop model, which can be derived from a structured model (see [35]), is that it accounts for the observation that phytoplankton can store nutrient over and above that needed for immediate needs. In the case that the nutrient is assumed to be supplied at constant concentration, the system has been studied in [51], [50] where it is shown that there are two possible outcomes. Either there is insufficient nutrient or the dilution rate is too high causing the washout of the organism from the chemostat, or it can survive, eventually reaching a constant concentration in the chemostat. Following [45], we will outline here the main ideas required to show that the same result holds even when the nutrient is supplied at a periodically varying concentration in the feed except in the case of survival the organisms concentration oscillates periodically. See also [37], a study which motivated our own.

The Droop model, sometimes called the variable yield model, takes the following form:

$$\begin{aligned} N' &= N(\mu(Q) - D) \\ Q' &= \rho(S, Q) - \mu(Q)Q \\ S' &= D(S^0(t) - S) - N\rho(S, Q) \end{aligned} \tag{9}$$

N represents biomass concentration of phytoplankton, S the nutrient concentration in the chemostat and Q is the stored nutrient per unit phytoplankton biomass. It is also called the cell quota. D is the dilution rate and $S^0(t)$ is the time-dependent feed nutrient concentration. The phytoplankton growth rate $\mu(Q)$ is assumed to be strictly increasing with cell quota. More precisely, we assume that μ is defined and continuously differentiable for all Q larger than some positive minimum value Q_m and that

$$\mu(Q_m) = 0, \quad \mu'(Q) > 0, \quad Q \geq Q_m \ ,$$

holds. Note that the cell quota must exceed the threshold Q_m for growth.

The nutrient uptake rate $\rho(S, Q)$ is assumed to be strictly increasing with nutrient concentration S, nonincreasing with cell quota Q, and to vanish in the absence of nutrient. In other words, an increase in the ambient nutrient concentration with no change in cell quota leads to a greater uptake rate, while an increase in the internal nutrient pool with no change in ambient nutrient concentration can only decrease the uptake rate. Thus, ρ is continuously differentiable for $S \geq 0$ and $Q \geq Q_m$ and satisfies:

$$\rho(0, Q) = 0, \quad \frac{\partial \rho}{\partial S} > 0, \quad \frac{\partial \rho}{\partial Q} \leq 0 \ .$$

The latter inequality allows for the possibility that ρ is independent of Q.

A principal assumption here is that the nutrient concentration in the feed $S^0(t)$ varies periodically with time with period ω:

$$S^0(t) = S^0(t + \omega) \geq 0 \ .$$

We assume further that $S^0(t)$ is positive for some, but not necessarily all, values of t. We remark that the domain $N \geq 0$, $Q \geq Q_m$, $S \geq 0$ is positively invariant for (9).

The second of equations (9) is easily derived from the relation

$$(S + NQ)' = D(S^0(t) - S - NQ)$$

which basically says that total nutrient $S + NQ$ changes only due to the input of fresh nutrient S^0 and the outflow due to washout.

There is a trivial periodic solution of (9) corresponding to the absence of phytoplankton in the chemostat which we now describe. Putting $N = 0$ in the last of the equations (9) results in

$$S' = D(S^0(t) - S) \ . \tag{10}$$

This linear equation has a unique positive periodic solution $S = S^*(t) = S^*(t + \omega) > 0$ which describes the available nutrient in a phytoplankton-free chemostat. There remains the rather artificial second equation in (9) for the (virtual) cell quota which must be taken into account even though there are no cells present. Putting $S = S^*(t)$ in this equation results in

$$Q' = \rho(S^*(t), Q) - \mu(Q)Q \ .$$

It may be shown that this scalar equation has a unique periodic solution $Q = Q^*(t) = Q^*(t+\omega) > Q_m$ which is globally attracting. Therefore, the trivial periodic solution is given by:

$$N = 0, \quad Q = Q^*(t), \quad S = S^*(t) \ .$$

Our main result says that either this trivial periodic solution is stable, in which case it is a global attractor, or it is unstable, in which case there is a non-trivial periodic solution which is globally attracting. Before stating our main result, it is useful to create some notation for the time-average of a periodic function, defined as follows

$$\langle f(t) \rangle \equiv \omega^{-1} \int_0^\omega f(t)dt \ .$$

Theorem 5 *If*

$$\langle \mu(Q^*(t)) \rangle - D < 0 \ , \tag{11}$$

then the plankton population is washed out of the chemostat:

$$N(t) \to 0, \quad t \to \infty \ .$$

If

$$\langle \mu(Q^*(t)) \rangle - D > 0 \ , \tag{12}$$

then there is a unique periodic solution

$$N = \bar{N}(t) = \bar{N}(t+\omega), \quad Q = \bar{Q}(t) = \bar{Q}(t+\omega), \quad S = \bar{S}(t) = \bar{S}(t+\omega)$$

of (9) satisfying $\bar{N}(t) > 0$ to which all solutions of (9) with $N(0) > 0$ are attracted asymptotically:

$$(N(t), Q(t), S(t)) - (\bar{N}(t), \bar{Q}(t), \bar{S}(t)) \to 0, \quad t \to \infty \ .$$

It is not our intention to give a complete proof of this result here (see [45] for this) but rather to show how the results of the previous chapter apply, even though the system (9) is neither competitive nor two-dimensional. There are three main ingredients to the proof. First is the reduction of the system to a two-dimensional system by making use of a conservation principle common to most chemostat models. Second, we must find appropriate variables in which this two-dimensional system is competitive. Finally, the uniqueness of the non-trivial periodic solution must be established. We will show each of these steps.

The total amount of nutrient T in the chemostat includes both free nutrient S and stored nutrient NQ:

$$T \equiv S + QN \ .$$

A simple calculation shows that T satisfies (10) and consequently T approaches the unique periodic solution $S^*(t)$ of (10):

$$T(t) = S^*(t) + O(e^{-Dt}), \quad t \to \infty \ .$$

It is therefore natural to take advantage of the fact that the solutions of (9) approach the invariant time-periodic surface

$$S^*(t) = S + QN$$

at an exponential rate by dropping the equation for S and replacing S in (9) by $S = S^*(t) - QN$ in the second equation. This results in the (limiting) two-dimensional system:

$$\begin{aligned} N' &= N(\mu(Q) - D) \\ Q' &= \rho(S^*(t) - QN, Q) - \mu(Q)Q \ . \end{aligned} \tag{13}$$

As S must be nonnegative, initial conditions for (13) must satisfy $Q(0)N(0) \leq S^*(0)$. It is convenient to change variables in (13) by letting $U = QN$, the total stored nutrient, which leads to the system

$$\begin{aligned} U' &= U[Q^{-1}\rho(S^*(t) - U, Q) - D] \\ Q' &= \rho(S^*(t) - U, Q) - \mu(Q)Q \ . \end{aligned} \tag{14}$$

Here too, we must choose initial conditions such that $U(0) \leq S^*(0)$. Since $\rho(0,Q) = 0$, it follows from the first equation of (14) and (10) that $(S^* - U)' \geq 0$ whenever $S^* = U$, so $U(t) \leq S^*(t)$ for $t \geq 0$.

System (14) is competitive since

$$\frac{\partial U'}{\partial Q} \leq 0 \quad \text{and} \quad \frac{\partial Q'}{\partial U} < 0 \ .$$

Consequently, Thm. 4 settles the asymptotic behavior of the system (14), as well as the equivalent system (13): every solution is asymptotic to a periodic solution. Hence our problem is reduced to determining the periodic solutions of (14) and their stability properties. There is, of course, the trivial periodic solution $(0, Q^*(t))$. In the next result, we prove that there can be at most one non-trivial periodic solution.

Lemma 4 *Equation (13) has at most one periodic solution with the property that $N(t) > 0$. Equivalently, (14) has at most one periodic solution with the property that $U(t) > 0$.*

Proof: The two equations (13) and (14) are equivalent so there is a one-to-one correspondence between their periodic solutions. In fact, it will prove convenient to argue using both equations since each has useful properties. Suppose $(N(t), Q(t))$ and $(\bar{N}(t), \bar{Q}(t))$ are distinct periodic solutions with $N(t), \bar{N}(t) > 0$. If $U(t) = N(t)Q(t), \bar{U}(t) = \bar{N}(t)\bar{Q}(t)$, then $(U(t), Q(t))$ and $(\bar{U}(t), \bar{Q}(t))$ are periodic solutions of (14) with $U(t), \bar{U}(t) > 0$. The first equation in (13) implies that

$$\langle \mu(Q(t)) \rangle = D = \langle \mu(\bar{Q}(t)) \rangle$$

and as μ is strictly increasing, it follows that $Q(t_0) = \bar{Q}(t_0)$ for some t_0. Without loss of generality, we assume that $N(t_0) < \bar{N}(t_0)$, or equivalently for (14), $U(t_0) < \bar{U}(t_0)$. The monotonicity of the right-hand side of the second equation of (14) in U implies that $Q'(t_0) > \bar{Q}'(t_0)$. Consequently, for $t_1 > t_0$ and t_1 sufficiently near to t_0, we have $U(t_1) < \bar{U}(t_1)$ and $Q(t_1) > \bar{Q}(t_1)$. Equivalently, $(U(t_1), Q(t_1)) \ll_K (\bar{U}(t_1), \bar{Q}(t_1))$, so this relation must be preserved for $t \geq t_1$ by Remark 1. Thus, we conclude that $U(t) < \bar{U}(t)$ and $Q(t) > \bar{Q}(t)$ for all $t > t_1$, and hence for all t by periodicity. This contradiction to the equality displayed above proves the lemma. □

It is easy to see that there can be no non-trivial periodic solution if (11) holds. Indeed, dividing the first of (13) by N and integrating over a period leads to:

$$\ln(N(t+\omega)/N(t)) = \int_t^{t+\omega} (\mu(Q(s)) - D)ds$$

which must vanish if (N, Q) is to be periodic. But from the second of (13) we have

$$Q' \leq \rho(S^*(t), Q) - \mu(Q)Q$$

from which it follows that any periodic solution must satisfy $Q(t) \leq Q^*(t)$. The monotonicity of μ then implies that $\langle \mu(Q(t)) \rangle \leq \langle \mu(Q^*(t)) \rangle$ from which a contradiction is apparent. Thus, if (11) holds, every solution is asymptotic to the trivial periodic solution as there are no others, proving the first assertion of Thm. 5.

If (12) holds, it is not difficult to see that the trivial periodic solution of (13) has a positive Floquet exponent and consequently $(0, Q^*(0))$ is a saddle fixed point of the period map whose stable manifold consists of the Q-axis in the $N - Q$ plane. The unstable manifold points into the interior of the first quadrant and therefore one can show that the orbit under the period map of a point on this manifold must converge to a non-trivial fixed point, thus establishing the existence of a non-trivial fixed point when (12) holds. The latter is globally attracting for solutions of (14) starting with $U(0) > 0$ because the trivial periodic solution only attracts solutions with $U(0) = 0$. This completes our sketch of the proof of Thm. 5.

One message from this chapter is that competitive systems need not arise from mathematical models of competition. It is conceded that the change of variables by which a competitive system was obtained was not obvious.

5 Autonomous Competitive Systems

5.1 General Results

In this chapter, we consider the autonomous system:

$$x' = f(x) \tag{15}$$

which is assumed to be defined for x in a domain $D \subset \mathbb{R}^n$. We adopt standard notation for the flow generated by (15):

$$\phi_t(x_0) = x(t, 0, x_0) \ .$$

Recall that (15) is cooperative in the convex domain D if the Jacobian of f is a quasi-positive matrix at each $x \in D$ and is competitive if $-f$ has this property. If (15) is competitive and $x(t)$ is a solution, then $y(t) = x(-t)$ is a solution of the time-reversed cooperative system $y' = -f(y)$, and vice-versa. We will exploit this fact repeatedly. We use the standard notation $\omega(p)$ and $\alpha(p)$ for the omega and alpha limit sets of the orbit through p.

Let $x(t)$ be a solution of the system (15) on an interval I. A subinterval $[a, b]$ of I is called a *rising interval* if $x(a) < x(b)$ and a *falling interval* if $x(b) < x(a)$. Our first result says that cooperative (and hence also competitive) systems cannot have solutions that oscillate back and forth between "larger" and "smaller" values, where large and small refer to the order relation $\ll$. It was first stated by Hirsch in [17] who attributes the proof to Ito. The proof is not difficult (see also [43]) but we take it for granted here and construct the theory based on it.

Lemma 5 *Let $x(t)$ be a solution of the cooperative system (15) on the interval I. Then $x(t)$ cannot have a rising interval and a falling interval that are disjoint.*

An immediate corollary is fundamental. Limit sets (α or ω) of competitive and cooperative systems are unordered.

Theorem 6 *A compact limit set of a competitive or cooperative system cannot contain two points x, y with $x \ll y$.*

Proof Sketch: First, we may, by time reversal if necessary, assume that the system is cooperative (but note that time-reversal carries an omega limit set to an alpha limit set and vice-versa). If the omega limit set of the solution $x(t)$ did contain two such points, x and y, then there would exist t_1 such that $x(t_1)$ is so near x that $x(t_1) \ll y$. Similarly, we can find $t_2 > t_1$ such that $x(t_2)$ is so near y that $x(t_1) \ll x(t_2)$ and we have produced a rising interval $[t_1, t_2]$. Clearly this can be repeated, starting near y and then near x to get a falling interval disjoint from $[t_1, t_2]$, contradicting Lemma 5. A similar argument applies to an alpha limit set. □

We are now in position to prove the fundamental result that the asymptotic behavior of a competitive or cooperative autonomous system in $\mathbb{R}^n$ can be no more complicated than that of a general system in $\mathbb{R}^{n-1}$. To make this more precise, we need a way to compare the flow of two different equations.

This can be done using the idea of topological equivalence of flows. The definition follows.

Let A be an invariant set for (15) with flow ϕ_t and let B be an invariant set for the system $y' = F(y)$ with flow ψ_t. Note that the state spaces of the two systems need not have the same dimension. We say that the flow ϕ_t on A is *topologically equivalent* to the flow ψ_t on B provided there is a homeomorphism $H : A \to B$ such that $H(\phi_t(x)) = \psi_t(H(x))$ for all $x \in A$ and all $t \in \mathbb{R}$. The relationship of topological equivalence is one of several equivalence relations on the set of all flows that say, roughly, that the dynamics of the two flows are the same.

Theorem 7 *The flow on a compact limit set of a competitive or cooperative system in $\mathbb{R}^n$ is topologically equivalent to a flow on a compact invariant set of a Lipschitz system of differential equations in $\mathbb{R}^{n-1}$.*

Proof Sketch: Let L be the limit set. Let v be a unit vector satisfying $0 \ll v$ and let $v^\perp$ be the hyperplane orthogonal to v. $v^\perp$ consists of vectors x such that $x \cdot v = 0$, where "$\cdot$" is the standard dot (or scalar) product in $\mathbb{R}^n$. Let H be the orthogonal projection onto $v^\perp$, that is $Hx = x - (x \cdot v)v$. By Thm. 6, H is one-to-one on L (this could fail only if L contains two points that differ by a multiple of v and hence are related by $\ll$). Therefore, H_L, the restriction of H to L, is a Lipschitz homeomorphism of L onto a compact subset of $v^\perp$. It is straightforward to argue, by contradicting Thm. 6, that there exists $m > 0$ such that $|H_L x_1 - H_L x_2| \geq m|x_1 - x_2|$ whenever $x_1 \neq x_2$ are points of L. Therefore, H_L^{-1} is Lipschitz on $H(L)$. Since L is a limit set, it is an invariant set for (15). It follows that the dynamical system restricted to L can be modeled on a dynamical system on $H(L)$. In fact, if $y \in H(L)$ then $y = H_L(x)$ for a unique $x \in L$ and $\psi_t(y) \equiv H_L(\phi_t(x))$ is a flow on $H(L)$ generated by the vector field

$$F(y) = H_L(f(H_L^{-1}(y)))$$

on $H(L)$. According to McShane(1934) [32], a Lipschitz vector field on an arbitrary subset of $v^\perp$ can be extended to a Lipschitz vector field on all of $v^\perp$, preserving the Lipschitz constant. It follows that F can be extended to all of $v^\perp$ as a Lipschitz vector field. It is easy to see that $H(L)$ is an invariant set for the latter vector field. We have established the topological equivalence of the flow ϕ_t on L with the flow ψ_t on $H(L)$. $H(L)$ is a compact invariant set for the $(n-1)$-dimensional dynamical system on $v^{\perp}$ generated by the extended vector field. □

As a consequence of Thm. 7, the flow on a compact limit set, L, of a competitive or cooperative system shares common dynamical properties with the flow of a system of differential equations in one less dimension, restricted to a compact invariant set, $H(L)$. However, $H(L)$ need not be a limit set of any orbit of the Lipschitz vector field. Obviously, $H(L)$ is compact and connected since L has these properties.

Thm. 7 provides a sort of converse to a construction of Smale [40] who showed that any vector field on the standard simplex $S = \{x \in \mathbb{R}^n_+ : \sum x_i = 1\}$ can be extended to a competitive system on $\mathbb{R}^n_+$ in such a way that S is an attracting set. His goal was to warn population biologists that essentially any $n-1$-dimensional dynamics can be imbedded in an n-dimensional competitive system.

5.2 3-Dimensional Competitive Systems

Given our present ignorance of dimension-dependent dynamical features of flows on limit sets, Thm. 7 may seem of limited value. However, it can be exploited for three dimensional competitive systems because we may use the Poincaré-Bendixson Thm. to determine the planar, compact, connected, invariant sets. In this way, we have the following:

Theorem 8 (The Poincaré-Bendixson Theorem) *A compact limit set of a* competitive *or* cooperative *system in* $\mathbb{R}^3$ *that contains no equilibrium points is a periodic orbit.*

Proof Sketch: Let L be the limit set. By Thm. 7, the flow ϕ_t on L is topologically equivalent to the flow ψ_t, generated by a Lipschitz planar vector field, restricted to the compact, connected, ψ-invariant set $H(L)$. Since L contains no equilibria neither does $H(L)$. The Poincaré-Bendixson Thm. implies that $H(L)$ consists of periodic orbits and, possibly, non-periodic orbits whose omega and alpha limit sets are periodic orbits contained in $H(L)$. If we could rule out orbits of the latter variety, then there would be only two cases. Either L is a single periodic orbit or L is an annulus foliated (or filled) by periodic orbits.

It is nontrivial to exclude the possibility that $H(L)$ contains a non-periodic orbit whose alpha and omega limit sets are periodic orbits contained in $H(L)$. Recall that $H(L)$ itself is not necessarily a limit set (L is) but it must share with L all dynamical properties that are preserved by topological equivalence. To see what is at issue in a slightly simpler way, suppose a planar system has two equilibria joined by a heteroclinic orbit-an orbit whose alpha limit set is one equilibrium and whose omega limit set is the other. Let A denote this set consisting of two equilibria and the heteroclinic orbit. Can such a set A be (or have the dynamical properties of) a limit set? The answer is "No!", although it possesses the usual properties of compactness, connectedness, and invariance. It is somehow intuitive that no orbit can accumulate on such a set but actually the dynamics on A preclude it from being a limit set. For exactly the same reasons, $H(L)$ cannot contain two distinct periodic orbits and a non-periodic orbit whose alpha limit set is one of these and whose omega limit set is the other.

A dynamical feature possessed by all limit sets which is not so well-known is *chain recurrence*, first described by Conley [4], [5]. The definition is as follows. Let A be a compact invariant set for the flow ψ_t. Given two points z and y in A and positive numbers ϵ and t, an (ϵ, t)-chain from z to y in A is an ordered set

$$\{z = x_1, x_2, \dots, x_{n+1} = y; t_1, t_2, \dots, t_n\}$$

of points $x_i \in A$ and times $t_i \geq t$ such that

$$|\psi_{t_i}(x_i) - x_{i+1}| < \epsilon, \qquad i = 1, 2, \dots, n \ .$$

In other words, an ϵ-chain starting at z consists of a long (if t is large) stretch of the forward orbit through z followed by a tiny mistake (if ϵ is small) to x_2 and then a long stretch of the forward orbit through x_2 followed by another small mistake, and so forth until, following a final small mistake we arrive at y. Of course, the mistakes are optional. A is said to be chain recurrent if for every $z \in A$ and for every $\epsilon > 0$ and $t > 0$, there is an (ϵ, t)-chain from z to z in A.

That all limit sets have the property of chain recurrence (see [4], [43] for a proof) is intuitively clear. Indeed, if L is a limit set of some semi-orbit (forward or backward) through point x and if L contains the point z then the semi-orbit through x must repeatedly pass arbitrarily near to z providing segments of an (ϵ, t)-chain. The only problem with this argument is that the chain may not belong to the limit set as required. To get around this, one must take limits of these orbit segments through x.

The reader should convince her/himself that the set A described above is not chain recurrent. If ϵ is small and t is large and if z lies close to the repelling equilibrium, then following the orbit through z for a large time puts one very close to the attracting equilibrium and since the allowed mistake is very small, after it we are still far away from the starting point z. Further long stretches of orbits and mistakes cannot bring us sufficiently close to z to complete the chain. Thus A cannot be a limit set.

It is easy to see that the flow ψ_t on the compact invariant set $H(L)$ inherits the property of chain recurrence since it is topologically equivalent to the flow ϕ_t on the chain recurrent limit set L. Now we see that $H(L)$ cannot contain distinct periodic orbits and a non-periodic orbit connecting the two and also be chain recurrent, for the very same reason that A above is not chain recurrent. Consequently, we conclude that every orbit of $H(L)$, and hence of L, is periodic. Since $H(L)$ is connected, it is either a single periodic orbit or an annulus consisting of periodic orbits. It follows that L is either a single periodic orbit or a "cylinder" of periodic orbits. We must rule out this last alternative.

In fact, the argument to rule out that L can be a "cylinder" foliated by periodic orbits (see [22]) was not presented by Hirsch until nine years after the original paper [17] had established the two alternatives and it is neither simple nor particularly intuitive, although it uses only Thm. 6. We give a proof due to the author [42] in a special (but the most interesting) case that L is an omega limit set of a competitive system under the additional hypothesis that (15) is irreducible. In applications, the irreducibility assumption is not a severe one. As it is easier to work with cooperative systems we reverse time and consider instead the equivalent case that L is an alpha limit set of the backward orbit through x of a cooperative and irreducible system. We assume that L consists of a "cylinder" consisting of periodic orbits (the homeomorphic image of an annulus). We will use the Perron-Frobenius Thm. to show that each periodic orbit of L is unstable, having one real Floquet multiplier exceeding one and the other two multipliers equal to one. The degeneracy is due to the fact the orbit belongs to the manifold L of periodic orbits which might be viewed as a center manifold corresponding to the given periodic orbit. Transversality arguments then imply that the union U of the two-dimensional local unstable manifolds of all the periodic orbits making up L has non-empty interior in $\mathbb{R}^3$. Since the interior of U contains points of L, $\phi_t(x)$ must enter the interior of U for large negative times t but then it must also intersect the unstable manifold of a periodic orbit, call it γ. The latter implies that $L = \gamma$, a contradiction to our assumption that L is a cylinder foliated by periodic orbits.

Finally, let $\gamma = \{\phi_t(p) : 0 \le t < T\}$ be a periodic orbit of a cooperative and irreducible system in $\mathbb{R}^3$ with minimal period $T > 0$. The stability of γ can be determined by considering the Floquet exponents of the periodic linearized system:

$$y' = D_x f(\phi_t(p))y \ .$$

If $Y(t)$ is the fundamental matrix satisfying $Y(0) = I$, then the Floquet multipliers are the eigenvalues of $Y(T)$. Now $Y(t) \gg 0$ for $t > 0$ by Thm. 2 (see also the remarks following it) so the Perron-Frobenius Thm. [2] implies that $Y(T)$ has a simple positive eigenvalue, μ, larger than the modulus of all other eigenvalues, with a corresponding eigenvector $v \gg 0$ and that any other eigenvector u of $Y(T)$ belonging to $\mathbb{R}^n_+$ must be a multiple of v. Of course, $Y(t)$ is just the Jacobian matrix $(D_x\phi_t)(p)$ and consequently $Y(T)$ has the trivial eigenvalue one: $Y(T)f(p) = f(p)$. Now $f(p) \gg 0$ cannot hold since $\phi_t(p) = p + tf(p) + o(t)$ would then imply that $\phi_\tau(p) \gg p$ for small $\tau > 0$ and therefore γ, which can be viewed as the limit set of itself, would violate Thm. 6. In a similar way, $-f(p) \gg 0$ cannot hold so $f(p)$ is not a multiple of v. Since μ is a simple eigenvalue larger than the modulus of all others, we conclude that $\mu > 1$, implying the instability of γ. If γ belongs to the cylinder of periodic orbits L, then γ is degenerate and has two independent eigenvectors corresponding to the Floquet exponent one. This completes our sketch of the proof. □

Thm. 8 is more difficult to apply than its two-dimensional counterpart because it is more difficult to exclude the possibility that a limit set contains an equilibrium for three-dimensional systems. If the region of interest contains a single equilibrium, the task is easier. The following result is quite useful.

Theorem 9 *Let (15) be a competitive system in $D \subset \mathbb{R}^3$ and suppose that D contains a unique equilibrium point p which is hyperbolic and assume that $D_x f(p)$ is irreducible. Suppose further that $W^s(p)$, the stable manifold of p, is one-dimensional. If $q \in D \setminus W^s(p)$ and $\{\phi_t(q) : t \ge 0\}$ has compact closure in D then its omega limit set is a nontrivial periodic orbit.*

Proof: The result will follow from Thm. 8 provided $p \notin \omega(q)$. Clearly, $\omega(q) \ne p$ since $q \notin W^s(p)$. If $p \in \omega(q)$ then the Butler-McGehee Lemma [50] implies that $\omega(q)$ contains a point y on $W^s(p)$ different from p. By the invariance of

$\omega(q)$ we may assume that y is a point of the local stable manifold, arbitrarily close to p. We will use the Perron-Frobenius Thm. to show that either $p \ll y$ or $y \ll p$, if y is sufficiently near p and belongs to $W^s(p)$. This contradicts Thm. 6 since both p and y are assumed to belong to $\omega(q)$.

The Jacobian matrix of f at p, $Df(p)$, has the property that $A = -Df(p) + sI \geq 0$ for all large values of s. If $Df(p)$ is irreducible then so is A and consequently, by the Perron-Frobenius Thm., A has an eigenvector $v \gg 0$ corresponding to the eigenvalue $r > 0$ having maximum modulus among all eigenvalues. Then, v is an eigenvector for $Df(p)$ corresponding to the eigenvalue $s - r$. The eigenvalues λ of $Df(p)$ give rise to the eigenvalues $s - \lambda$ of A, so, as r is the eigenvalue of A with maximal real part, $s - r$ is the eigenvalue of $Df(p)$ of minimal real part. The hypotheses of Thm. 9 imply that $s - r < 0$ and the corresponding eigenvector is v. Thus we see that the tangent space of $W^s(p)$ at p is the vector v and since $v \gg 0$, so $W^s(p) = p + \epsilon v + o(\epsilon)$, and we easily see that y and p are ordered as claimed in the previous paragraph. □

It is of interest to establish the existence of an orbitally stable or asymptotically stable periodic orbit since these are observable. For general planar systems one can prove the following (the reader is invited to supply the proof):

Theorem 10 *Let a continuously differentiable, dissipative, planar system of ordinary differential equations have a unique equilibrium point p. If $Df(p)$ is nonsingular then either p is stable or there exists an orbitally stable periodic orbit. If, in addition, the system is real analytic then there exists an orbitally asymptotically stable periodic orbit when p is unstable.*

Roughly, if the point at infinity is a repeller and p is unstable, then there should be a stable periodic orbit. The proof can be found in the thesis of H.-R. Zhu [57]. In her thesis (see also [58]), this result is carried over to three dimensional competitive systems. In order to state the result, it is convenient to introduce the following hypotheses.

(H1) D is an open and convex subset of $\mathbb{R}^3$.

(H2) (15) is competitive and irreducible in D.

(H3) System (15) is dissipative in D: for each $x \in D$, the forward orbit through x has compact closure in D. Moreover, there exists a compact subset B of D such that for each $x \in D$, there exists $T = T(x)$ so that $\phi_t(x) \in B$ for $t \geq T$.

(H4) D contains a unique equilibrium point p and $\det Df(p) < 0$.

Theorem 11 *Let (H1)-(H4) hold. Then either p is stable or there exists a non-trivial orbitally stable periodic orbit in D. If, in addition, f is real analytic in D, then there is an orbitally asymptotically stable periodic orbit when p is unstable.*

We remark that in both Thm. 10 and Thm. 11, when p is unstable and the system is real analytic, then there can be at most finitely many periodic orbits. Basically, the proof of Thm. 11 is similar to the proof of Thm. 10 using Thm. 8. The idea is to show that the compact attractor A exists and is unordered and therefore is Lipschitz-homeomorphic to a set in the plane exactly as was proved to hold for a limit set in Thm. 7. Once this is established, one can argue as in the proof of Thm. 10. The hypothesis that the determinant of the Jacobian at p is negative implies that if p is unstable then either it is hyperbolic and has a two dimensional unstable manifold or it is non-hyperbolic due to two purely imaginary eigenvalues and one negative eigenvalue. It is not difficult to see using the Brouwer degree that if D is homeomorphic to an open ball, then (H3) and $\det Df(p) \neq 0$ implies $\det Df(p) < 0$. See [57], [58] for a careful proof and additional remarks. Of course, even if p is stable, there may exist orbitally stable periodic orbits.

6 Examples of Competitive Systems in $\mathbb{R}^3$

The most well-known competitive system in $\mathbb{R}^3$ is the famous Lotka-Volterra system

$$x_i' = x_i(r_i - \sum_{j=1}^{3} c_{ij}x_j), \quad 1 \leq i \leq 3 \ , \tag{16}$$

where $c_{ij}, r_i > 0$. It has been studied by numerous authors but we would like to single out the work of M.L. Zeeman [59], which is based on a result of Hirsch [20] implying that all orbits approach a "carrying simplex", a Lipschitz two-dimensional manifold homeomorphic to the standard simplex in $\mathbb{R}^3_+$ and containing all nontrivial equilibria. By studying the geometric structure of the isoclines, Zeeman defines a combinatorial equivalence relation on the set of all systems (16) and identifies 33 stable equivalence classes. Of these, 25 equivalence classes exhibit convergence to equilibrium for all orbits while periodic orbits are possible for some of the remaining 8 classes. Open problems remain concerning the number of periodic orbits in the latter classes. Hofbauer and So [25] give an example with two limit cycles surrounding the interior equilibrium. Apparently, it is an open question as to whether or not there are at most finitely many periodic orbits on the carrying simplex [56]. These authors conjecture that there are at most two. See also [53] for a study of Kolmogorov-type systems modeling three-species competition.

There is essentially one other type of three dimensional system (15) which may be transformed to a cooperative (competitive) system by the change of variables $y = Q_m x$. This is because, if we allow permutation of variables, and take account that cone K and $-K$ generate equivalent partial orderings, there are essentially two distinct cones, the usual one $K = \mathbb{R}^3_+$ and $K_m = \{x \in \mathbb{R}^3 : x_1 \geq 0, x_2 \leq 0, x_3 \geq 0\}$ with order isomorphism $Q = Q_m = \text{diag}(+1, -1, +1)$. Thus, aside from the usual competitive systems, there is another class of vector fields whose Jacobian matrix at each point of the convex set $D \subset \mathbb{R}^3$ has the following sign structure (the diagonal entries are irrelevant):

$$Df(x) = \begin{pmatrix} * & + & - \\ + & * & + \\ - & + & * \end{pmatrix} .$$

We emphasize that a $+$ sign in an entry of the matrix means that the entry is non-negative for all $x \in D$ (possibly identically zero) and similarly for a $-$ sign. If the vector field f has this property then the change of variables $y = Qx$ leads to the system

$$y' = g(y) \equiv Qf(Qy)$$

which is competitive in the usual sense in the convex set QD since for $y \in QD$

$$Dg(y) = QDf(Qy)Q = \begin{pmatrix} * & - & - \\ - & * & - \\ - & - & * \end{pmatrix} .$$

Obviously, all the results of the previous section apply to these systems, which we will also refer to as competitive. We intend to show by way of example that this class of three dimensional competitive systems include many which are of importance to biology.

6.1 A Predator-Prey Model

Harrison [15] introduces various predator-prey models in order to fit the experimental data of Luckinbill [30] on *Didinium nasutum* (predator) and *Paramecium aurelia* (prey) in a methyl cellulose thickened cerophyl medium. A better fit required introduction of a third variable, z, the predator stored energy, in the usual planar predator-prey models, effectively creating a time delay in the predator response function. One of several models considered by Harrison (a similar analysis can be applied to the others) takes the following form (we have scaled the system in [15]).

$$\begin{aligned} x' &= \rho x(1 - x/K) - \frac{\omega xy}{\phi + x} \\ y' &= z - \gamma y \\ z' &= \frac{\omega xy}{\phi + x} - \delta z \end{aligned} \tag{17}$$

All parameters $\rho, K, \omega, \phi, \delta, \gamma$ are positive and we take $D = (\mathbf{R}^3_+)^0$. We leave to the reader to check that (17) is competitive and irreducible in D (re-order the variables (y, z, x)). Assuming that

$$\frac{\omega K}{\phi + K} > \delta\gamma \ , \tag{18}$$

there is a unique equilibrium $p = (x^*, y^*, z^*) \in D$ with $z^* = \gamma y^*$, $\frac{\omega x^*}{\phi + x^*} = \delta\gamma$ and $y^* = (\rho/\gamma\delta)(1 - x^*/K)x^*$, at which

$$\det Df(p) = -\frac{\omega\gamma\phi\gamma}{(\phi + x^*)^2} y^* < 0 \ .$$

Lemma 6 *If (18) holds, there is a compact subset $B \subset D$ which attracts all solutions starting in D.*

Proof: First, we show the system is dissipative. Obviously, $\limsup_{t\to\infty} x(t) \leq K$ follows from a simple differential inequality argument. Thus, there exists $T = T(x(0)) > 0$ such that $x(t) < 2K$ for $t \geq T$. Choose L so that $\delta(L - 2K) > K\rho/4$. If $x(t) + z(t) > L$ for some $t \geq T$ then $(x + z)' < K\rho/4 - \delta(L - 2K) < 0$ from which we conclude that $\limsup_{t\to\infty}(x(t) + z(t)) \leq L$. This in turn leads to $\limsup_{t\to\infty} y(t) \leq 2L$. The system (17) is dissipative.

We may use Thm. 4.5 of [54] or Thm. 3.2 of [14] to conclude that system (17) is ***uniformly persistent.*** That is, there exists $\epsilon > 0$ such that the limit inferior of each component of every solution starting in D exceeds ϵ. To do this, we must show the following: (1) the stable manifold of each equilibrium on the boundary of D does not intersect D, and (2) there are no heteroclinic cycles consisting of boundary equilibria and entire trajectories on the boundary whose α and ω limit sets are these equilibria. The stable manifold of the trivial equilibrium is the $y-z$ plane on the boundary of D and the stable manifold of the equilibrium $(K,0,0)$ is two dimensional but intersects $\bar{D}$ only in the x-axis. Neither of these stable manifolds intersect D. There is an entire orbit on the x-axis connecting the trivial equilibrium to $(K,0,0)$ but there is no heteroclinic orbit connecting the latter equilibrium to the former one so there are no heteroclinic cycles. Thus, we conclude that (17) is uniformly persistent. This completes our proof. □

Our main result for (17) can now be stated. The so-called "paradox of enrichment" holds: increasing the carrying capacity of the prey destabilizes the system.

Theorem 12 *If (18) holds then there exists K^* satisfying $x^* < K^* < K_0 \equiv \frac{\phi(\omega+\delta\gamma)}{\omega-\delta\gamma}$ such that for each $K > K^*$, p is unstable, there is an orbitally asymptotically stable periodic orbit and the omega limit set of every orbit of (17) starting in $D \backslash W^s(p)$ is a non-trivial periodic orbit. $W^s(p)$ is one-dimensional in this case. If $K < K^*$, then p is asymptotically stable.*

Proof: We have established all of the hypotheses of Thm. 9 and Thm. 11 except the instability of p, the hyperbolicity of p, and the fact that its stable manifold $W^s(p)$ is one-dimensional. A calculation of the characteristic polynomial for $Df(p)$ leads to:

$$\lambda^3 + a_1\lambda^2 + a_2\lambda + a_3 = 0 \ ,$$

where $a_1 = \gamma+\delta+b-a$, $a_2 = (b-a)(\gamma+\delta)$, $a_3 = \delta\gamma b$, $b = \frac{\phi\rho}{\phi+x^*}(1-x^*/K) > 0$ and $a = \rho(1-2x^*/K)$. We find that $b-a = \frac{\rho\delta\gamma}{\omega K}[K_0-K]$ where K_0 is defined above. If $K \geq K_0$ then $b-a \leq 0$ so $a_2 \leq 0$ and thus one of the Routh-Hurwitz criteria for all eigenvalues to satisfy $\Re\lambda < 0$ is violated so at least one eigenvalue satisfies $\Re\lambda \geq 0$. If $b-a > 0$ then all coefficients a_i are positive and the Routh-Hurwitz criteria $\Delta \equiv a_1a_2 - a_3 > 0$ must also be made to fail. We note that

$$\Delta(K) = (\gamma+\delta)(b-a)^2 + (\gamma+\delta)^2(b-a) - \delta\gamma b$$

is strictly decreasing in K so long as $b - a \geq 0$ (or, equivalently $K \leq K_0$). Furthermore, $\Delta(x^*+) = (\gamma + \delta + \rho)\rho(\gamma + \delta) > 0$ and $\Delta(K_0) = -\delta\gamma b < 0$ so there exists a unique value K^* such that $\Delta(K^*) = 0$. If $K > K^*$ then $\Delta < 0$ or $b - a \leq 0$. If $K < K^*$, then $b - a > 0$ and $\Delta > 0$ so p is asymptotically stable.

But $Df(p)$ has negative determinant so no eigenvalue can vanish and, setting $\lambda = i\eta$, with η a nonzero real number, in the characteristic polynomial and considering only the imaginary part, we find that there are no such eigenvalues if $b - a \leq 0$ or $\Delta < 0$. From this and the preceding paragraph we conclude that $Df(p)$ is hyperbolic and has precisely two eigenvalues with positive real part. The assertions then follow from the theorems mentioned above. □

6.2 Model of cAMP signaling in the Slime Mold

The scaled Goldbeter-Segel model of cAMP synthesis by the amoeba *Dictyostelium discoideum* is given by (see [39], [12], [11]):

$$\begin{aligned} \alpha' &= 1 - \sigma\phi(\alpha,\gamma) \\ \beta' &= q\sigma\phi(\alpha,\gamma) - k_1\beta \\ \gamma' &= k_1 h^{-1}\beta - k_2\gamma \end{aligned} \tag{19}$$

where α represents intra-cellular ATP concentration, β represents intra-cellular cAMP, and γ represents extra-cellular cAMP. All parameters L, k_i, σ, q, h are positive and the function ϕ is given by

$$\phi(\alpha,\gamma) = \frac{\alpha(1+\alpha)(1+\gamma)^2}{L + (1+\alpha)^2(1+\gamma)^2} \ .$$

It satisfies $0 \leq \phi \leq 1$ and its partial derivatives ϕ_α and ϕ_γ are positive.

It is easy to check that (19) is competitive and irreducible on $D = (\mathbb{R}^3_+)^0$ and, if the cAMP production parameter σ satisfies $\sigma > 1$ then there is a unique equilibrium $p = (\alpha^*, \beta^*, \gamma^*) \in D$ with $\beta^* = q/k_1, \gamma^* = q/hk_2$ and α^* is the positive root of

$$0 = (\sigma - 1)\alpha^2 + (\sigma - 2)\alpha - [1 + (1+\gamma^*)^{-2}L] \ .$$

There are no other equilibria. Furthermore

$$\det Df(p) = -k_1 k_2 \sigma \phi_\alpha < 0 \ .$$

It is not difficult to see that if $\sigma > 1$ then the system (19) is uniformly persistent (see [54]) in the sense that there exists $\epsilon > 0$ such that $\liminf_{t\to\infty} \alpha(t) > \epsilon$, and similarly for the other coordinates, for every solution $(\alpha(t), \beta(t), \gamma(t))$ starting in D. Also, the system is easily seen to be dissipative so there is a set $B = [\epsilon, \bar{\alpha}] \times [\epsilon, \bar{\beta}] \times [\epsilon, \bar{\gamma}]$ which attracts all orbits starting in D. The details are similar to those in the example above. Thm. 9 and Thm. 11 imply the following result.

Theorem 13 *Assume $\sigma > 1$ in (19). If p is unstable then there exists an orbitally asymptotically stable periodic orbit. If, in addition, p is hyperbolic, the omega limit set of every orbit starting off the one-dimensional stable manifold of p is a non-trivial periodic orbit.*

Proof: Note the vector field is real analytic so Thm. 11 gives the first assertion. Because the Jacobian at p is irreducible and has a negative determinant and p is assumed to be hyperbolic and unstable, it follows that there is exactly one eigenvalue with negative real part. Thus Thm. 9 applies to give the second assertion. □

We refer the reader to the monographs [12], [39] for regions in parameter space where the hypotheses of the theorem are satisfied and for further discussions of the interesting dynamics exhibited by system (19).

6.3 Model of a Mitotic Oscillator

The Goldbeter model [12] of a mitotic oscillator controlling the cell cycle of an embryonic cell is given by

$$\begin{aligned} C' &= \nu_i - \nu_d X \frac{C}{K_d + C} - k_d C \\ M' &= V_{M1} \frac{C}{K_c + C} \frac{1 - M}{K_1 + 1 - M} - V_2 \frac{M}{K_2 + M} \\ X' &= V_{M3} M \frac{1 - X}{K_3 + 1 - X} - V_4 \frac{X}{K_4 + X} \end{aligned} \tag{20}$$

C is the cyclin concentration, $M \in [0, 1]$ is the fraction of active cdc2 kinase, and $X \in [0, 1]$ is the active fraction of cyclin protease. As usual, all parameters are positive. The reader may easily check that (20) is competitive and irreducible in $D = (0, \infty) \times (0, 1)^2$. An entirely similar analysis to that of the previous example may be carried out to obtain a result like the previous theorem. See [12] for a description of the parameter region where periodic solutions are observed.

7 Abstract Dynamics of Competition

The amount of research devoted to mathematical models of two competing populations is enormous. Most such models consist of ordinary differential equations or difference equations, but more recently models consisting of delay differential equations, partial differential equations, and even partial differential equations with delays have been studied. Despite the fact that there are techniques of analysis that are common to a large body of this research, it was not until very recently that an abstract approach to competition was taken. The work of Hess and Lazer [16] seems to be the first in this direction. They consider competitive dynamics generated by order preserving smooth (C^1) maps, making essential use of the well-known Krein-Rutman theorem, the infinite dimensional analog of the Perron-Frobenius theorem. This chapter is based on the more topological approach taken in [27], where smoothness assumptions are avoided.

In nearly all models of two-population competition, the state space is a cartesian product of two spaces, each the state space for a single organism. The single-population state space consists of that population density, which might be a scalar quantity, simply a measure of the number of organisms, or it might be a vector, say the number of organisms in habitat patch i, for $1 \leq i \leq n$, or it might be the density of organisms at position x for each x in a domain Ω. Of course, each population density must be non-negative in a suitable sense. In order to capture all these cases, we are led to take a cone X_i^+ in the Banach space X_i as the single-population state space and $X^+ = X_1^+ \times X_2^+$ as the natural state space for two-population competition. We assume that X_i^+ has non-empty interior $(X_i^+)^0$ and, for simplicity, we assume that the cones X_i^+ are *normal*, that is, there exists $k_i > 0$ such that $x_i, y_i \in X_i$ and $0 \leq x_i \leq y_i$ implies $\|x_i\| \leq k_i \|y_i\|$. The latter assumption implies that order intervals $[x_i, y_i]$ are bounded sets in X_i. See [27] for a more general treatment; the cone $C_+(\bar{\Omega})$ is a normal cone in $C(\bar{\Omega})$ with $k_i = 1$. We use the same symbol for the partial orders generated by the cones X_i^+. If $x_i, \bar{x}_i \in X_i$, then we write $x_i \leq \bar{x}_i$ if $\bar{x}_i - x_i \in X_i^+$, $x_i < \bar{x}_i$ if $x_i \leq \bar{x}_i$ and $x_i \neq \bar{x}_i$, and $x_i \ll \bar{x}_i$ if $\bar{x}_i - x_i \in (X_i^+)^0$. If $x_i, y_i \in X_i$ satisfies $x_i < y_i$, then the order interval $[x_i, y_i]$ is defined by $[x_i, y_i] = \{u \in X_i : x_i \leq u \leq y_i\}$.

Let $X = X_1 \times X_2$, $X^+ = X_1^+ \times X_2^+$, and $K = X_1^+ \times (-X_2^+)$. X^+ is a cone in X with nonempty interior given by $(X^+)^0 = (X_1^+)^0 \times (X_2^+)^0$. It generates the order relations $\leq, <, \ll$ in the usual way. In particular, if $x = (x_1, x_2)$ and $\bar{x} = (\bar{x}_1, \bar{x}_2)$, then $x \leq \bar{x}$ if and only if $x_i \leq \bar{x}_i$, for $i = 1, 2$. For our purposes, the more important cone is K which also has nonempty interior given by $K^0 = (X_1^+)^0 \times (-(X_2^+)^0)$. It generates the partial order relations $\leq_K, <_K, \ll_K$. In this case,

$$x \leq_K \bar{x} \iff x_1 \leq \bar{x}_1 \quad \text{and} \quad \bar{x}_2 \leq x_2.$$

A similar statement holds with $\ll_K$ replacing $\leq_K$ and $\ll$ replacing $\leq$. If $x, y \in X$ satisfy $x <_K y$, then we define the order interval $[x, y]_K \equiv \{z \in X : x \leq_K z \leq_K y\}$. Obviously, $z = (z_1, z_2) \in [x, y]_K$ if and only if $x_1 \leq z_1 \leq y_1$ and $y_2 \leq z_2 \leq x_2$. The ordering $\leq_K$ is a generalization of the south-east ordering used in Section 3.

Consider first the case of discrete-time models of competition generated by a map (e.g. the period map of the periodic Lotka-Volterra system (7) or the map P induced by (8)):

$$x_{n+1} = T(x_n) \quad n \geq 0.$$

Let $T : X^+ \to X^+$ be continuous and denote by T^n the n-fold composition of T. The following hypotheses on T are meant to capture the essence of competition between two adequate competitors (ones which can survive in the absence of competition).

(H1) T is completely continuous and strictly order-preserving with respect to $<_K$. That is, $x <_K \bar{x}$ implies $T(x) <_K T(\bar{x})$.

(H2) $T(0) = 0$ and 0 is a repelling fixed point: there exists a neighborhood U of 0 in X^+ such that for each $x \in U$, $x \neq 0$, there is an integer $n = n(x)$ such that $T^n(x) \notin U$.

(H3) $T(X_1^+ \times \{0\}) \subset X_1^+ \times \{0\}$. There exists $\hat{x}_1$ satisfying $0 \ll \hat{x}_1$ such that $T((\hat{x}_1, 0)) = (\hat{x}_1, 0)$. Furthermore, $T^n((x_1, 0)) \to (\hat{x}_1, 0)$ for every x_1 satisfying $0 < x_1$. The symmetric conditions hold for T on $\{0\} \times X_2$. The fixed point is denoted by $(0, \tilde{x}_2)$.

(H4) If $x, y \in X^+$ satisfy $x <_K y$ and either x or y belongs to $(X^+)^0$, then $T(x) \ll_K T(y)$. If $x = (x_1, x_2) \in X^+$ satisfies $x_i \neq 0, i = 1, 2$, then $T(x) \gg 0$.

Recall that T is *completely continuous* if for every bounded set B, it follows that $T(B)$ has compact closure in X.

The strict order preserving property described in (H1) is the signature of a competitive system. It is biologically intuitive. The two related states $x = (x_1, x_2)$ and $\bar{x} = (\bar{x}_1, \bar{x}_2)$, where $x <_K \bar{x}$, represent initial conditions in which the state of the first population is given by the first component and the state of the second population is given by the second component. The relation says that the second population has an advantage over the first in the state x relative to the state $\bar{x}$ since the second population is greater in state x and its competitors population is smaller. Viewed differently, population one has the advantage over population two in state $\bar{x}$. The order preservation property says merely that the relative advantage of one state over the other is preserved into the future.

We introduce the following notation for the "boundary" fixed points of T:

$$E_0 = (0,0), \quad E_1 = (\hat{x}_1, 0), \quad E_2 = (0, \tilde{x}_2) \ .$$

We say that a fixed point E_* of T is positive if it belongs to the interior of X^+. Such a fixed point represents the coexistence of the two populations. The order interval I defined by

$$I \equiv [0, \hat{x}_1] \times [0, \tilde{x}_2]$$

will play an important role.

Given $x \in X^+$ we write $O(x) = \{T^n(x) : n \geq 0\}$ for the positive orbit of x. Its omega limit set is defined in the usual way as

$$\omega(x) = \{y \in X^+ : T^{n_i}(x) \to y, \ \text{ some sequence } \{n_i\} \text{ satisfying } n_i \to \infty\} \ .$$

Our main result says that for a competitive system, either there is a positive fixed point of T, representing coexistence of the two populations, or one population drives the other to extinction.

Theorem 14 *Let (H1)-(H4) hold. Then the omega limit set of every orbit is contained in I and exactly one of the following holds:*

(a) There exists a positive fixed point E_* of T in I.

(b) $T^n(x) \to E_1$ as $n \to \infty$ for every $x = (x_1, x_2) \in I$ with $x_i \neq 0$, $i = 1, 2$.

(c) $T^n(x) \to E_2$ as $n \to \infty$ for every $x = (x_1, x_2) \in I$ with $x_i \neq 0$, $i = 1, 2$.

Finally, if (b) or (c) hold and $x \in X^+ \setminus I$, *then either* $T^n(x) \to E_1$ *or* $T^n(x) \to E_2$.

In case (b), our result may seem a bit unsatisfactory in the sense that we do not conclude $T^n(x) \to E_1$ for all $x = (x_1, x_2)$ with $x_1 \neq 0$. More precisely, we cannot rule out that there is such an $x \in X^+ \setminus I$ such that $T^n(x) \to E_2$. In fact, we give an example of such behavior below. If such a point $x = (x_1, x_2)$ with $x_i \neq 0$ exists, then it is easy to see that E_2 attracts the set $\{y \in X^+ : y \leq_K T(x)\}$ which has nonempty interior in X^+. As E_2 also repels a relatively open set in I in case (b), E_2 would certainly be non-hyperbolic if T were a smooth map.

An immediate corollary of Thm. 14 is that a coexistence equilibrium exists whenever both E_1 and E_2 are stable or when both are unstable for the restriction of the map T to I. The reason is that neither (b) nor (c) of the theorem are compatible with either of these conditions. See [49] for an application of Thm. 14 to the Lotka-Volterra difference equations (8).

We now formulate a continuous-time version of Thm. 14. Assume that $T : [0, \infty) \times X^+ \to X^+$ is a continuous *semiflow*. We write $T_t(x) = T(t, x)$. The semiflow properties are (i) $T_0 = id_X$, and (ii) $T_t \circ T_s = T_{t+s}$ for $t, s \geq 0$. The analogous hypotheses to (H1)-(H4) above are given below.

(H1) T is strictly order-preserving with respect to $<_K$. That is, $x <_K \bar{x}$ implies $T_t(x) <_K T_t(\bar{x})$. For each $t > 0$, $T_t : X^+ \to X^+$ is completely continuous.

(H2) $T_t(0) = 0$ for all $t \geq 0$ and 0 is a repelling equilibrium: there exists a neighborhood U of 0 in X^+ such that for each $x \in U$, $x \neq 0$, there is a $t_0 > 0$ such that $T_{t_0}(x) \notin U$.

(H3) $T_t(X_1^+ \times \{0\}) \subset X_1^+ \times \{0\}$ for all $t \geq 0$. There exists $\hat{x}_1 \gg 0$ such that $T_t((\hat{x}_1, 0)) = (\hat{x}_1, 0)$ for all $t \geq 0$. Furthermore, $T_t((x_1, 0)) \to (\hat{x}_1, 0)$ as $t \to \infty$ for all $x_1 \neq 0$. The symmetric conditions hold for T on $\{0\} \times X_2^+$ with equilibrium point $(0, \tilde{x}_2)$.

(H4) If $x = (x_1, x_2) \in X^+$ satisfies $x_i \neq 0, i = 1, 2$, then $T_t(x) \gg 0$ for $t > 0$. If $x, y \in X^+$ satisfy $x <_K y$ and either x or y belongs to $(X^+)^0$, then $T_t(x) \ll_K T_t(y)$ for $t > 0$.

As in the discrete case, we use the notation E_i for the boundary equilibria. We say that E_* is a positive equilibrium of T if it belongs to the interior of

X^+. If $x \in X^+$ then $O(x) = \{T_t(x) : t \geq 0\}$ is called the positive orbit of T. Its omega limit set is defined in the usual way.

Theorem 15 *Let (H1)-(H4) hold. Then the omega limit set of every orbit is contained in I and exactly one of the following holds:*

(a) There exists a positive equilibrium E_* of T in I.

(b) $T_t(x) \to E_1$ as $t \to \infty$ for every $x = (x_1, x_2) \in I$ with $x_i \neq 0, i = 1, 2$.

(c) $T_t(x) \to E_2$ as $t \to \infty$ for every $x = (x_1, x_2) \in I$ with $x_i \neq 0, i = 1, 2$.

Finally, if (b) or (c) hold, $x \in X^+ \setminus I$, then either $T_t(x) \to E_1$ or $T_t(x) \to E_2$ as $t \to \infty$.

The following example shows that case (c) of Thm. 15 may hold yet some open set of initial data outside I is attracted to E_1. Consider the planar system

$$\begin{aligned} x_1' &= x_1(1 - x_1 - \mu x_2)^3 \\ x_2' &= x_2(1 - x_1 - x_2) \end{aligned}$$

where $\mu > 1$. It is possible to verify that all positive solutions beginning in $I = [0,1] \times [0,1]$ are attracted to $E_2 = (0,1)$ but that solutions starting at (x_1, x_2) near $E_1 = (1,0)$ and satisfying $x_1 > 1$, $0 < x_2 < (x_1 - 1)^2$ are attracted to E_1.

It is not our intention to give a detailed proof of Thm. 14 here but we do want to give some of the main ideas. It is easy to see that the order interval I attracts all orbits. First, observe that I can be viewed as $I = [E_2, E_1]_K = \{x \in X : E_2 \leq_K x \leq_K E_1\}$ and is positively invariant for if $x \in I \setminus \{E_1, E_2\}$ then $E_2 <_K x <_K E_1$ so, on applying T and using (H1), we find $E_2 <_K Tx <_K E_1$. If $x = (x_1, x_2) \in X^+ \setminus I$ with $x_i \neq 0$, then $(0, x_2) <_K x <_K (x_1, 0)$, from which it follows that $T^n(0, x_2) <_K T^n x <_K T^n(x_1, 0)$ for $n \geq 1$. Since $T^n(0, x_2) \to E_2$ and $T^n(x_1, 0) \to E_1$, we conclude that $O(x)$ is bounded and since T is completely continuous, it has compact closure. But $T^n(0, x_2) \to E_2$ and $T^n(x_1, 0) \to E_1$ as $n \to \infty$ and so we may conclude that $\omega(x)$ is a non-empty subset of I. For if $T^{n_i}x \to z$ as $i \to \infty$, then $E_2 \leq_K z \leq_K E_1$.

Further progress requires some new ideas. Let Y be a Banach space and Y^+ be a cone in Y. As usual, we denote by $\leq$ and $<$ the partial order relation generated by Y^+. Recall that if C is a convex subset of Y and $e \in C$, then we say that e is an extreme point of C if there do not exist points $x, y \in C \setminus \{e\}$ such that $e = 1/2(x+y)$.

Let $U \subset Y$ and $S : U \to U$ be a continuous function. S is strictly order-preserving if $x < y \Rightarrow S(x) < S(y)$.

A fixed point u of S is said to be an *ejective fixed point* if there is an open subset V of U containing u such that for every $x \in V \setminus \{u\}$ there is an integer m for which $S^m(x) \notin V$. Let Z denote the set of integers. A sequence $\{x_n\}_{n \in Z}$ in U is called an *entire orbit* of S if $x_{n+1} = S(x_n)$ for all $n \in Z$. If there exist fixed points u and v of S in U such that $x_n \to v$ as $n \to \infty$ and $x_n \to u$ as $n \to -\infty$, then we say that the entire orbit joins u to v.

The following result, a modification of Prop. 1 of Dancer and Hess [16] we call the order interval trichotomy, is the key to our proof of Thm. 14.

Proposition 1 *Let $u_1 < u_2$ be fixed points of the strictly order-preserving continuous map $S : U \to U$, let $I \equiv [u_1, u_2] \subset U$, and suppose that $S(I)$ has compact closure in I. Suppose further that S has an ejective fixed point $e \in I \setminus \{u_1, u_2\}$ which is an extreme point of I. Then at least one of the following holds:*

(a) S has a fixed point distinct from u_1, u_2, e in I.

(b) There is an entire orbit $\{x_n\}_{n \in Z}$ of S joining u_1 to u_2 and satisfying $x_{n+1} > x_n, \quad n \in Z$.

(c) There is an entire orbit $\{y_n\}_{n \in Z}$ of S joining u_2 to u_1 and satisfying $y_{n+1} < y_n, \quad n \in Z$.

Dancer and Hess proved Prop. 1 in the case that the fixed point e is not assumed to exist and where it is omitted from (a). See [27] for the proof, which is only a slight modification of the original proof of Dancer and Hess. We will indicate how it can be used to prove Thm. 14. Take $Y = X$ with cone K, $U = X^+$, $u_1 = E_2$, $u_2 = E_1$, and note that $I = [E_2, E_1]_K = \{x \in X : E_2 \leq_K x \leq_K E_1\}$. Setting $S = T$, we observe that by (H1) and (H2), all hypotheses of Prop. 1 hold (E_0 is an extreme point of I). The three alternatives of the order interval trichotomy correspond to the three alternatives of Thm. 14. We indicate briefly how (b) of the order interval trichotomy leads to (b) of the theorem. First, it can be shown that the hypothesis (H4) of Thm. 14 implies that exactly one of the alternatives of

Prop. 1 hold and furthermore, the entire orbit $\{x_n\}$, in case (b), satisfies $E_2 \ll_K x_n \ll_K x_{n+1} \ll_K E_1$. Now, suppose that $x = (x_1, x_2) \in I$ satisfies $x_i \neq 0$. By (H4), $0 \ll Tx \in I$ implying that $E_2 \ll_K Tx \ll_K E_1$. Since $x_n \to E_2$ as $n \to -\infty$, there is an $n \in Z$ such that $E_2 \ll_K x_n \ll_K Tx$ and therefore, $x_{n+k} = T^k x_n <_K T^{k+1}x <_K E_1$ for $k \geq 0$. As $x_n \to E_1$ as $n \to \infty$, we conclude that $T^k x \to E_1$ as well.

Continuing our demonstration in case (b), suppose that $x \in X_+ \setminus I$. The omega limit set $\omega(x) \subset I$ is nonempty, compact, invariant ($T(\omega(x)) = \omega(x)$), and *invariantly connected*. The latter means that $\omega(x)$ is not the disjoint union of two nonempty, closed, invariant subsets. Let $z \in \omega(x)$ and $z = (z_1, z_2)$. If $z_i \neq 0$ for $i = 1, 2$, then $T(z) \gg 0$ and therefore $E_2 \ll_K T^2(z) \ll_K E_1$ by (H4). In this case, since $T^2(z) \in \omega(x)$, there exists m such that $E_2 \ll_K T^m(x) \ll_K E_1$. But then $T^n(x) \to E_1$ as in the previous paragraph. Therefore, we may now assume that for every $z \in \omega(x)$, $z_i = 0$ for exactly one index i, either $i = 1$ or $i = 2$ ($0 \notin \omega(x)$ or else we are in the previous case). Now $\omega(x)$ is invariantly connected so by (H3), either $z_1 = 0$ for all $z \in \omega(x)$ or $z_2 = 0$ for all $z \in \omega(x)$. By (H3), the only compact invariant subset of $X_1^+ \times \{0\}$ which does not include $(0,0)$ is E_1. A symmetric conclusion holds for the other 'axis' $\{0\} \times X_2^+$. We conclude that either $\omega(x) = \{E_2\}$ or $\omega(x) = \{E_1\}$. This concludes our sketch.

8 Competition and Diffusion

As an application of the results of the previous section, we consider a Lotka-Volterra type reaction-diffusion system modeling the competition between two species which can disperse throughout a domain $\Omega \subset \mathbb{R}^m$. The population densities $u_i(x,t)$ of the competing populations are assumed to satisfy:

$$\begin{aligned} \frac{\partial u_1}{\partial t} &= d_1 \Delta u_1 + u_1[1 - u_1 - c_{12}u_2] \\ \frac{\partial u_2}{\partial t} &= d_2 \Delta u_2 + \rho u_2[1 - c_{21}u_1 - u_2], \quad x \in \Omega \end{aligned} \tag{21}$$

where $d_i, \rho, c_{ij} > 0$. Assuming no flux of organisms out of Ω through its boundary, $\partial\Omega$, is allowed, the appropriate boundary conditions are given by:

$$\frac{\partial u_i}{\partial n} = 0, \quad x \in \partial\Omega, \quad i = 1, 2. \tag{22}$$

Here, $\frac{\partial u_i}{\partial n}$ denotes the directional derivative of u_i in the direction of the outward pointing unit normal vector field $n(x)$ for $x \in \partial\Omega$. We assume that $\partial\Omega$ is smooth. Nonnegative, continuous initial conditions are also specified:

$$u_i(x,0) = u_{i0}(x) \geq 0, \quad x \in \Omega, \quad i = 1,2 \tag{23}$$

Observe that solutions of the ordinary differential equations

$$\begin{aligned} u_1' &= u_1[1 - u_1 - c_{12}u_2] \\ u_2' &= \rho u_2[1 - c_{21}u_1 - u_2] \end{aligned} \tag{24}$$

are solutions of (21)-(23). We use the notation $\phi_t(u) = (\phi_t^1(u), \phi_t^2(u))$, where $u = (u_1, u_2)$, for the flow associated with this equation.

An appropriate state space for (21)-(23) is given by the normal cone

$$X^+ = X_1^+ \times X_2^+, \quad X_i^+ = C_+(\bar{\Omega})$$

in the Banach space $X = C(\bar{\Omega}) \times C(\bar{\Omega})$. We will often view a real number z as an element of $C(\bar{\Omega})$, the constant function identically equal to z. This should cause no confusion.

The following well-posedness result for the system (21)-(23) is well-known and we take it for granted here. It says that (21)-(23) generates a semiflow on X^+. See [43] for a proof.

Theorem 16 *For each $u_0 \in X^+$, there is a unique solution $u(x,t) = (u_1(x,t), u_2(x,t))$ of (21)-(23), defined on $\bar{\Omega} \times [0,\infty)$ and $u(x,t) \geq 0$ on its domain. Furthermore*

$$T_t(u_0) \equiv u(\bullet, t)$$

defines a (nonlinear) semiflow on X^+. If $B \subset X^+$, $t_0 > 0$ and $\cup_{0 \leq t \leq t_0} T_t(B)$ is bounded, then $T_{t_0}B$ has compact closure in X^+.

Recall from the previous section that the important cone for competitive systems is $K = X_1^+ \times (-X_2^+)$ and its associated (south-east) partial order relation $\leq_K$. Let's review this ordering briefly. If $u = (u_1, u_2)$ and $v = (v_1, v_2)$ are elements of X (each component a continuous function on $\bar{\Omega}$) then $u \leq_K v$ is equivalent to $u_1(x) \leq v_1(x)$ and $u_2(x) \geq v_2(x)$ for $x \in \bar{\Omega}$. The inequality $u <_K v$ means, in addition, $u \neq v$. The strong inequality $u \ll_K v$ means that strict inequality holds in both inequalities for every x.

Our semiflow T_t preserves this order relation.

Theorem 17 *The following are true:*

(1) If $u_{i0} \neq 0$, then $u_i(x,t) > 0$ for all $(x,t) \in \bar{\Omega} \times (0,\infty)$.

(2) If $u_0, v_0 \in X^+$ satisfy $u_0 <_K v_0$, then $T_t(u_0) <_K T_t(v_0)$ for $t \geq 0$.

(3) If u_0, v_0 are as in (2) and either $u_{i0} \neq 0$ for $i = 1,2$ or $v_{i0} \neq 0$ for $i = 1,2$, then $T_t(u_0) \ll_K T_t(v_0)$ for $t > 0$.

Proof: u_1 satisfies the differential inequality

$$\frac{\partial u}{\partial t} - d\Delta u + c(x,t)u \geq 0$$

where $c(x,t) = -(1 - u_1(x,t) - c_{12}u_2(x,t))$. Thus the assertion (1) follows from Lemma 1. Make the change of variables $w = Qu = (u_1, -u_2)$ in (21) to obtain the system

$$\begin{aligned} \frac{\partial w_1}{\partial t} &= d_1\Delta w_1 + w_1[1 - w_1 + c_{12}w_2] \\ \frac{\partial w_2}{\partial t} &= d_2\Delta w_2 + \rho w_2[1 - c_{21}w_1 + w_2], \quad x \in \Omega \end{aligned} \tag{25}$$

for $w(x,t) \leq_K 0$. Define $f(w) = (w_1[1 - w_1 + c_{12}w_2], \rho w_2[1 - c_{21}w_1 + w_2])$ for $w \leq_K 0$ and observe that f is cooperative in $Q\mathbb{R}^2_+$. Set $f^+ = f$ and $v^+(\bullet,t) = QT_t(v_0)$ be the solution of the system above with boundary conditions given by (22) and $v^+(x,0) = Qv_0(x)$. Now apply Thm. 3 and the remarks following it with $\phi = Qu_0$ and $w = QT_t(u_0)$, to obtain $w(x,t) \leq v^+(x,t)$ and $w(\bullet,t) \neq v^+(\bullet,t)$ for $t > 0$. Therefore, $T_t(u_0) \leq_K T_t(v_0)$ and equality does not hold. This proves (2). Assertion (3) follows from the final assertion of Thm. 3. □

If only one population is present initially then the system (21)-(23) behaves exactly as the ordinary differential equation system (24). Solutions asymptotically approach the single-population equilibria (1,0) and (0,1).

Proposition 2 *If $u_{10} \neq 0$ and $u_{20} = 0$, then $T_t(u_0) = (u_1(\bullet,t), 0)$ and $u_1(\bullet,t) \to 1$ as $t \to \infty$, uniformly in $x \in \bar{\Omega}$. If $u_{10} = 0$ and $u_{20} \neq 0$, then $T_t(u_0) = (0, u_2(\bullet,t))$ and $u_2(\bullet,t) \to 1$ as $t \to \infty$, uniformly in $x \in \bar{\Omega}$.*

Proof: It suffices to prove the first assertion. Obviously, the solution has the form $(u_1(x,t), 0)$ if $u_{20} = 0$ and, by Thm. 17 (1), there exists $m, M > 0$ such that $m < u_1(x,1) < M$ for all $x \in \bar{\Omega}$. As $(m,0) <_K (u_1(\bullet,1),0) <_K$

$(M,0)$, we may conclude from the semiflow property and Thm. 17 (2) that $T_t(m,0) <_K T_{t+1}(u_0) <_K T_t(M,0)$, or equivalently, $\phi_t^1(m,0) \le u_1(\bullet, t+1) \le \phi_t^1(M,0)$. But $\phi_t^1(z,0) \to 1$ as $t \to \infty$ for $z > 0$ and, together with the previous inequality, implies our result. □

As in the previous section, we introduce the notation

$$E_0 = (0,0), \quad E_1 = (1,0), \quad E_2 = (0,1)$$

for the trivial equilibria and set

$$I = [0,1] \times [0,1] \ .$$

In addition to these trivial equilibria, there is a positive, spatially homogeneous, equilibrium solution of (21),(22)

$$E_* = (u_1^*, u_2^*) = \left(\frac{1-c_{12}}{1-c_{12}c_{21}}, \frac{1-c_{21}}{1-c_{12}c_{21}}\right) \tag{26}$$

when $c_{12}, c_{21} > 1$ or when $c_{12}, c_{21} < 1$. The interesting question is whether these account for all equilibria of (21)-(22), or equivalently, do nonconstant equilibria exist.

For the statement of our main result, it is useful to recall results for the ordinary differential system (24) in the case that $c_{12}, c_{21} > 1$ where E_* is an unstable saddle point. Its stable manifold

$$\Gamma \equiv W^s(E_*)$$

consists of E_* and two orbits of (24), a heteroclinic orbit connecting E_0 to E_* and an orbit connecting E_* to the point at infinity. The time-reversed system to (24) is cooperative and irreducible in $\mathbb{R}_+^2$ so the Perron-Frobenius theorem and the strong order preserving property of the flow imply that Γ is totally ordered by $\ll$ in the sense that if x, y are distinct points of Γ, then either $x \ll y$ or $y \ll x$ (see [43] Thm. 4.3.3). $\Gamma \cup \{E_0\}$ separates $\mathbb{R}_+^2$ into two connected components, B_1 and B_2, which are the basins of attraction of E_1 and E_2, both being asymptotically stable. Thm. 15 and the hyperbolicity of the E_i imply parts of the following well-known result (see e.g. deMottoni [9], Zhou and Pao [60], [36]).

Theorem 18 *The following hold for (21):*

(i) If $c_{12} < 1$ and $c_{21} < 1$, then E_* attracts all solutions with $u_{i0} \neq 0$, $i = 1,2$.

(ii) If $c_{12} > 1$ *and* $c_{21} > 1$*, then* E_* *is unstable and* E_1 *and* E_2 *are asymptotically stable. Furthermore,* E_1 *attracts all orbits for which there exists* $q \in \Gamma$ *such that* $q <_K u_0(x)$ *for all* $x \in \bar{\Omega}$ *and* E_2 *attracts all orbits for which there exists* $q \in \Gamma$ *such that* $u_0(x) <_K q$ *for all* $x \in \bar{\Omega}$*.*

(iii) If $c_{12} < 1$ *and* $c_{21} > 1$*, then* E_1 *attracts all solutions with* $u_{10} \neq 0$*.*

(iv) If $c_{12} > 1$ *and* $c_{21} < 1$*, then* E_2 *attracts all solutions with* $u_{20} \neq 0$*.*

Proof: Theorem 16, Thm. 17, and Prop. 2 supply most of the hypotheses (H1), (H3), and (H4) of Thm. 15. The complete continuity of T_t for $t > 0$ follows from the final assertion of Thm. 16 and the fact that bounded sets have bounded orbits. The latter is a consequence of Thm. 17 and the arguments used in the previous section to show that the set I attracts all orbits. We must show that (H2) holds, that is, that E_0 is repelling. We can find a neighborhood U of E_0 such that if $u_0 \in U \setminus \{E_0\}$, then so long as $u(\bullet, t) \in \overline{U}$, we have

$$\begin{aligned} \frac{\partial u_1}{\partial t} &\geq d_1 \Delta u_1 + (1/2) u_1 \\ \frac{\partial u_2}{\partial t} &\geq d_2 \Delta u_2 + (\rho/2) u_2, \quad x \in \Omega \end{aligned}$$

In view of Thm. 17 (1), we may as well assume that $u_i(x, 0) > 0$ for all x for some i (see the proof of Prop. 2). Without loss of generality, suppose that $i = 1$; then there exists $\epsilon > 0$ such that $u_1(x, 0) \geq \epsilon$ for all x. But then $u_1(x, t) \geq \epsilon \exp(t/2)$ so long as $u(\bullet, t) \in \overline{U}$, by Lemma 1 (set $u = u_1(x,t) - \epsilon \exp(t/2)$). It follows that there exists $t > 0$ such that $T_t(u_0) \neq U$, implying that (H2) holds.

Theorem 15 implies assertions (iii) and (iv). We provide details for case (iii). In this case, the assertion holds as stated for the ordinary differential equations (24). If we show there are no positive equilibria for (21), then (iii) will follow from the previously mentioned theorem because the alternative (c) of that theorem is incompatible with the behavior of solutions of (24). If $\hat{u} = (\hat{u}_1, \hat{u}_2)$ is a positive equilibrium then $E_2 <_K \hat{u} <_K E_1$ so on applying Thm. 17(3) we conclude that $E_2 \ll_K \hat{u} \ll_K E_1$. We can find $(a, b) \in \mathbb{R}^2$ so near to $(0, 1)$ such that $E_2 \ll_K (a, b) <_K \hat{u}$. Applying Thm. 17, we conclude that $\phi_t(a, b) <_K \hat{u}$ for $t \geq 0$, and since $\phi_t(a, b) \to E_1$ as $t \to \infty$, we have a contradiction.

Now consider case (ii). The instability of E_* follows since it is unstable for (24). The asymptotic stability of the E_i follows from a simple comparison argument and the fact the E_i are asymptotically stable for (24). However, we will show much more. Suppose that there exists $q \in \Gamma$ such that $q <_K u_0(x)$ holds for all x. From Thm. 17 we conclude that $\phi_t(q) \ll_K T_t(u_0)$ for $t \geq 0$. As $\phi_t(q) \in \Gamma$, we can find b in the basin of attraction B_1 for E_1 for (24) such that $\phi_1(q) \ll_K b \ll_K T_1(u_0)$. Applying Thm. 17, we conclude that $\phi_t(b) \ll_K T_{t+1}(u_0)$ for $t \geq 0$. Now $\phi_t(b) \to E_1$ as $t \to \infty$ from which it follows that any omega limit point z of the orbit $O(u_0)$ satisfies $E_1 \leq_K z$. But $z \in I$ by Thm. 15 so $z \leq_K E_1$. Therefore, the omega limit set is E_1 as asserted.

We now turn to the assertion (i), which of course holds for (24). Now consider a solution $T_t(u_0)$ where u_0 satisfies $u_{i0} \neq 0$ for $i = 1, 2$. Without loss of generality, we may assume that $u_{i0}(x) > 0$ for all x and i since, by Thm. 17(1), this will hold for $T_t(u_0)$ for $t > 0$. Let $m_i = \inf_{x \in \bar{\Omega}} u_{i0}(x)$ and $M_i = \sup_{x \in \bar{\Omega}} u_{i0}(x)$. Then $m_i > 0$ and $(m_1, M_2) <_K u_0 <_K (M_1, m_2)$ so by Thm. 17(2) we have $\phi_t(m_1, M_2) <_K T_t(u_0) <_K \phi_t(M_1, m_2)$ for $t \geq 0$. As the two solutions of (24) converge to E_*, the same holds for $T_t(u_0)$. □

The dynamics and the possibility of nonconstant equilibria are not completely resolved in the interesting case (ii), where inter-specific competition is stronger than intra-specific competition. In fact, there may exist nonconstant equilibria in this case, though not if both diffusion constants are sufficiently large, or equivalently, if the diameter of Ω is sufficiently small. In that case, results of Conway, Hoff and Smoller [6] show that the L^2-norm of the gradient of any solution starting in I decays to zero at an exponential rate. See [6] or [41], Thm. 14.17 for quantitative estimates for this result to apply. In this case, all solutions must approach one of the equilibria E_1, E_2 or E_*. Furthermore, there is a remarkable difference between the case that Ω is convex and when it is nonconvex. A general result of Kishimoto and Weinberger [28] implies that any nonconstant equilibrium is necessarily unstable in the linear approximation if Ω is convex. A sort of converse has been established by Matano and Mimura [31].

Theorem 19 (Matano & Mimura) *Assume that $c_{12} > 1$ and $c_{21} > 1$ hold. Then, for $m > 1$ and any $d_1, d_2 > 0$ there is a (nonconvex) domain $\Omega \in \mathbb{R}^m$ such that (21)-(22) has a stable, nonconstant, positive equilibrium.*

There is no conflict between this result and the previously quoted result of Conway, Hoff and Smoller because in the theorem of Matano and Mimura the domain will depend on the diffusivities.

A numerical example described in [31] depicts a domain $\Omega \subset \mathbb{R}^2$ consisting of two disjoint, equal size squares joined together by a very thin strip. The stable equilibrium has the feature that u_1 dominates u_2 in one square while u_2 dominates u_1 in the other square. The thin strip allows a transition between a high level of u_1 in one square and a low value in the other. This stable equilibrium represents the spatial segregation of the two populations. It is a sort of pattern formation more subtle than the Turing instability. Should a fenced wildlife refuge have a nonconvex shape?

Remark 2 *If there exists a nonconstant equilibrium $\hat{E}$ for (21) in case (ii) of Thm. 18, then we may apply Prop. 1, as originally formulated in [8] (without e), to study the dynamics of (21) restricted to the positively invariant order intervals $[\hat{E}, E_1]_K$ and $[E_2, \hat{E}]_K$. We consider only the former interval. Prop. 1 gives three alternatives ($u_1 = \hat{E}$ and $u_2 = E_1$) but (c) is impossible since $u_2 = E_1$ is asymptotically stable. Thus either there is another equilibrium E in $[\hat{E}, E_1]_K$ distinct from $\hat{E}$ and E_1 or there is a connecting orbit from $\hat{E}$ to E_1 implying that $\hat{E}$ is unstable. In the former case, $E \neq E_*$. For if $E = E_*$ then $E_2 \ll_K \hat{E} \ll_K E_*$ by Thm. 17 (3). But then by Thm. 18 (ii) with $q = E_*$ and $u_0 = \hat{E}$ we obtain the contradiction $\hat{E} = E_2$. Therefore, $E \neq E_*$ and so E is also a nonconstant equilibrium. If it is known that $\hat{E}$ is stable, then $E \in [\hat{E}, E_1]_K$ exists, distinct from $\hat{E}$ and E_1 and is a nonconstant equilibrium. Arguing similarly with the other order interval, we conclude that there are at least three nonconstant equilibria under the circumstances of Matano and Mimura's result.*

Equation (21) can have an inertial manifold under suitable hypotheses and the dynamics on this finite dimensional invariant manifold can be described by an ordinary differential equation similar in form to (24). See Morita [34] for the description of this approach which is further exploited by Mimura et al in [33]. Ahmad and Lazer [1] prove Thm. 18 in the case of temporally periodic coefficients in (21). See also [16].

For more significant applications of Thm.15 to models of microbial competition in bio-reactors, see [43], [44], [52], [26].

References

[1] Ahmad, S., Lazer, A.C.: Asymptotic behavior of solutions of periodic competition-diffusion system, Nonlinear Analysis **13** (1993) 263–284.

[2] Berman, A., Plemmons, R.: *Nonnegative matrices in the mathematical sciences*, Academic Press, New York, 1979.

[3] Capasso, V.: *Mathematical structures of epidemic systems*, Lecture Notes in Biomathematics, **97**, Springer-Verlag, New York.

[4] Conley, C.: The gradient structure of a flow :I, IBM Research, RC 3939 (17806) Yorktown Heights, NY, 1972. Also, Ergodic Theory and Dynamical Systems **8** (1988) 11–26.

[5] Conley, C.: Isolated Invariant Sets and the Morse Index, CBMS 38, Amer. Math. Soc., Providence, R.I., 1978.

[6] Conway, E., Hoff, D., Smoller, J.: Large-time behavior of solutions of systems of reaction-diffusion equations, Siam J. Appl. Math. **35** (1978) 1–16.

[7] Coppel, W.A.: *Stability and Asymptotic Behavior of Differential Equations*, Heath, Boston, 1965.

[8] Dancer, E., Hess, P.: Stability of fixed points for order preserving discrete time dynamical systems, J. reine angew. Math. **419** (1991) 125–139.

[9] deMottoni, P.: Qualitative analysis for some quasi-linear parabolic systems, Inst. Math. Pol. Acad. Sci. Zam **190** (1979).

[10] deMottoni, P., Schiaffino, A.: Competition systems with periodic coefficients: a geometric approach, J. Math. Biology **11** (1982) 319–335.

[11] Goldbeter, A., Segel, L.: Unified mechanism for relay and oscillation of cyclic AMP in Distyostelium discoideum, Proc. Nat. Acad. Sci. U.S.A. **74** (1977) 1543–1547.

[12] Goldbeter, A.: *Biochemical Oscillations and Cellular Rhythms, the molecular bases of periodic and chaotic behavior*, Cambridge Univ. Press, London, 1996.

[13] Hale, J., Somolinas, A.: Competition for fluctuating nutrient, J. Math. Biology **18** (1983), 255–280.

[14] Hale, J., Waltman, P.: Persistence in infinite-dimensional systems, SIAM J. Math. Anal. **20** (1989), 388–395.

[15] Harrison, G.: Comparing predator-prey models to Luckinbill's experiment with Didinium and Paramecium, Ecology **76** (1995) 357-374.

[16] Hess, P., Lazer, A.C.: On an abstract competition model and applications, Nonlinear Analysis T.M.A. **16** (1991) 917–940.

[17] Hirsch, M.: Systems of differential equations which are competitive or cooperative 1: limit sets, SIAM J. Appl. Math. **13** (1982) 167–179.

[18] Hirsch, M.: Systems of differential equations which are competitive or cooperative II: convergence almost everywhere, SIAM J. Math. Anal. **16** (1985) 423–439.

[19] Hirsch, M.: The dynamical systems approach to differential equations, Bull. A.M.S. **11** (1984) 1–64.

[20] Hirsch, M.: Systems of differential equations which are competitive or cooperative III. Competing species. Nonlinearity **1** (1988a) 51–71.

[21] Hirsch, M.: Stability and Convergence in Strongly Monotone dynamical systems, J. reine angew. Math. **383** (1988b) 1–53.

[22] Hirsch, M.: Systems of differential equations that are competitive or cooperative. IV: Structural stability in three dimensional systems. SIAM J. Math. Anal. **21** (1990) 1225–1234.

[23] Hirsch, M.: Systems of differential equations that are competitive or cooperative. V: Convergence in 3-dimensional systems, J. Diff. Eqns. **80** (1989) 94–106.

[24] Hirsch, M.: Systems of differential equations that are competitive or cooperative. VI: A local C^r closing lemma for 3-dimensional systems, Ergod. Th. Dynamical Sys. **11** (1991) 443–454.

[25] Hofbauer, J. and So, J.W.-H.: Multiple limit cycles for three dimensional Lotka-Volterra equations, Appl. Math. Lett. **7** (1994) 65–70.

[26] Hsu, S.-B., Smith, H., Waltman, P.: Dynamics of competition in the unstirred chemostat, Canadian Applied Math. Quart. **2** (1994) 461–483.

[27] Hsu, S.-B., Smith, H., Waltman, P.: Competitive exclusion and coexistence for competitive systems on ordered Banach spaces, Trans. Amer. Math. Soc. **348** (1996) 4083–4094.

[28] Kishimoto, K., Weinberger, H.: The spatial homogeneity of stable equilibria of some reaction-diffusion systems on convex domains, J. Diff. Eqns. **58** (1985) 15–21.

[29] Leung, A.: *Systems of Nonlinear Partial Differential Equations*, Kluwer Academic Publishers, Boston, 1989.

[30] Luckinbill, L.: Coexistence in laboratory populations of Paramecium aurelia and its predator Didinium nasutum, Ecology **54** (1973) 1320–1327.

[31] Matano, H., Mimura, M.: Pattern formation in competition-diffusion systems in nonconvex domains, Pub. Res. Inst. Math. Sci. Kyoto Univ. **19** (1983) 1050–1079.

[32] McShane, E.J.: Extension of range of functions, Bull. Amer. Math. Soc. **40** (1934) 837–842.

[33] Mimura, M., Ei, S.-I., Fang, Q.: Effect of domain shape on coexistence problems in a competition-diffusion system, J. Math. Biol. **29** (1991) 219–237.

[34] Morita, Y.: Reaction-Diffusion systems in nonconvex domains: invariant manifold and reduced form, J. Dyn. and Diff. Eqns. **2** (1990) 69–115.

[35] Nisbet, R.M., Gurney, W.S.C.: *Modelling Fluctuating Populations*, New York, Wiley, 1982.

[36] Pao, C.V.: *Nonlinear parabolic and elliptic equations*, Plenum Press, New York, 1992.

[37] Pascual, M.: Periodic response to periodic forcing of the Droop equations for phytoplankton growth. J. Math. Biol. **32** (1994) 743–759.

[38] Protter, M.H., Weinberger, H.F.: *Maximum Principles in Differential Equations*, Prentice Hall, N.J., 1967.

[39] Segel, L.: *Modeling dynamic phenomena in molecular and cellular biology*, Cambridge, London, 1984.

[40] Smale, S.: On the differential equations of species in competition, J. Math. Biol. **3** (1976) 5–7.

[41] Smoller, J.: *Shock Waves and Reaction Diffusion Equations*, Springer-Verlag, New York, 1983.

[42] Smith, H.L.: Periodic orbits of competitive and cooperative systems, J. Diff. Eqns. **65** (1986) 361–373.

[43] Smith, H.L.: *Monotone Dynamical Systems: An introduction to the Theory of Competitive and Cooperative Systems* AMS Math. Surv.& Monographs **41**, Providence, R.I, 1995.

[44] Smith, H.L.: An application of monotone dynamical systems theory to a model of microbial competition, in *Differential Equations and Control Theory*, Proc. of Int. Conf. on Differential Equations and Control Theory, Wuhan, China, ed. Z. Deng et al, Marcel Dekker, Inc., New York, 1996.

[45] Smith, H.L.: The periodically forced Droop model for phytoplankton growth in a chemostat, J. Math. Biol. **35** (1997) 545–556.

[46] Smith, H.L.: Periodic competitive differential equations and the discrete dynamics of competitive maps, J. Diff. Eqns. **64** (1986) 165–194.

[47] Smith, H.L.: Periodic solutions of periodic competitive and cooperative systems, Siam J. Math. Anal. **17** (1986) 1289–1318.

[48] Smith, H.L.: Complicated dynamics for low-dimensional strongly monotone maps, to appear, Proc. WCNA, (1996).

[49] Smith, H.L.: Planar Competitive and Cooperative Difference Equations, to appear, J. Difference Equations.

[50] Smith, H., Waltman, P.: *The Theory of the Chemostat*, Cambridge Univ. Press, London, 1995.

[51] Smith, H.L., Waltman, P.: Competition for a single limiting resource in continuous culture: the variable yield model, SIAM J. Appl. Math. **54** (1994) 1113–1131.

[52] Smith, H., Waltman, P.: Competition in an unstirred multi-dimensional chemostat, in *Differential Equations and Applications to Biology and Industry, Proceedings of Claremont International Conference Dedicated to the Memory of Stavros Busenberg*, (M. Martelli et al, eds.), World Scientific, Singapore, 1996.

[53] Smith, H., Waltman, P.: A classification theorem for three dimensional competitive systems, J. Diff. Eq. **70** (1987) 325–332.

[54] Thieme, H.R.: Persistence under relaxed point-dissipativity (with application to an epidemic model), SIAM J. Math. Anal. **24** (1993) 407–435.

[55] Tereščák, I.: Dynamics of C^1 smooth strongly monotone discrete-time dynamical systems, preprint.

[56] Xiao, D., Li, W.: Limit cycles for competitive three-dimensional Lotka-Volterra system, preprint.

[57] Zhu, H.-R.: The Existence of Stable Periodic Orbits for Systems of Three Dimensional Differential Equations that are Competitive, Ph.D. thesis, Arizona State University, 1991.

[58] Zhu, H.-R., Smith, H.L.: Stable periodic orbits for a class of three dimensional competitive systems, J. Diff. Eqns. **110** (1994) 143–156.

[59] Zeeman, M.L.: Hopf bifurcation in competitive three-dimensional Lotka-Volterra systems, Dynamics and Stability of Systems **8** (1993) 189–217.

[60] Zhou, L., Pao, C.V.: Asymptotic behavior of a competitive-diffusion system in population dynamics, J. Nonlinear Analysis **6** (1982) 1163–1184.

LIST OF PARTICIPANTS

Arlotti Luisa	ARLOTTI@dic.uniud.it
Arrigoni Francesca	arrigoni@alpha.science.unitn.it
Baker Jennifer	jabaker@wins.uva.nl
Buonomo Bruno	buonomo@matna2.dma.unina.it
Calaud Valere	edval@infomie.fr
Capasso Vincenzo	capasso@miriam.mat.unimi.it
Cellina Arrigo	cellina@elanor.mat.unimi.it
Clother David Robin	rob@amsta.leeds.ac.uk
Conde Susana	sconde@pie.xtec.es
Cuadrado Silvia	silvia@manwe.mat.uab.es
D'Ambrogio Enos	dambrogi@univ.trieste.it
De Cesare Luigi	irmald01@area.ba.cnr.it
Deutsch Andreas	andreas.deutsch@uni-bonn.de
Diekmann Odo	O.Diekmann@math.ruu.nl
Di Liddo Andrea	irmaad02@area.ba.cnr.it
Durret Richard	rtd1@cornell.edu
El Idrissi Omar	idrissi@mat.uab.es
Fasangova Eva	fasanga@karlin.mff.cuni.cz
Giacomin Giambattista	gbg@amath.unizh.ch
Giberti Claudio	giberti@unimo.it
Gyllenberg Mats	mats.gyllenberg@utu.fi
Guatteri Giuseppina	guatteri@alpha.science.unitn.it
Hadeler Karl Peter	k.p.hadeler@uni-tuebingen.de
Huiskes Mark	mark.huiskes@ztw.wk.wau.nl
Kirkilionis Markus	markus@cwi.nl
Kohli Raymond	kohli@dma.epfl.ch
Krivan Vastimil	krivan@entu.cas.cz
Kuehnemund Franziska	frku@michelangelo.mathematik.uni-tubingen.de
Lacitignola Deborah	deborahl@tin.it
Lawniczk Anna	alawnicz@fields.utoronto.ca
Lopez Brito Belen	blopez@dma.ulpgc.es
Lunardini Francesca	lunardin@prmat.math.unipr.it
Maini Philip	maini@maths.ox.ac.uk
Matucci Serena	matucci@udini.math.unifi.it
Micheletti Alessandra	micheletti@elanor.mat.unimi.it
Milota Jaroslav	milota@karlin.mff.cuni.cz
Morale Daniela	morale@ares.mat.unimi.it
Muller Johannes	johannes.mueller@uni-tuebingen.de
Noikova Nelly	issim@bgcict.acad.bg
Pappalardo Luca	pappa@csi.unimi.it
Parvinen Kalle	kalparvi@utu.fi

Pels Bas	pels@bio.uva.nl
Perera Ferrer Luis Gonzalo	perera@stats.math.u-psud.fr
Piazzera Susanna	supi@michelangelo.mathematik.uni-tuebingen.de
Salvatori Maria Cesarina	salva@dipmat.unipg.it
Sanfelici Simona	sanfelici@prmat.math.unipr.it
Seno Hiromi	seno@ics.nara-wu.ac.jp
Smith Hal	halsmith@asu.edu
Stoecker Sabine	stoecker@calvino.polito.it
Tolic Iva	iva.tolic@public.srce.hr
Tonetto Lorenza	tonetto@alpha.science.unitn.it
Valensin Silvana	valensin@myosotis.unimo.it
Vernia Cecilia	vernia@unimo.it

LIST OF C.I.M.E. SEMINARS		Publisher
1954 -	1. Analisi funzionale	C.I.M.E.
2. Quadratura delle superficie e questioni connesse	"
3. Equazioni differenziali non lineari	"
1955 -	4. Teorema di Riemann-Roch e questioni connesse	"
5. Teoria dei numeri	"
6. Topologia	"
7. Teorie non linearizzate in elasticità, idrodinamica,aerodinamica	"
8. Geometria proiettivo-differenziale	"
1956 -	9. Equazioni alle derivate parziali a caratteristiche reali	"
10. Propagazione delle onde elettromagnetiche	"
11. Teoria della funzioni di più variabili complesse e delle funzioni automorfe	"
1957 -	12. Geometria aritmetica e algebrica (2 vol.)	"
13. Integrali singolari e questioni connesse	"
14. Teoria della turbolenza (2 vol.)	"
1958 -	15. Vedute e problemi attuali in relatività generale	"
16. Problemi di geometria differenziale in grande	"
17. Il principio di minimo e le sue applicazioni alle equazioni funzionali	"
1959 -	18. Induzione e statistica	"
19. Teoria algebrica dei meccanismi automatici (2 vol.)	"
20. Gruppi, anelli di Lie e teoria della coomologia	"
1960 -	21. Sistemi dinamici e teoremi ergodici	"
22. Forme differenziali e loro integrali	"
1961 -	23. Geometria del calcolo delle variazioni (2 vol.)	"
24. Teoria delle distribuzioni	"
25. Onde superficiali	"
1962 -	26. Topologia differenziale	"
27. Autovalori e autosoluzioni	"
28. Magnetofluidodinamica	"

1963 - 29. Equazioni differenziali astratte "
30. Funzioni e varietà complesse "
31. Proprietà di media e teoremi di confronto in Fisica Matematica "

1964 - 32. Relatività generale "
33. Dinamica dei gas rarefatti "
34. Alcune questioni di analisi numerica "
35. Equazioni differenziali non lineari "

1965 - 36. Non-linear continuum theories "
37. Some aspects of ring theory "
38. Mathematical optimization in economics "

1966 - 39. Calculus of variations Ed. Cremonese, Firenze
40. Economia matematica "
41. Classi caratteristiche e questioni connesse "
42. Some aspects of diffusion theory "

1967 - 43. Modern questions of celestial mechanics "
44. Numerical analysis of partial differential equations "
45. Geometry of homogeneous bounded domains "

1968 - 46. Controllability and observability "
47. Pseudo-differential operators "
48. Aspects of mathematical logic "

1969 - 49. Potential theory "
50. Non-linear continuum theories in mechanics and physics and their applications "
51. Questions of algebraic varieties "

1970 - 52. Relativistic fluid dynamics "
53. Theory of group representations and Fourier analysis "
54. Functional equations and inequalities "
55. Problems in non-linear analysis "

1971 - 56. Stereodynamics "
57. Constructive aspects of functional analysis (2 vol.) "
58. Categories and commutative algebra "

1972 - 59. Non-linear mechanics "
60. Finite geometric structures and their applications "
61. Geometric measure theory and minimal surfaces "

1973 - 62. Complex analysis "
63. New variational techniques in mathematical physics "
64. Spectral analysis "

1974 - 65. Stability problems "
66. Singularities of analytic spaces "
67. Eigenvalues of non linear problems "

1975 - 68. Theoretical computer sciences "
69. Model theory and applications "
70. Differential operators and manifolds "

1976 - 71. Statistical Mechanics Ed Liguori, Napoli
72. Hyperbolicity "
73. Differential topology "

1977 - 74. Materials with memory "
75. Pseudodifferential operators with applications "
76. Algebraic surfaces "

1978 - 77. Stochastic differential equations "
78. Dynamical systems Ed Liguori, Napoli and Birhäuser Verlag

1979 - 79. Recursion theory and computational complexity "
80. Mathematics of biology "

1980 - 81. Wave propagation "
82. Harmonic analysis and group representations "
83. Matroid theory and its applications "

1981 - 84. Kinetic Theories and the Boltzmann Equation (LNM 1048) Springer-Verlag
85. Algebraic Threefolds (LNM 947) "
86. Nonlinear Filtering and Stochastic Control (LNM 972) "

1982 - 87. Invariant Theory (LNM 996) "
88. Thermodynamics and Constitutive Equations (LN Physics 228) "
89. Fluid Dynamics (LNM 1047) "

1983 -	90.	Complete Intersections	(LNM 1092)	Springer-Verlag
	91.	Bifurcation Theory and Applications	(LNM 1057)	"
	92.	Numerical Methods in Fluid Dynamics	(LNM 1127)	"
1984 -	93.	Harmonic Mappings and Minimal Immersions	(LNM 1161)	"
	94.	Schrödinger Operators	(LNM 1159)	"
	95.	Buildings and the Geometry of Diagrams	(LNM 1181)	"
1985 -	96.	Probability and Analysis	(LNM 1206)	"
	97.	Some Problems in Nonlinear Diffusion	(LNM 1224)	"
	98.	Theory of Moduli	(LNM 1337)	"
1986 -	99.	Inverse Problems	(LNM 1225)	"
	100.	Mathematical Economics	(LNM 1330)	"
	101.	Combinatorial Optimization	(LNM 1403)	"
1987 -	102.	Relativistic Fluid Dynamics	(LNM 1385)	"
	103.	Topics in Calculus of Variations	(LNM 1365)	"
1988 -	104.	Logic and Computer Science	(LNM 1429)	"
	105.	Global Geometry and Mathematical Physics	(LNM 1451)	"
1989 -	106.	Methods of nonconvex analysis	(LNM 1446)	"
	107.	Microlocal Analysis and Applications	(LNM 1495)	"
1990 -	108.	Geometric Topology: Recent Developments	(LNM 1504)	"
	109.	H_{∞} Control Theory	(LNM 1496)	"
	110.	Mathematical Modelling of Industrial Processes	(LNM 1521)	"
1991 -	111.	Topological Methods for Ordinary Differential Equations	(LNM 1537)	"
	112.	Arithmetic Algebraic Geometry	(LNM 1553)	"
	113.	Transition to Chaos in Classical and Quantum Mechanics	(LNM 1589)	"
1992 -	114.	Dirichlet Forms	(LNM 1563)	"
	115.	D-Modules, Representation Theory, and Quantum Groups	(LNM 1565)	"
	116.	Nonequilibrium Problems in Many-Particle Systems	(LNM 1551)	"

1993 -	117.	Integrable Systems and Quantum Groups	(LNM 1620)	Springer-Verlag
	118.	Algebraic Cycles and Hodge Theory	(LNM 1594)	
	119.	Phase Transitions and Hysteresis	(LNM 1584)	"
1994 -	120.	Recent Mathematical Methods in Nonlinear Wave Propagation	(LNM 1640)	"
	121.	Dynamical Systems	(LNM 1609)	"
	122.	Transcendental Methods in Algebraic Geometry	(LNM 1646)	"
1995 -	123.	Probabilistic Models for Nonlinear PDE's	(LNM 1627)	"
	124.	Viscosity Solutions and Applications	(LNM 1660)	"
	125.	Vector Bundles on Curves. New Directions	(LNM 1649)	"
1996 -	126.	Integral Geometry, Radon Transforms and Complex Analysis	(LNM 1684)	"
	127.	Calculus of Variations and Geometric Evolution Problems	LNM 1713	
	128.	Financial Mathematics	LNM 1656	"
1997 -	129.	Mathematics Inspired by Biology	LNM 1714	
	130.	Advanced Numerical Approximation of Nonlinear Hyperbolic Equations	LNM 1697	
	131.	Arithmetic Theory of Elliptic Curves	LNM 1716r	"
	132.	Quantum Cohomology	to appear	"
1998 -	133.	Optimal Shape Design	to appear	
	134.	Dynamical Systems and Small Divisors	to appear	
	135.	Mathematical Problems in Semiconductor Physics	to appear	
	136.	Stochastic PDE's and Kolmogorov Equations in Infinite Dimension	LNM 1715	
	137.	Filtration in Porous Media and Industrial Applications	to appear	
1999 -	138.	Computional Mathematics driven by Industrual Applicationa	to appear	
	139.	Iwahori-Hecke Algebras and Representation Theory	to appear	
	140.	Theory and Applications of Hamiltonian Dynamics	to appear	

141. Global Theory of Minimal Surfaces in Flat Spaces — to appear

142. Direct and Inverse Methods in Solving Nonlinear Evolution Equations — to appear

FONDAZIONE C.I.M.E.
CENTRO INTERNAZIONALE MATEMATICO ESTIVO
INTERNATIONAL MATHEMATICAL SUMMER CENTER

"Computational Mathematics driven by Industrial Applications"

is the subject of the first 1999 C.I.M.E. Session.

The session, sponsored by the Consiglio Nazionale delle Ricerche (C.N.R.), the Ministero dell'Università e della Ricerca Scientifica e Tecnologica (M.U.R.S.T.) and the European Community, will take place, under the scientific direction of Professors Vincenzo CAPASSO (Università di Milano), Heinz W. ENGL (Johannes Kepler Universitaet, Linz) and Doct. Jacques PERIAUX (Dassault Aviation) at the Ducal Palace of Martina Franca (Taranto), from 21 to 27 June, 1999.

Courses

a) Paths, trees and flows: graph optimisation problems with industrial applications (5 lectures in English) Prof. Rainer BURKARD (Technische Universität Graz)

Abstract

Graph optimisation problems play a crucial role in telecommunication, production, transportation, and many other industrial areas. This series of lectures shall give an overview about exact and heuristic solution approaches and their inherent difficulties. In particular the essential algorithmic paradigms such as greedy algorithms, shortest path computation, network flow algorithms, branch and bound as well as branch and cut, and dynamic programming will be outlined by means of examples stemming from applications.

References

1) R. K. Ahuja, T. L. Magnanti & J. B. Orlin, *Network Flows: Theory, Algorithms and Applications*, Prentice Hall, 1993

2) R. K. Ahuja, T. L. Magnanti, J.B.Orlin & M. R. Reddy, *Applications of Network Optimization.* Chapter 1 in: Network Models (Handbooks of Operations Research and Management Science, Vol. 7), ed. by M. O. Ball et al., North Holland 1995, pp. 1-83

3) R. E. Burkard & E. Cela, *Linear Assignment Problems and Extensions*, Report 127, June 1998 (to appear in Handbook of Combinatorial Optimization, Kluwer, 1999).

Can be downloaded by anonymous ftp from

ftp.tu-graz.ac.at, directory/pub/papers/math

4) R. E. Burkard, E. Cela, P. M. Pardalos & L. S. Pitsoulis, *The Quadratic Assignment Problem*, Report 126 May 1998 (to appear in Handbook of Combinatorial Optimization, Kluwer, 1999). Can be downloaded by anonymous ftp from ftp.tu-graz.ac.at, directory /pub/papers/math.

5) E. L. Lawler, J. K. Lenstra, A. H. G.Rinnooy Kan & D. B. Shmoys (Eds.), *The Travelling Salesman Problem*, Wiley, Chichester, 1985.

b) New Computational Concepts, Adaptive Differential Equations Solvers and Virtual Labs (5 lectures in English) Prof. Peter DEUFLHARD (Konrad Zuse Zentrum, Berlin).

Abstract

The series of lectures will address computational mathematical projects that have been tackled by the speaker and his group. In all the topics to be presented novel mathematical modelling, advanced algorithm developments. and efficient visualisation play a joint role to solve problems of practical relevance. Among the applications to be exemplified are:

1) Adaptive multilevel FEM in clinical cancer therapy planning;

2) Adaptive multilevel FEM in optical chip design;

3) Adaptive discrete Galerkin methods for countable ODEs in polymer chemistry;

4) Essential molecular dynamics in RNA drug design.

References

1) P. Deuflhard & A Hohmann, *Numerical Analysis. A first Course in Scientific Computation*, Verlag de Gruyter, Berlin, 1995

2) P. Deuflhard et al *A nonlinear multigrid eigenproblem solver for the complex Helmoltz equation*, Konrad Zuse Zentrum Berlin SC 97-55 (1997)

3) P. Deuflhard et al. *Recent developments in chemical computing*, Computers in Chemical Engineering, **14**, (1990),pp.1249-1258.

4) P. Deuflhard et al. (eds) *Computational molecular dynamics: challenges, methods, ideas*, Lecture Notes in Computational Sciences and Engineering, vol.4 Springer Verlag, Heidelberg, 1998.

5) P.Deuflhard & M. Weiser, *Global inexact Newton multilevel FEM for nonlinear elliptic problems*, Konrad Zuse Zentrum SC 96-33, 1996.

c) Computational Methods for Aerodynamic Analysis and Design. (5 lectures in English) Prof. Antony JAMESON (Stanford University, Stanford).

Abstract

The topics to be discussed will include: - Analysis of shock capturing schemes, and fast solution algorithms for compressible flow; - Formulation of aerodynamic shape optimisation based on control theory; - Derivation of the adjoint equations for compressible flow modelled by the potential Euler and Navies-Stokes equations; - Analysis of alternative numerical search procedures; - Discussion of geometry control and mesh perturbation methods; - Discussion of numerical implementation and practical applications to aerodynamic design.

d) Mathematical Problems in Industry (5 lectures in English) Prof. Jacques-Louis LIONS (Collège de France and Dassault Aviation, France).

Abstract

1. Interfaces and scales. The industrial systems are such that for questions of reliability, safety, cost no subsystem can be underestimated. Hence the need to address problems of scales, both in space variables and in time and the crucial importance of modelling and numerical methods.

2. Examples in Aerospace Examples in Aeronautics and in Spatial Industries. Optimum design.

3. Comparison of problems in Aerospace and in Meteorology. Analogies and differences

4 Real time control. Many methods can be thought of. Universal decomposition methods will be presented.

References

1) J. L. Lions, *Parallel stabilization hyperbolic and Petrowsky systems*, WCCM4 Conference, CDROM Proceedings, Buenos Aires, June 29- July 2, 1998.

2) W. Annacchiarico & M. Cerolaza, *Structural shape optimization of 2-D finite elements models using Beta-splines and genetic algorithms*, WCCM4 Conference, CDROM Proceedings, Buenos Aires, June 29- July 2, 1998.

3) J. Periaux, M. Sefrioui & B. Mantel, *Multi-objective strategies for complex optimization problems in aerodynamics using genetic algorithms*, ICAS '98 Conference, Melbourne, September '98, ICAS paper 98-2.9.1

e) Wavelet transforms and Cosine Transform in Signal and Image Processing (5 lectures in English) Prof. Gilbert STRANG (MIT, Boston).

Abstract

In a series of lectures we will describe how a linear transform is applied to the sampled data in signal processing, and the transformed data is compressed (and quantized to a string of bits). The quantized signal is transmitted and then the inverse transform reconstructs a very good approximation to the original signal. Our analysis concentrates on the construction of the transform. There are several important constructions and we emphasise two: 1) the discrete cosine transform (DCT); 2) discrete wavelet transform (DWT). The DCT is an orthogonal transform (for which we will give a new proof). The DWT may be orthogonal, as for the Daubechies family of wavelets. In other cases it may be biorthogonal - so the reconstructing transform is the inverse but not the transpose of the analysing transform. The reason for this possibility is that orthogonal wavelets cannot also be symmetric, and symmetry is essential property in image processing (because our visual system objects to lack of symmetry). The wavelet construction is based on a "bank" of filters - often a low pass and high pass filter. By iterating the low pass filter we decompose the input space into "scales" to produce a multiresolution. An infinite iteration yields in the limit the scaling function and a wavelet: the crucial equation for the theory is the refinement equation or dilatation equation that yields the scaling function. We discuss the mathematics of the refinement equation: the existence and the smoothness of the solution, and the construction by the cascade algorithm. Throughout these lectures we will be developing the mathematical ideas, but always for a purpose. The insights of wavelets have led to new bases for function spaces and there is no doubt that other ideas are waiting to be developed. This is applied mathematics.

References

1) I. Daubechies, *Ten lectures on wavelets*, SIAM, 1992.

2) G. Strang & T. Nguyen, *Wavelets and filter banks*, Wellesley-Cambridge, 1996.

3) Y. Meyer, *Wavelets: Algorithms and Applications*, SIAM, 1993.

Seminars

Two hour seminars will be held by the Scientific Directors and Professor R. Mattheij.

1) **Mathematics of the crystallisation process of polymers.** Prof. Vincenzo CAPASSO (Un. di Milano).

2) **Inverse Problems: Regularization methods, Application in Industry.** Prof. H. W. ENGL (Johannes Kepler Un., Linz).

3) **Mathematics of Glass.** Prof. R. MATTHEIJ (TU Eindhoven).

4) **Combining game theory and genetic algorithms for solving multi-objective shape optimization problems in Aerodynamics Engineering.** Doct. J. PERIAUX (Dassault Aviation).

Applications

Those who want to attend the Session should fill in an application to C.I.M.E Foundation at the address below, **not later than April 30**, 1999. An important consideration in the acceptance of applications is the scientific relevance of the Session to the field of interest of the applicant. Applicants are requested, therefore, to submit, along with their application, a scientific curriculum and a letter of recommendation. Participation will only be allowed to persons who have applied in due time and have had their application accepted. CIME will be able to partially support some of the youngest participants. Those who plan to apply for support have to mention it explicitely in the application form.

Attendance

No registration fee is requested. Lectures will be held at Martina Franca on June 21, 22, 23, 24, 25, 26, 27. Participants are requested to register on June 20, 1999.

Site and lodging

Martina Franca is a delightful baroque town of white houses of Apulian spontaneous architecture. Martina Franca is the major and most aristocratic centre of the "Murgia dei Trulli" standing on an hill which dominates the well known Itria valley spotted with "Trulli" conical dry stone houses which go back to the 15th century. A masterpiece of baroque architecture is the Ducal palace where the workshop will be hosted. Martina Franca is part of the province of Taranto, one of the major centres of Magna Grecia, particularly devoted to mathematics. Taranto houses an outstanding museum of Magna Grecia with fabulous collections of gold manufactures.

Lecture Notes

Lecture notes will be published as soon as possible after the Session.

Arrigo CELLINA
CIME Director

Vincenzo VESPRI
CIME Secretary

Fondazione C.I.M.E. c/o Dipartimento di Matematica ?U. Dini? Viale Morgagni, 67/A - 50134 FIRENZE (ITALY) Tel. +39-55-434975 / +39-55-4237123 FAX +39-55-434975 / +39-55-4222695 E-mail CIME@UDINI.MATH.UNIFI.IT

Information on CIME can be obtained on the system World-Wide-Web on the file HTTP: //WWW.MATH.UNIFI.IT/CIME/WELCOME.TO.CIME

FONDAZIONE C.I.M.E.
CENTRO INTERNAZIONALE MATEMATICO ESTIVO
INTERNATIONAL MATHEMATICAL SUMMER CENTER

"Iwahori-Hecke Algebras and Representation Theory"

is the subject of the second 1999 C.I.M.E. Session.

The session, sponsored by the Consiglio Nazionale delle Ricerche (C.N.R.), the Ministero dell'Università e della Ricerca Scientifica e Tecnologica (M.U.R.S.T.) and the European Community, will take place, under the scientific direction of Professors Velleda BALDONI (Università di Roma "Tor Vergata") and Dan BARBASCH (Cornell University) at the Ducal Palace of Martina Franca (Taranto), from June 28 to July 6, 1999.

Courses

a) Double HECKE algebras and applications (6 lectures in English)
Prof. Ivan CHEREDNIK (Un. of North Carolina at Chapel Hill, USA)

Abstract:

The starting point of many theories in the range from arithmetic and harmonic analysis to path integrals and matrix models is the formula:

$$\Gamma(k+1/2) = 2\int_0^\infty e^{-x^2}x^{2k}dx.$$

Recently a q-generalization was found based on the Hecke algebra technique, which completes the 15 year old Macdonald program.

The course will be about applications of the double affine Hecke algebras (mainly one-dimensional) to the Macdonald polynomials, Verlinde algebras, Gauss integrals and sums. It will be understandable for those who are not familiar with Hecke algebras and (hopefully) interesting to the specialists.

1) *q-Gauss integrals.* We will introduce a q-analogue of the classical integral formula for the gamma-function and use it to generalize the Gaussian sums at roots of unity.

2) *Ultraspherical polynomials.* A connection of the q-ultraspherical polynomials (the Rogers polynomials) with the one-dimensional double affine Hecke algebra will be established.

3) *Duality.* The duality for these polynomials (which has no classical counterpart) will be proved via the double Hecke algebras in full details.

4) *Verlinde algebras.* We will study the polynomial representation of the 1-dim. DHA at roots of unity, which leads to a generalization and a simplification of the Verlinde algebras.

5) *$PSL_2(\mathbf{Z})$-action.* The projective action of the $PSL_2(\mathbf{Z})$ on DHA and the generalized Verlinde algebras will be considered for A_1 and arbitrary root systems.

6) *Fourier transform of the q-Gaussian.* The invariance of the q-Gaussian with respect to the q-Fourier transform and some applications will be discussed.

References:

1) *From double Hecke algebra to analysis*, Proceedings of ICM98, Documenta Mathematica (1998).

2) *Difference Macdonald–Mehta conjecture*, IMRN:10, 449–467 (1997).

3) *Lectures on Knizhnik-Zamolodchikov equations and Hecke algebras*, MSJ Memoirs (1997).

b) Representation theory of affine Hecke algebras

Prof. Gert HECKMAN (Catholic Un., Nijmegen, Netherlands)

Abstract.

1. The Gauss hypergeometric equation.
2. Algebraic aspects of the hypergeometric system for root systems.
3. The hypergeometric function for root systems.
4. The Plancherel formula in the hypergeometric context.
5. The Lauricella hypergeometric function.
6. A root system analogue of 5.

I will assume that the audience is familiar with the classical theory of ordinary differential equations in the complex plane, in particular the concept of regular singular points and monodromy (although in my first lecture I will give a brief review of the Gauss hypergeometric function). This material can be found in many text books, for example E.L. Ince, Ordinary differential equations, Dover Publ, 1956. E.T. Whittaker and G.N. Watson, A course of modern analysis, Cambridge University Press, 1927.

I will also assume that the audience is familiar with the theory of root systems and reflection groups, as can be found in N. Bourbaki, Groupes et algèbres de Lie, Ch. 4,5 et 6, Masson, 1981. J. E. Humphreys, Reflection groups and Coxeter groups, Cambridge University Press, 1990. or in one of the text books on semisimple groups.

For the material covered in my lectures references are W.J. Couwenberg, Complex reflection groups and hypergeometric functions, Thesis Nijmegen, 1994. G.J. Heckman, Dunkl operators, Sem Bourbaki no 828, 1997. E.M. Opdam, Lectures on Dunkl operators, preprint 1998.

c) Representations of affine Hecke algebras.

Prof. George LUSZTIG (MIT, Cambridge, USA)

Abstract

Affine Hecke algebras appear naturally in the representation theory of p-adic groups. In these lectures we will discuss the representation theory of affine Hecke algebras and their graded version using geometric methods such as equivariant K-theory or perverse sheaves.

References.

1. V. Ginzburg, *Lagrangian construction of representations of Hecke algebras*, Adv. in Math. 63 (1987), 100-112.

2. D. Kazhdan and G. Lusztig, *Proof of the Deligne-Langlands conjecture for Hecke algebras.*, Inv. Math. 87 (1987), 153-215.

3. G. Lusztig, *Cuspidal local systems and graded Hecke algebras, I*, IHES Publ. Math. 67 (1988),145-202; II, in "Representation of groups" (ed. B. Allison and G. Cliff), Conf. Proc. Canad. Math. Soc.. 16, Amer. Math. Soc. 1995, 217-275.

4. G. Lusztig, *Bases in equivariant K-theory, Represent. Th.*, 2 (1998).

d) Affine-like Hecke Algebras and p-adic representation theory

Prof. Roger HOWE (Yale Un., New Haven, USA)

Abstract

Affine Hecke algebras first appeared in the study of a special class of representations (the spherical principal series) of reductive groups with coefficients in *p*-adic fields. Because of their connections with this and other topics, the structure and representation theory of affine Hecke algebras has been intensively studied by a variety of authors. In the meantime, it has gradually emerged that affine Hecke algebras, or slight generalizations of them, allow one to understand far more of the representations of *p*-adic groups than just the spherical principal series. Indeed, it seems possible that such algebras will allow one to understand all representations of *p*-adic groups. These lectures will survey progress in this approach to *p*-adic representation theory.

Topics:

1) Generalities on spherical function algebras on *p*-adic groups.

2) Iwahori Hecke algebras and generalizations.

3) - 4) Affine Hecke algebras and harmonic analysis

5) - 8) Affine-like Hecke algebras and representations of higher level.

References:

J. Adler, *Refined minimal K-types and supercuspidal representations*, Ph.D. Thesis, University of Chicago.

D. Barbasch, *The spherical dual for p-adic groups*, in Geometry and Representation Theory of Real and p-adic Groups, J. Tirao, D. Vogan, and J. Wolf, eds, Prog. In Math. 158, Birkhauser Verlag, Boston, 1998, 1 - 20.

D. Barbasch and A. Moy, *A unitarity criterion for p-adic groups*, Inv. Math. 98 (1989), 19 - 38.

D. Barbasch and A. Moy, *Reduction to real infinitesimal character in affine Hecke algebras*, J. A. M. S.6 (1993), 611- 635.

D. Barbasch, *Unitary spherical spectrum for p-adic classical groups*, Acta. Appl. Math. 44 (1996), 1 - 37.

C. Bushnell and P. Kutzko, *The admissible dual of GL(N) via open subgroups*, Ann. of Math. Stud. 129, Princeton University Press, Princeton, NJ, 1993.

C. Bushnell and P. Kutzko, *Smooth representations of reductive p-adic groups*: Structure theory via types, D. Goldstein, *Hecke algebra isomorphisms for tamely ramified characters*, R. Howe and A. Moy, *Harish-Chandra Homomorphisms for p-adic Groups*, CBMS Reg. Conf. Ser. 59, American Mathematical Society, Providence, RI, 1985.

R. Howe and A. Moy, *Hecke algebra isomorphisms for GL(N) over a p-adic field*, J. Alg. 131 (1990), 388 - 424.

J-L. Kim, *Hecke algebras of classical groups over p-adic fields and supercuspidal representations,I, II, III*, preprints, 1998.

G. Lusztig, *Classification of unipotent representations of simple p-adic groups*, IMRN 11 (1995), 517 - 589.

G. Lusztig, *Affine Hecke algebras and their graded version*, J. A. M. S. 2 (1989), 599 - 635.

L. Morris, *Tamely ramified supercuspidal representations of classical groups, I, II*, Ann. Ec. Norm. Sup 24, (1991) 705 - 738; 25 (1992), 639 - 667.

L. Morris, *Tamely ramified intertwining algebras*, Inv. Math. 114 (1994), 1 - 54.

A. Roche, *Types and Hecke algebras for principal series representations of split reductive p-adic groups*, preprint, (1996).

J-L. Waldspurger, *Algebres de Hecke et induites de representations cuspidales pour GLn*, J. reine u. angew. Math. 370 (1986), 27 - 191.

J-K. Yu, *Tame construction of supercuspidal representations*, preprint, 1998.

Applications

Those who want to attend the Session should fill in an application to the Director of C.I.M.E at the address below, **not later than** April 30, 1999.

An important consideration in the acceptance of applications is the scientific relevance of the Session to the field of interest of the applicant.

Applicants are requested, therefore, to submit, along with their application, a scientific curriculum and a letter of recommendation.

Participation will only be allowed to persons who have applied in due time and have had their application accepted.

CIME will be able to partially support some of the youngest participants. Those who plan to apply for support have to mention it explicitely in the application form.

Attendance

No registration fee is requested. Lectures will be held at Martina Franca on June 28, 29, 30, July 1, 2, 3, 4, 5, 6. Participants are requested to register on June 27, 1999.

Site and lodging

Martina Franca is a delightful baroque town of white houses of Apulian spontaneous architecture. Martina Franca is the major and most aristocratic centre of the Murgia dei Trulli standing on an hill which dominates the well known Itria valley spotted with Trulli conical dry stone houses which go back to the 15th century. A masterpiece of baroque architecture is the Ducal palace where the workshop will be hosted. Martina Franca is part of the province of Taranto, one of the major centres of Magna Grecia, particularly devoted to mathematics. Taranto houses an outstanding museum of Magna Grecia with fabulous collections of gold manufactures.

Lecture Notes

Lecture notes will be published as soon as possible after the Session.

Arrigo CELLINA
CIME Director

Vincenzo VESPRI
CIME Secretary

Fondazione C.I.M.E. c/o Dipartimento di Matematica U. Dini Viale Morgagni, 67/A - 50134 FIRENZE (ITALY) Tel. +39-55-434975 / +39-55-4237123 FAX +39-55-434975 / +39-55-4222695 E-mail CIME@UDINI.MATH.UNIFI.IT

Information on CIME can be obtained on the system World-Wide-Web on the file HTTP: //WWW.MATH.UNIFI.IT/CIME/WELCOME.TO.CIME.

FONDAZIONE C.I.M.E.
CENTRO INTERNAZIONALE MATEMATICO ESTIVO
INTERNATIONAL MATHEMATICAL SUMMER CENTER

"Theory and Applications of Hamiltonian Dynamics"

is the subject of the third 1999 C.I.M.E. Session.

The session, sponsored by the Consiglio Nazionale delle Ricerche (C.N.R.), the Ministero dell'Università e della Ricerca Scientifica e Tecnologica (M.U.R.S.T.) and the European Community, will take place, under the scientific direction of Professor Antonio GIORGILLI (Un. di Milano), at Grand Hotel San Michele, Cetraro (Cosenza), from July 1 to July 10, 1999.

Courses

a) Physical applications of Nekhoroshev theorem and exponential estimates (6 lectures in English)

Prof. Giancarlo BENETTIN (Un. di Padova, Italy)

Abstract

The purpose of the lectures is to introduce exponential estimates (i.e., construction of normal forms up to an exponentially small remainder) and Nekhoroshev theorem (exponential estimates plus geometry of the action space) as the key to understand the behavior of several physical systems, from the Celestial mechanics to microphysics.

Among the applications of the exponential estimates, we shall consider problems of adiabatic invariance for systems with one or two frequencies coming from molecular dynamics. We shall compare the traditional rigorous approach via canonical transformations, the heuristic approach of Jeans and of Landau–Teller, and its possible rigorous implementation via Lindstet series. An old conjecture of Boltzmann and Jeans, concerning the possible presence of very long equilibrium times in classical gases (the classical analog of "quantum freezing") will be reconsidered. Rigorous and heuristic results will be compared with numerical results, to test their level of optimality.

Among the applications of Nekhoroshev theorem, we shall study the fast rotations of the rigid body, which is a rather complete problem, including degeneracy and singularities. Other applications include the stability of elliptic equilibria, with special emphasis on the stability of triangular Lagrangian points in the spatial restricted three body problem.

References:

For a general introduction to the subject, one can look at chapter 5 of V.I. Arnold, VV. Kozlov and A.I. Neoshtadt, in *Dynamical Systems III,* V.I. Arnold Editor (Springer, Berlin 1988). An introduction to physical applications of Nekhorshev theorem and exponential estimates is in the proceeding of the Noto School "Non-Linear Evolution and Chaotic Phenomena", G. Gallavotti and P.W. Zweifel Editors (Plenum Press, New York, 1988), see the contributions by G. Benettin, L. Galgani and A. Giorgilli.

General references on Nekhoroshev theorem and exponential estimates: N.N. Nekhoroshev, Usp. Mat. Nauk. **32:6**, 5-66 (1977) [Russ. Math. Surv. **32:6**, 1-65

(1977)]; G. Benettin, L. Galgani, A. Giorgilli, Cel. Mech. **37**, 1 (1985); A. Giorgilli and L. Galgani, Cel. Mech. **37**, 95 (1985); G. Benettin and G. Gallavotti, Journ. Stat. Phys. **44**, 293-338 (1986); P. Lochak, Russ. Math. Surv. **47**, 57-133 (1992); J. Pöschel, Math. Z. **213**, 187–216 (1993).

Applications to statistical mechanics: G. Benettin, in: *Boltzmann's legacy 150 years afrer his birth*, Atti Accad. Nazionale dei Lincei **131**, 89-105 (1997); G. Benettin, A. Carati and P. Sempio, Journ. Stat. Phys. **73**, 175–192 (1993); G. Benettin, A. Carati and G. Gallavotti, Nonlinearity **10**, 479-505 (1997); G. Benettin, A. Carati e F. Fassò, Physica D **104**, 253-268 (1997); G. Benettin, P. Hjorth and P. Sempio, *Exponentially long equilibrium times in a one dimensional collisional model of a classical gas*, in print in Journ. Stat. Phys.

Applications to the rigid body: G. Benettin and F. Fassò, Nonlinearity **9**, 137-186 (1996); G. Benettin, F. Fassò e M. Guzzo, Nonlinearity **10**, 1695-1717 (1997).

Applications to elliptic equilibria (recent nonisochronous approach): F. Fassò, M. Guzzo e G. Benettin, Comm. Math. Phys. **197**, 347-360 (1998); L. Niederman, *Nonlinear stability around an elliptic equilibrium point in an Hamiltonian system*, preprint (1997). M. Guzzo, F. Fasso' e G. Benettin, Math. Phys. Electronic Journal, Vol. **4**, paper 1 (1998); G. Benettin, F. Fassò e M. Guzzo, *Nekhoroshev–stability of L4 and L5 in the spatial restricted three–body problem*, in print in Regular and Chaotic Dynamics.

b) KAM-theory (6 lectures in English)

Prof. Hakan ELIASSON (Royal Institute of Technology, Stockholm, Sweden)

Abstract

Quasi-periodic motions (or invariant tori) occur naturally when systems with periodic motions are coupled. The perturbation problem for these motions involves small divisors and the most natural way to handle this difficulty is by the quadratic convergence given by Newton's method. A basic problem is how to implement this method in a particular perturbative situation. We shall describe this difficulty, its relation to linear quasi-periodic systems and the way given by KAM-theory to overcome it in the most generic case. Additional difficulties occur for systems with elliptic lower dimensional tori and even more for systems with weak non-degeneracy.

We shall also discuss the difference between initial value and boundary value problems and their relation to the Lindstedt and the Poincaré-Lindstedt series.

The classical books Lectures in Celestial Mechanics by Siegel and Moser (Springer 1971) and Stable and Random Motions in Dynamical Systems by Moser (Princeton University Press 1973) are perhaps still the best introductions to KAM-theory. The development up to middle 80's is described by Bost in a Bourbaki Seminar (no. 6 1986). After middle 80's a lot of work have been devoted to elliptic lower dimensional tori, and to the study of systems with weak non-degeneracy starting with the work of Cheng and Sun (for example "*Existence of KAM-tori in Degenerate Hamiltonian systems*", J. Diff. Eq. 114, 1994). Also on linear quasi-periodic systems there has been some progress which is described in my article "*Reducibility and point spectrum for quasi-periodic skew-products*", Proceedings of the ICM, Berlin volume II 1998.

c) The Adiabatic Invariant in Classical Dynamics: Theory and applications (6 lectures in English).

Prof. Jacques HENRARD (Facultés Universitaires Notre Dame de la Paix, Namur, Belgique).

Abstract

The adiabatic invariant theory applies essentially to oscillating non-autonomous Hamiltonian systems when the time dependance is considerably slower than the oscillation periods. It describes "easy to compute" and "dynamicaly meaningful" quasi-invariants by which on can predict the approximate evolution of the system on very large time scales. The theory makes use and may serve as an illustration of several classical results of Hamiltonian theory.

1) Classical Adiabatic Invariant Theory (Including an introduction to angle-action variables)

2) Classical Adiabatic Invariant Theory (continued) and some applications (including an introduction to the "magnetic bottle")

3) Adiabatic Invariant and Separatrix Crossing (Neo-adiabatic theory)

4) Applications of Neo-Adiabatic Theory: Resonance Sweeping in the Solar System

5) The chaotic layer of the "Slowly Modulated Standard Map"

References:

J.R. Cary, D.F. Escande, J.L. Tennison: Phys.Rev. A, 34, 1986, 3256-4275

J. Henrard, in "*Dynamics reported*" (n=B02- newseries), Springer Verlag 1993; pp 117-235)

J. Henrard: in "*Les méthodes moderne de la mécanique céleste*" (Benest et Hroeschle eds), Edition Frontieres, 1990, 213-247

J. Henrard and A. Morbidelli: Physica D, 68, 1993, 187-200.

d) Some aspects of qualitative theory of Hamiltonian PDEs (6 lectures in English).

Prof. Sergei B. KUKSIN (Heriot-Watt University, Edinburgh, and Steklov Institute, Moscow)

Abstract.

I) Basic properties of Hamiltonian PDEs. Symplectic structures in scales of Hilbert spaces, the notion of a Hamiltonian PDE, properties of flow-maps, etc.

II) Around Gromov's non-squeezing property. Discussions of the finite-dimensional Gromov's theorem, its version for PDEs and its relevance for mathematical physics, infinite-dimensional symplectic capacities.

III) Damped Hamiltonian PDEs and the turbulence-limit. Here we establish some qualitative properties of PDEs of the form <non-linear Hamiltonian PDE>+<small linear damping> and discuss their relations with theory of decaying turbulence

Parts I)-II) will occupy the first three lectures, Part III - the last two.

References

[1] S.K., *Nearly Integrable Infinite-dimensional Hamiltonian Systems.* LNM 1556, Springer 1993.

[2] S.K., *Infinite-dimensional symplectic capacities and a squeezing theorem for Hamiltonian PDE's.* Comm. Math. Phys. 167 (1995), 531-552.

[3] Hofer H., Zehnder E., *Symplectic invariants and Hamiltonian dynamics.* Birkhauser, 1994.

[4] S.K. *Oscillations in space-periodic nonlinear Schroedinger equations.* Geometric and Functional Analysis 7 (1997), 338-363.

For I) see [1] (Part 1); for II) see [2,3]; for III) see [4]."

e) An overview on some problems in Celestial Mechanics (6 lectures in English)

Prof. Carles SIMO' (Universidad de Barcelona, Spagna)

Abstract

1. Introduction. The N-body problem. Relative equilibria. Collisions.

2. The 3D restricted three-body problem. Libration points and local stability analysis.

3. Periodic orbits and invariant tori. Numerical and symbolical computation.

4. Stability and practical stability. Central manifolds and the related stable/unstable manifolds. Practical confiners.

5. The motion of spacecrafts in the vicinity of the Earth-Moon system. Results for improved models. Results for full JPL models.

References:

C. Simò, *An overview of some problems in Celestial Mechanics*, available at http://www-mal.upc.es/escorial .

Click of "curso completo" of Prof. Carles Simó

Applications

Deadline for application: **May 15, 1999.**

Applicants are requested to submit, along with their application, a scientific curriculum and a letter of recommendation.

CIME will be able to partially support some of the youngest participants. Those who plan to apply for support have to mention it explicitely in the application form.

Attendance

No registration fee is requested. Lectures will be held at Cetraro on July 1, 2, 3, 4, 5, 6, 7, 8, 9, 10. Participants are requested to register on June 30, 1999.

Site and lodging

The session will be held at Grand Hotel S. Michele at Cetraro (Cosenza), Italy. Prices for full board (bed and meals) are roughly 150.000 italian liras p.p. day in a single room, 130.000 italian liras in a double room. Cheaper arrangements for multiple lodging in a residence are avalaible. More detailed information may be obtained from the Direction of the hotel (tel. +39-098291012, Fax +39-098291430, email: sanmichele@antares.it.

Further information on the hotel at the web page www.sanmichele.it

Arrigo CELLINA
CIME Director

Vincenzo VESPRI
CIME Secretary

Fondazione C.I.M.E. c/o Dipartimento di Matematica U. Dini Viale Morgagni, 67/A - 50134 FIRENZE (ITALY) Tel. +39-55-434975 / +39-55-4237123 FAX +39-55-434975 / +39-55-4222695 E-mail CIME@UDINI.MATH.UNIFI.IT

Information on CIME can be obtained on the system World-Wide-Web on the file HTTP: //WWW.MATH.UNIFI.IT/CIME/WELCOME.TO.CIME.

FONDAZIONE C.I.M.E.
CENTRO INTERNAZIONALE MATEMATICO ESTIVO
INTERNATIONAL MATHEMATICAL SUMMER CENTER

"Global Theory of Minimal Surfaces in Flat Spaces"

is the subject of the fourth 1999 C.I.M.E. Session.

The session, sponsored by the Consiglio Nazionale delle Ricerche (C.N.R.), the Ministero dell'Università e della Ricerca Scientifica e Tecnologica (M.U.R.S.T.) and the European Community, will take place, under the scientific direction of Professor Gian Pietro PIROLA (Un. di Pavia), at Ducal Palace of Martina Franca (Taranto), from July 7 to July 15, 1999.

Courses

a) Asymptotic geometry of properly embedded minimal surfaces (6 lecture in English)

Prof. William H. MEEKS, III (Un. of Massachusetts, Amherst, USA).

Abstract:

In recent years great progress has been made in understanding the asymptotic geometry of properly embedded minimal surfaces. The first major result of this type was the solution of the generalized Nitsch conjecture by P. Collin, based on earlier work by Meeks and Rosenberg. It follows from the resolution of this conjecture that whenever M is a properly embedded minimal surface with more than one end and $E \subset M$ is an annular end representative, then E has finite total curvature and is asymptotic to an end of a plan or catenoid. Having finite total curvature in the case of an annular end is equivalent to proving the end has quadratic area growth with respect to the radial function r. Recently Collin, Kusner, Meeks and Rosenberg have been able to prove that any middle end of M, even one with infinite genus, has quadratic area growth. It follows from this result that middle ends are never limit ends and hence M can only have one or two limit ends which must be top or bottom ends. With more work it is shown that the middle ends of M stay a bounded distance from a plane or an end of a catenoid.

The goal of my lectures will be to introduce the audience to the concepts in the theory o f properly embedded minimal surfaces needed to understand the above results and to understand some recent classification theorems on proper minimal surfaces of genus 0 in flat three-manifolds.

References

1) H. Rosenberg, *Some recent developments in the theory of properly embedded minimal surfaces in E*, Asterisque **206**, (19929, pp. 463-535;

2) W. Meeks & H. Rosenberg, *The geometry and conformal type of properly embedded minimal surfaces in E*, Invent.Math. **114**, (1993), pp. 625-639;

3) W. Meeks, J. Perez & A. Ros, *Uniqueness of the Riemann minimal examples*, Invent. Math. **131**, (1998), pp. 107-132;

4) W. Meeks & H. Rosenberg, *The geometry of periodic minimal surfaces*, Comm. Math. Helv. **68**, (1993), pp. 255-270;

5) P. Collin, *Topologie et courbure des surfaces minimales proprement plongees dans E*, Annals of Math. **145**, (1997), pp. 1-31;

6) H. Rosenberg, *Minimal surfaces of finite type*, Bull. Soc. Math. France **123**, (1995), pp. 351-359;

7) Rodriquez & H. Rosenberg, *Minimal surfaces in E with one end and bounded curvature*, Manusc. Math. **96**, (1998), pp. 3-9.

b) Properly embedded minimal surfaces with finite total curvature (6 lectures in English)

Prof. Antonio ROS (Universidad de Granada, Spain)

Abstact:

Among properly embedded minimal surfaces in Euclidean 3-space, those that have finite total curvature form a natural and important subclass. These surfaces have finitely many ends which are all parallel and asymptotic to planes or catenoids. Although the structure of the space $\mathcal{M}$ of surfaces of this type which have a fixed topology is not well understood, we have a certain number of partial results and some of them will be explained in the lectures we will give.

The first nontrivial examples, other than the plane and the catenoid, were constructed only ten years ago by Costa, Hoffman and Meeks. Schoen showed that if the surface has two ends, then it must be a catenoid and López and Ros proved that the only surfaces of genus zero are the plane and the catenoid. These results give partial answers to an interesting open problem: decide which topologies are supported by this kind of surfaces. Ros obtained certain compactness properties of $\mathcal{M}$. In general this space is known to be noncompact but he showed that $\mathcal{M}$ is compact for some fixed topologies. Pérez and Ros studied the local structure of $\mathcal{M}$ around a nondegenerate surface and they proved that around these points the moduli space can be naturally viewed as a Lagrangian submanifold of the complex Euclidean space.

In spite of that analytic and algebraic methods compete to solve the main problems in this theory, at this moment we do not have a satisfactory idea of the behaviour of the moduli space $\mathcal{M}$. Thus the above is a good research field for young geometers interested in minimal surfaces.

References

1) C. Costa, *Example of a compete minimal immersion in* $\mathbb{R}^3$ *of genus one and three embedded ends,* Bull. SOc. Bras. Math. **15, (1984), pp. 47-54;**

2) D. Hoffman & H. Karcher, *Complete embedded minimal surfaces of finite total curvature*, R. Osserman ed., Encyclopedia of Math., vol. of Minimal Surfaces, **5-90**, Springer 1997;

3) D. Hoffman & W. H. Meeks III, *Embedded minimal surfaces of finite topology,* Ann. Math. **131**, (1990), pp. 1-34;

4) F. J. Lòpez & A. Ros, *On embedded minimal surfaces of genus zero*, J. Differential Geometry **33**, (1991), pp. 293-300;

5) J. P. Perez & A. Ros, *Some uniqueness and nonexistence theorems for embedded minimal surfaces*, Math. Ann. **295** (3), (1993), pp. 513-525;

6) J. P. Perez & A. Ros, *The space of properly embedded minimal surfaces with finite total curvature*, Indiana Univ. Math. J. **45** 1, (1996), pp.177-204.

c) Minimal surfaces of finite topology properly embedded in E (Euclidean 3-space).(6 lectures in English)

Prof. Harold ROSENBERG (Univ. Paris VII, Paris, France)

Abstract:

We will prove that a properly embedded minimal surface in E of finite topology and at least two ends has finite total curvature. To establish this we first prove that each annular end of such a surface *M* can be made transverse to the horizontal planes

(after a possible rotation in space), [Meeks-Rosenberg]. Then we will prove that such an end has finite total curvature [Pascal Collin]. We next study properly embedded minimal surfaces in E with finite topology and one end. The basic unsolved problem is to determine if such a surface is a plane or helicoid when simply connected. We will describe partial results. We will prove that a properly immersed minimal surface of finite topology that meets some plane in a finite number of connected components, with at most a finite number of singularities, is of finite conformal type. If in addition the curvature is bounded, then the surface is of finite type. This means M can be parametrized by meromorphic data on a compact Riemann surface. In particular, under the above hypothesis, M is a plane or helicoid when M is also simply connected and embedded. This is work of Rodriquez- Rosenberg, and Xavier. If time permits we will discuss the geometry and topology of constant mean curvature surfaces properly embedded in E.

References

1) H. Rosenberg, *Some recent developments in the theory of properly embedded minimal surfaces in E*, Asterique **206**, (1992), pp. 463-535;

2) W.Meeks & H. Rosenberg, *The geometry and conformal type of properly embedded minimal surfaces in E*, Invent. **114**, (1993), pp.625-639;

3) P. Collin, *Topologie et courbure des surfaces minimales proprement plongées dans E*, Annals of Math. **145**, (1997), pp. 1-31

4) H. Rosenberg, *Minimal surfaces of finite type*, Bull. Soc. Math. France **123**, (1995), pp. 351-359;

5) Rodriquez & H. Rosenberg, *Minimal surfaces in E with one end and bounded curvature*, Manusc. Math. **96**, (1998), pp. 3-9.

Applications

Those who want to attend the Session should fill in an application to the C.I.M.E Foundation at the address below, **not later than** May 15, 1999.

An important consideration in the acceptance of applications is the scientific relevance of the Session to the field of interest of the applicant.

Applicants are requested, therefore, to submit, along with their application, a scientific curriculum and a letter of recommendation.

Participation will only be allowed to persons who have applied in due time and have had their application accepted.

CIME will be able to partially support some of the youngest participants. Those who plan to apply for support have to mention it explicitely in the application form

Attendance

No registration fee is requested. Lectures will be held at Martina Franca on July 7, 8, 9, 10, 11, 12, 13, 14, 15. Participants are requested to register on July 6, 1999.

Site and lodging

Martina Franca is a delightful baroque town of white houses of Apulian spontaneous architecture. Martina Franca is the major and most aristocratic centre of the Murgia dei Trulli standing on an hill which dominates the well known Itria valley spotted with Trulli conical dry stone houses which go back to the 15th century. A masterpiece of baroque architecture is the Ducal palace where the workshop will be hosted. Martina Franca is part of the province of Taranto, one of the major centres of Magna Grecia, particularly devoted to mathematics. Taranto houses an outstanding museum of Magna Grecia with fabulous collections of gold manufactures.

Lecture Notes

Lecture notes will be published as soon as possible after the Session.

Arrigo CELLINA
CIME Director

Vincenzo VESPRI
CIME Secretary

Fondazione C.I.M.E. c/o Dipartimento di Matematica U. Dini Viale Morgagni, 67/A - 50134 FIRENZE (ITALY) Tel. +39-55-434975 / +39-55-4237123 FAX +39-55-434975 / +39-55-4222695 E-mail CIME@UDINI.MATH.UNIFI.IT

Information on CIME can be obtained on the system World-Wide-Web on the file HTTP: //WWW.MATH.UNIFI.IT/CIME/WELCOME.TO.CIME.

FONDAZIONE C.I.M.E.
CENTRO INTERNAZIONALE MATEMATICO ESTIVO
INTERNATIONAL MATHEMATICAL SUMMER CENTER

"Direct and Inverse Methods in Solving Nonlinear Evolution Equations"

is the subject of the fifth 1999 C.I.M.E. Session.

The session, sponsored by the Consiglio Nazionale delle Ricerche (C.N.R.), the Ministero dell'Università e della Ricerca Scientifica e Tecnologica (M.U.R.S.T.) and the European Community, will take place, under the scientific direction of Professor Antonio M. Greco (Università di Palermo), at Grand Hotel San Michele,Cetraro (Cosenza), from September 8 to September 15, 1999.

a) Exact solutions of nonlinear PDEs by singularity analysis (6 lectures in English)

Prof. Robert CONTE (Service de physique de l'état condensé, CEA Saclay, Gif-sur-Yvette Cedex, France)

Abstract

1) Criteria of integrability : Lax pair, Darboux and Bäcklund transformations. Partial integrability, examples. Importance of involutions.

2) The Painlevé test for PDEs in its invariant version.

3) The "truncation method" as a Darboux transformation, ODE and PDE situations.

4) The one-family truncation method (WTC), integrable (Korteweg-de Vries, Boussinesq, Hirota-Satsuma, Sawada-Kotera) and partially integrable (Kuramoto-Sivashinsky) cases.

5) The two-family truncation method, integrable (sine-Gordon, mKdV, Broer-Kaup) and partially integrable (complex Ginzburg-Landau and degeneracies) cases.

6) The one-family truncation method based on the scattering problems of Gambier: BT of Kaup-Kupershmidt and Tzitzéica equations.

References

References are divided into three subsets: prerequisite (assumed known by the attendant to the school), general (not assumed known, pedagogical texts which would greatly benefit the attendant if they were read before the school), research (research papers whose content will be exposed from a synthetic point of view during the course).

Prerequisite bibliography.

The following subjects will be assumed to be known : the Painlevé property for nonlinear ordinary differential equations, and the associated Painlevé test.

Prerequisite recommended texts treating these subjects are

[P.1] E. Hille, *Ordinary differential equations in the complex domain* (J. Wiley and sons, New York, 1976).

[P.2] R. Conte, *The Painlevé approach to nonlinear ordinary differential equations, The Painlevé property, one century later*, 112 pages, ed. R. Conte, CRM series in mathematical physics (Springer, Berlin, 1999). Solv-int/9710020.

The interested reader can find many applications in the following review, which should not be read before [P.2] :

[P.3] A. Ramani, B. Grammaticos, and T. Bountis, *The Painlevé property and singularity analysis of integrable and nonintegrable systems*, Physics Reports 180 (1989) 159–245.

A text to be avoided by the beginner is Ince's book, the ideas are much clearer in Hille's book.

There exist very few pedagogical texts on the subject of this school.

A general reference, covering all the above program, is the course delivered at a Cargèse school in 1996 :

[G.1] M. Musette, *Painlevé analysis for nonlinear partial differential equations, The Painlevé property, one century later*, 65 pages, ed. R. Conte, CRM series in mathematical physics (Springer, Berlin, 1999). Solv-int/9804003.

A short subset of [G.1], with emphasis on the ideas, is the conference report

[G.2] R. Conte, *Various truncations in Painlevé analysis of partial differential equations*, 16 pages, Nonlinear dynamics : integrability and chaos, ed. M. Daniel, to appear (Springer? World Scientific?). Solv-int/9812008. Preprint S98/047.

Research papers.

[R.2] J. Weiss, M. Tabor and G. Carnevale, *The Painlevé property for partial differential equations*, J. Math. Phys. 24 (1983) 522–526.

[R.3] Numerous articles of Weiss, from 1983 to 1989, all in J. Math. Phys. [singular manifold method].

[R.4] M. Musette and R. Conte, *Algorithmic method for deriving Lax pairs from the invariant Painlevé analysis of nonlinear partial differential equations*, J. Math. Phys. 32 (1991) 1450–1457 [invariant singular manifold method].

[R.5] R. Conte and M. Musette, *Linearity inside nonlinearity: exact solutions to the complex Ginz-burg-Landau equation*, Physica D 69 (1993) 1–17 [Ginzburg-Landau].

[R.6] M. Musette and R. Conte, *The two-singular manifold method, I. Modified KdV and sine-Gordon equations*, J. Phys. A 27 (1994) 3895–3913 [Two-singular manifold method].

[R.7] R. Conte, M. Musette and A. Pickering, *The two-singular manifold method, II.* Classical Boussinesq system, J. Phys. A 28 (1995) 179–185 [Two-singular manifold method].

[R.8] A. Pickering, *The singular manifold method revisited*, J. Math. Phys. 37 (1996) 1894–1927 [Two-singular manifold method].

[R.9] M. Musette and R. Conte, *Bäcklund transformation of partial differential equations from the Painlevé-Gambier classification, I. Kaup-Kupershmidt equation*, J. Math. Phys. 39 (1998) 5617–5630. [Lecture 6].

[R.10] R. Conte, M. Musette and A. M. Grundland, *Bäcklund transformation of partial differential equations from the Painlevé-Gambier classification, II. Tzitzéica equation*, J. Math. Phys. 40 (1999) to appear. [Lecture 6].

b) Integrable Systems and Bi-Hamiltonian Manifolds (6 lectures in English)

Prof. Franco MAGRI (Università di Milano, Milano, Italy)

Abstract

1) Integrable systems and bi-hamiltonian manifolds according to Gelfand and Zakharevich.

2) Examples: KdV, KP and Sato's equations.

3) The rational solutions of KP equation.
4) Bi-hamiltonian reductions and completely algebraically integrable systems.
5) Connections with the separabilty theory.
6) The τ function and the Hirota's identities from a bi-hamiltonian point of view.

References

1) R. Abraham, J.E. Marsden, *Foundations of Mechanics*,Benjamin/Cummings, 1978

2) P. Libermann, C. M. Marle, *Symplectic Geometry and Analytical Mechanics*, Reidel Dordrecht, 1987

3) L. A. Dickey, *Soliton Equations and Hamiltonian Systems*, World Scientific, Singapore, 1991, Adv. Series in Math. Phys Vol. 12

4) I. Vaisman, *Lectures on the Geometry of Poisson Manifolds*, Progress in Math., Birkhäuser, 1994

5) P. Casati, G. Falqui, F. Magri, M. Pedroni (1996), *The KP theory revisited. I,II,III,IV.* Technical Reports, SISSA/2,3,4,5/96/FM, SISSA/ISAS, Trieste, 1995

c) Hirota Methods for non Linear Differential and Difference Equations (6 lectures in English)

Prof. Junkichi SATSUMA (University of Tokyo, Tokyo, Japan)

Abstract

1) Introduction;
2) Nonlinear differential systems;
3) Nonlinear differential-difference systems;
4) Nonlinear difference systems;
5) Sato theory;
6) Ultra-discrete systems.

References.

1) M.J.Ablowitz and H.Segur, *Solitons and the Inverse Scattering Transform*, (SIAM, Philadelphia, 1981).

2) Y.Ohta, J.Satsuma, D.Takahashi and T.Tokihiro, " Prog. Theor. Phys. Suppl. No.94, p.210-241 (1988)

3) J.Satsuma, *Bilinear Formalism in Soliton Theory*, Lecture Notes in Physics No.495, Integrability of Nonlinear Systems, ed. by Y.Kosmann-Schwarzbach, B.Grammaticos and K.M.Tamizhmani p.297-313 (Springer, Berlin, 1997).

d) Lie Groups and Exact Solutions of non Linear Differential and Difference Equations (6 lectures in English)

Prof. Pavel WINTERNITZ (Universitè de Montreal, Montreal, Canada) 3J7

Abstract

1) Algorithms for calculating the symmetry group of a system of ordinary or partial differential equations. Examples of equations with finite and infinite Lie point symmetry groups;

2) Applications of symmetries. The method of symmetry reduction for partial differential equations. Group classification of differential equations;

3) Classification and identification of Lie algebras given by their structure constants. Classification of subalgebras of Lie algebras. Examples and applications;

4) Solutions of ordinary differential equations. Lowering the order of the equation. First integrals. Painlevè analysis and the singularity structure of solutions;

5) Conditional symmetries. Partially invariant solutions.

6) Lie symmetries of difference equations.

References.

1) P. J. Olver, *Applications of Lie Groups to Differential Equations*, Springer,1993,

2) P. Winternitz, *Group Theory and Exact Solutions of Partially Integrable Differential Systems*, in Partially Integrable Evolution Equations in Physics, Kluwer, Dordrecht, 1990, (Editors R.Conte and N.Boccara).

3) P. Winternitz, in "*Integrable Systems, Quantum Groups and Quantum Field Theories*", Kluwer, 1993 (Editors L .A. Ibort and M. A. Rodriguez).

Applications

Those who want to attend the Session should fill in an application to the C.I.M.E Foundation at the address below, **not later than May 30**, 1999.

An important consideration in the acceptance of applications is the scientific relevance of the Session to the field of interest of the applicant.

Applicants are requested, therefore, to submit, along with their application, a scientific curriculum and a letter of recommendation.

Participation will only be allowed to persons who have applied in due time and have had their application accepted.

CIME will be able to partially support some of the youngest participants. Those who plan to apply for support have to mention it explicitely in the application form.

Attendance

No registration fee is requested. Lectures will be held at Cetraro on September 8, 9, 10, 11, 12, 13, 14, 15. Participants are requested to register on September 7, 1999.

Site and lodging

The session will be held at Grand Hotel S. Michele at Cetraro (Cosenza), Italy. Prices for full board (bed and meals) are roughly 150.000 italian liras p.p. day in a single room, 130.000 italian liras in a double room. Cheaper arrangements for multiple lodging in a residence are avalaible. More detailed informations may be obtained from the Direction of the hotel (tel. +39-098291012, Fax +39-098291430, email: sanmichele@antares.it.

Further information on the hotel at the web page www.sanmichele.it

Lecture Notes

Lecture notes will be published as soon as possible after the Session.

Arrigo CELLINA
CIME Director

Vincenzo VESPRI
CIME Secretary

Fondazione C.I.M.E. c/o Dipartimento di Matematica U. Dini Viale Morgagni, 67/A - 50134 FIRENZE (ITALY) Tel. +39-55-434975 / +39-55-4237123 FAX +39-55-434975 / +39-55-4222695 E-mail CIME@UDINI.MATH.UNIFI.IT

Information on CIME can be obtained on the system World-Wide-Web on the file HTTP: //WWW.MATH.UNIFI.IT/CIME/WELCOME.TO.CIME.

Lecture Notes in Mathematics

For information about Vols. 1–1525
please contact your bookseller or Springer-Verlag

Vol. 1526: J. Azéma, P. A. Meyer, M. Yor (Eds.), Séminaire de Probabilités XXVI. X, 633 pages. 1992.

Vol. 1527: M. I. Freidlin, J.-F. Le Gall, Ecole d'Eté de Probabilités de Saint-Flour XX – 1990. Editor: P. L. Hennequin. VIII, 244 pages. 1992.

Vol. 1528: G. Isac, Complementarity Problems. VI, 297 pages. 1992.

Vol. 1529: J. van Neerven, The Adjoint of a Semigroup of Linear Operators. X, 195 pages. 1992.

Vol. 1530: J. G. Heywood, K. Masuda, R. Rautmann, S. A. Solonnikov (Eds.), The Navier-Stokes Equations II – Theory and Numerical Methods. IX, 322 pages. 1992.

Vol. 1531: M. Stoer, Design of Survivable Networks. IV, 206 pages. 1992.

Vol. 1532: J. F. Colombeau, Multiplication of Distributions. X, 184 pages. 1992.

Vol. 1533: P. Jipsen, H. Rose, Varieties of Lattices. X, 162 pages. 1992.

Vol. 1534: C. Greither, Cyclic Galois Extensions of Commutative Rings. X, 145 pages. 1992.

Vol. 1535: A. B. Evans, Orthomorphism Graphs of Groups. VIII, 114 pages. 1992.

Vol. 1536: M. K. Kwong, A. Zettl, Norm Inequalities for Derivatives and Differences. VII, 150 pages. 1992.

Vol. 1537: P. Fitzpatrick, M. Martelli, J. Mawhin, R. Nussbaum, Topological Methods for Ordinary Differential Equations. Montecatini Terme, 1991. Editors: M. Furi, P. Zecca. VII, 218 pages. 1993.

Vol. 1538: P.-A. Meyer, Quantum Probability for Probabilists. X, 287 pages. 1993.

Vol. 1539: M. Coornaert, A. Papadopoulos, Symbolic Dynamics and Hyperbolic Groups. VIII, 138 pages. 1993.

Vol. 1540: H. Komatsu (Ed.), Functional Analysis and Related Topics, 1991. Proceedings. XXI, 413 pages. 1993.

Vol. 1541: D. A. Dawson, B. Maisonneuve, J. Spencer, Ecole d' Eté de Probabilités de Saint-Flour XXI - 1991. Editor: P. L. Hennequin. VIII, 356 pages. 1993.

Vol. 1542: J.Fröhlich, Th.Kerler, Quantum Groups, Quantum Categories and Quantum Field Theory. VII, 431 pages. 1993.

Vol. 1543: A. L. Dontchev, T. Zolezzi, Well-Posed Optimization Problems. XII, 421 pages. 1993.

Vol. 1544: M.Schürmann, White Noise on Bialgebras. VII, 146 pages. 1993.

Vol. 1545: J. Morgan, K. O'Grady, Differential Topology of Complex Surfaces. VIII, 224 pages. 1993.

Vol. 1546: V. V. Kalashnikov, V. M. Zolotarev (Eds.), Stability Problems for Stochastic Models. Proceedings, 1991. VIII, 229 pages. 1993.

Vol. 1547: P. Harmand, D. Werner, W. Werner, M-ideals in Banach Spaces and Banach Algebras. VIII, 387 pages. 1993.

Vol. 1548: T. Urabe, Dynkin Graphs and Quadrilateral Singularities. VI, 233 pages. 1993.

Vol. 1549: G. Vainikko, Multidimensional Weakly Singular Integral Equations. XI, 159 pages. 1993.

Vol. 1550: A. A. Gonchar, E. B. Saff (Eds.), Methods of Approximation Theory in Complex Analysis and Mathematical Physics IV, 222 pages, 1993.

Vol. 1551: L. Arkeryd, P. L. Lions, P.A. Markowich, S.R. S. Varadhan. Nonequilibrium Problems in Many-Particle Systems. Montecatini, 1992. Editors: C. Cercignani, M. Pulvirenti. VII, 158 pages 1993.

Vol. 1552: J. Hilgert, K.-H. Neeb, Lie Semigroups and their Applications. XII, 315 pages. 1993.

Vol. 1553: J.-L- Colliot-Thélène, J. Kato, P. Vojta. Arithmetic Algebraic Geometry. Trento, 1991. Editor: E. Ballico. VII, 223 pages. 1993.

Vol. 1554: A. K. Lenstra, H. W. Lenstra, Jr. (Eds.), The Development of the Number Field Sieve. VIII, 131 pages. 1993.

Vol. 1555: O. Liess, Conical Refraction and Higher Microlocalization. X, 389 pages. 1993.

Vol. 1556: S. B. Kuksin, Nearly Integrable Infinite-Dimensional Hamiltonian Systems. XXVII, 101 pages. 1993.

Vol. 1557: J. Azéma, P. A. Meyer, M. Yor (Eds.), Séminaire de Probabilités XXVII. VI, 327 pages. 1993.

Vol. 1558: T. J. Bridges, J. E. Furter, Singularity Theory and Equivariant Symplectic Maps. VI, 226 pages. 1993.

Vol. 1559: V. G. Sprindžuk, Classical Diophantine Equations. XII, 228 pages. 1993.

Vol. 1560: T. Bartsch, Topological Methods for Variational Problems with Symmetries. X, 152 pages. 1993.

Vol. 1561: I. S. Molchanov, Limit Theorems for Unions of Random Closed Sets. X, 157 pages. 1993.

Vol. 1562: G. Harder, Eisensteinkohomologie und die Konstruktion gemischter Motive. XX, 184 pages. 1993.

Vol. 1563: E. Fabes, M. Fukushima, L. Gross, C. Kenig, M. Röckner, D. W. Stroock, Dirichlet Forms. Varenna, 1992. Editors: G. Dell'Antonio, U. Mosco. VII, 245 pages. 1993.

Vol. 1564: J. Jorgenson, S. Lang, Basic Analysis of Regularized Series and Products. IX, 122 pages. 1993.

Vol. 1565: L. Boutet de Monvel, C. De Concini, C. Procesi, P. Schapira, M. Vergne. D-modules, Representation Theory, and Quantum Groups. Venezia, 1992. Editors: G. Zampieri, A. D'Agnolo. VII, 217 pages. 1993.

Vol. 1566: B. Edixhoven, J.-H. Evertse (Eds.), Diophantine Approximation and Abelian Varieties. XIII, 127 pages. 1993.

Vol. 1567: R. L. Dobrushin, S. Kusuoka, Statistical Mechanics and Fractals. VII, 98 pages. 1993.

Vol. 1568: F. Weisz, Martingale Hardy Spaces and their Application in Fourier Analysis. VIII, 217 pages. 1994.

Vol. 1569: V. Totik, Weighted Approximation with Varying Weight. VI, 117 pages. 1994.

Vol. 1570: R. deLaubenfels, Existence Families, Functional Calculi and Evolution Equations. XV, 234 pages. 1994.

Vol. 1571: S. Yu. Pilyugin, The Space of Dynamical Systems with the C^0-Topology. X, 188 pages. 1994.

Vol. 1572: L. Göttsche, Hilbert Schemes of Zero-Dimensional Subschemes of Smooth Varieties. IX, 196 pages. 1994.

Vol. 1573: V. P. Havin, N. K. Nikolski (Eds.), Linear and Complex Analysis – Problem Book 3 – Part I. XXII, 489 pages. 1994.

Vol. 1574: V. P. Havin, N. K. Nikolski (Eds.), Linear and Complex Analysis – Problem Book 3 – Part II. XXII, 507 pages. 1994.

Vol. 1575: M. Mitrea, Clifford Wavelets, Singular Integrals, and Hardy Spaces. XI, 116 pages. 1994.

Vol. 1576: K. Kitahara, Spaces of Approximating Functions with Haar-Like Conditions. X, 110 pages. 1994.

Vol. 1577: N. Obata, White Noise Calculus and Fock Space. X, 183 pages. 1994.

Vol. 1578: J. Bernstein, V. Lunts, Equivariant Sheaves and Functors. V, 139 pages. 1994.

Vol. 1579: N. Kazamaki, Continuous Exponential Martingales and *BMO*. VII, 91 pages. 1994.

Vol. 1580: M. Milman, Extrapolation and Optimal Decompositions with Applications to Analysis. XI, 161 pages. 1994.

Vol. 1581: D. Bakry, R. D. Gill, S. A. Molchanov, Lectures on Probability Theory. Editor: P. Bernard. VIII, 420 pages. 1994.

Vol. 1582: W. Balser, From Divergent Power Series to Analytic Functions. X, 108 pages. 1994.

Vol. 1583: J. Azéma, P. A. Meyer, M. Yor (Eds.), Séminaire de Probabilités XXVIII. VI, 334 pages. 1994.

Vol. 1584: M. Brokate, N. Kenmochi, I. Müller, J. F. Rodriguez, C. Verdi, Phase Transitions and Hysteresis. Montecatini Terme, 1993. Editor: A. Visintin. VII. 291 pages. 1994.

Vol. 1585: G. Frey (Ed.), On Artin's Conjecture for Odd 2-dimensional Representations. VIII, 148 pages. 1994.

Vol. 1586: R. Nillsen, Difference Spaces and Invariant Linear Forms. XII, 186 pages. 1994.

Vol. 1587: N. Xi, Representations of Affine Hecke Algebras. VIII, 137 pages. 1994.

Vol. 1588: C. Scheiderer, Real and Étale Cohomology. XXIV, 273 pages. 1994.

Vol. 1589: J. Bellissard, M. Degli Esposti, G. Forni, S. Graffi, S. Isola, J. N. Mather, Transition to Chaos in Classical and Quantum Mechanics. Montecatini Terme, 1991. Editor: 2S. Graffi. VII, 192 pages. 1994.

Vol. 1590: P. M. Soardi, Potential Theory on Infinite Networks. VIII, 187 pages. 1994.

Vol. 1591: M. Abate, G. Patrizio, Finsler Metrics – A Global Approach. IX, 180 pages. 1994.

Vol. 1592: K. W. Breitung, Asymptotic Approximations for Probability Integrals. IX, 146 pages. 1994.

Vol. 1593: J. Jorgenson & S. Lang, D. Goldfeld, Explicit Formulas for Regularized Products and Series. VIII, 154 pages. 1994.

Vol. 1594: M. Green, J. Murre, C. Voisin, Algebraic Cycles and Hodge Theory. Torino, 1993. Editors: A. Albano, F. Bardelli. VII, 275 pages. 1994.

Vol. 1595: R.D.M. Accola, Topics in the Theory of Riemann Surfaces. IX, 105 pages. 1994.

Vol. 1596: L. Heindorf, L. B. Shapiro, Nearly Projective Boolean Algebras. X, 202 pages. 1994.

Vol. 1597: B. Herzog, Kodaira-Spencer Maps in Local Algebra. XVII, 176 pages. 1994.

Vol. 1598: J. Berndt, F. Tricerri, L. Vanhecke, Generalized Heisenberg Groups and Damek-Ricci Harmonic Spaces. VIII, 125 pages. 1995.

Vol. 1599: K. Johannson, Topology and Combinatorics of 3-Manifolds. XVIII, 446 pages. 1995.

Vol. 1600: W. Narkiewicz, Polynomial Mappings. VII, 130 pages. 1995.

Vol. 1601: A. Pott, Finite Geometry and Character Theory. VII, 181 pages. 1995.

Vol. 1602: J. Winkelmann, The Classification of Three-dimensional Homogeneous Complex Manifolds. XI, 230 pages. 1995.

Vol. 1603: V. Ene, Real Functions – Current Topics. XIII, 310 pages. 1995.

Vol. 1604: A. Huber, Mixed Motives and their Realization in Derived Categories. XV, 207 pages. 1995.

Vol. 1605: L. B. Wahlbin, Superconvergence in Galerkin Finite Element Methods. XI, 166 pages. 1995.

Vol. 1606: P.-D. Liu, M. Qian, Smooth Ergodic Theory of Random Dynamical Systems. XI, 221 pages. 1995.

Vol. 1607: G. Schwarz, Hodge Decomposition – A Method for Solving Boundary Value Problems. VII, 155 pages. 1995.

Vol. 1608: P. Biane, R. Durrett, Lectures on Probability Theory. Editor: P. Bernard. VII, 210 pages. 1995.

Vol. 1609: L. Arnold, C. Jones, K. Mischaikow, G. Raugel, Dynamical Systems. Montecatini Terme, 1994. Editor: R. Johnson. VIII, 329 pages. 1995.

Vol. 1610: A. S. Üstünel, An Introduction to Analysis on Wiener Space. X, 95 pages. 1995.

Vol. 1611: N. Knarr, Translation Planes. VI, 112 pages. 1995.

Vol. 1612: W. Kühnel, Tight Polyhedral Submanifolds and Tight Triangulations. VII, 122 pages. 1995.

Vol. 1613: J. Azéma, M. Emery, P. A. Meyer, M. Yor (Eds.), Séminaire de Probabilités XXIX. VI, 326 pages. 1995.

Vol. 1614: A. Koshelev, Regularity Problem for Quasilinear Elliptic and Parabolic Systems. XXI, 255 pages. 1995.

Vol. 1615: D. B. Massey, Le Cycles and Hypersurface Singularities. XI, 131 pages. 1995.

Vol. 1616: I. Moerdijk, Classifying Spaces and Classifying Topoi. VII, 94 pages. 1995.

Vol. 1617: V. Yurinsky, Sums and Gaussian Vectors. XI, 305 pages. 1995.

Vol. 1618: G. Pisier, Similarity Problems and Completely Bounded Maps. VII, 156 pages. 1996.

Vol. 1619: E. Landvogt, A Compactification of the Bruhat-Tits Building. VII, 152 pages. 1996.

Vol. 1620: R. Donagi, B. Dubrovin, E. Frenkel, E. Previato, Integrable Systems and Quantum Groups. Montecatini Terme, 1993. Editors:M. Francaviglia, S. Greco. VIII, 488 pages. 1996.

Vol. 1621: H. Bass, M. V. Otero-Espinar, D. N. Rockmore, C. P. L. Tresser, Cyclic Renormalization and Auto-morphism Groups of Rooted Trees. XXI, 136 pages. 1996.

Vol. 1622: E. D. Farjoun, Cellular Spaces, Null Spaces and Homotopy Localization. XIV, 199 pages. 1996.

Vol. 1623: H.P. Yap, Total Colourings of Graphs. VIII, 131 pages. 1996.

Vol. 1624: V. Brınzanescu, Holomorphic Vector Bundles over Compact Complex Surfaces. X, 170 pages. 1996.

Vol.1625: S. Lang, Topics in Cohomology of Groups. VII, 226 pages. 1996.

Vol. 1626: J. Azéma, M. Emery, M. Yor (Eds.), Séminaire de Probabilités XXX. VIII, 382 pages. 1996.

Vol. 1627: C. Graham, Th. G. Kurtz, S. Méléard, Ph. E. Protter, M. Pulvirenti, D. Talay, Probabilistic Models for Nonlinear Partial Differential Equations. Montecatini Terme, 1995. Editors: D. Talay, L. Tubaro. X, 301 pages. 1996.

Vol. 1628: P.-H. Zieschang, An Algebraic Approach to Association Schemes. XII, 189 pages. 1996.

Vol. 1629: J. D. Moore, Lectures on Seiberg-Witten Invariants. VII, 105 pages. 1996.

Vol. 1630: D. Neuenschwander, Probabilities on the Heisenberg Group: Limit Theorems and Brownian Motion. VIII, 139 pages. 1996.

Vol. 1631: K. Nishioka, Mahler Functions and Transcendence.VIII, 185 pages.1996.

Vol. 1632: A. Kushkuley, Z. Balanov, Geometric Methods in Degree Theory for Equivariant Maps. VII, 136 pages. 1996.

Vol.1633: H. Aikawa, M. Essén, Potential Theory – Selected Topics. IX, 200 pages.1996.

Vol. 1634: J. Xu, Flat Covers of Modules. IX, 161 pages. 1996.

Vol. 1635: E. Hebey, Sobolev Spaces on Riemannian Manifolds. X, 116 pages. 1996.

Vol. 1636: M. A. Marshall, Spaces of Orderings and Abstract Real Spectra. VI, 190 pages. 1996.

Vol. 1637: B. Hunt, The Geometry of some special Arithmetic Quotients. XIII, 332 pages. 1996.

Vol. 1638: P. Vanhaecke, Integrable Systems in the realm of Algebraic Geometry. VIII, 218 pages. 1996.

Vol. 1639: K. Dekimpe, Almost-Bieberbach Groups: Affine and Polynomial Structures. X, 259 pages. 1996.

Vol. 1640: G. Boillat, C. M. Dafermos, P. D. Lax, T. P. Liu, Recent Mathematical Methods in Nonlinear Wave Propagation. Montecatini Terme, 1994. Editor: T. Ruggeri. VII, 142 pages. 1996.

Vol. 1641: P. Abramenko, Twin Buildings and Applications to S-Arithmetic Groups. IX, 123 pages. 1996.

Vol. 1642: M. Puschnigg, Asymptotic Cyclic Cohomology. XXII, 138 pages. 1996.

Vol. 1643: J. Richter-Gebert, Realization Spaces of Polytopes. XI, 187 pages. 1996.

Vol. 1644: A. Adler, S. Ramanan, Moduli of Abelian Varieties. VI, 196 pages. 1996.

Vol. 1645: H. W. Broer, G. B. Huitema, M. B. Sevryuk, Quasi-Periodic Motions in Families of Dynamical Systems. XI, 195 pages. 1996.

Vol. 1646: J.-P. Demailly, T. Peternell, G. Tian, A. N. Tyurin, Transcendental Methods in Algebraic Geometry. Cetraro, 1994. Editors: F. Catanese, C. Ciliberto. VII, 257 pages. 1996.

Vol. 1647: D. Dias, P. Le Barz, Configuration Spaces over Hilbert Schemes and Applications. VII. 143 pages. 1996.

Vol. 1648: R. Dobrushin, P. Groeneboom, M. Ledoux, Lectures on Probability Theory and Statistics. Editor: P. Bernard. VIII, 300 pages. 1996.

Vol. 1649: S. Kumar, G. Laumon, U. Stuhler, Vector Bundles on Curves – New Directions. Cetraro, 1995. Editor: M. S. Narasimhan. VII, 193 pages. 1997.

Vol. 1650: J. Wildeshaus, Realizations of Polylogarithms. XI, 343 pages. 1997.

Vol. 1651: M. Drmota, R. F. Tichy, Sequences, Discrepancies and Applications. XIII, 503 pages. 1997.

Vol. 1652: S. Todorcevic, Topics in Topology. VIII, 153 pages. 1997.

Vol. 1653: R. Benedetti, C. Petronio, Branched Standard Spines of 3-manifolds. VIII, 132 pages. 1997.

Vol. 1654: R. W. Ghrist, P. J. Holmes, M. C. Sullivan, Knots and Links in Three-Dimensional Flows. X, 208 pages. 1997.

Vol. 1655: J. Azéma, M. Emery, M. Yor (Eds.), Séminaire de Probabilités XXXI. VIII, 329 pages. 1997.

Vol. 1656: B. Biais, T. Björk, J. Cvitanic, N. El Karoui, E. Jouini, J. C. Rochet, Financial Mathematics. Bressanone, 1996. Editor: W. J. Runggaldier. VII, 316 pages. 1997.

Vol. 1657: H. Reimann, The semi-simple zeta function of quaternionic Shimura varieties. IX, 143 pages. 1997.

Vol. 1658: A. Pumarino, J. A. Rodrıguez, Coexistence and Persistence of Strange Attractors. VIII, 195 pages. 1997.

Vol. 1659: V, Kozlov, V. Maz'ya, Theory of a Higher-Order Sturm-Liouville Equation. XI, 140 pages. 1997.

Vol. 1660: M. Bardi, M. G. Crandall, L. C. Evans, H. M. Soner, P. E. Souganidis, Viscosity Solutions and Applications. Montecatini Terme, 1995. Editors: I. Capuzzo Dolcetta, P. L. Lions. IX, 259 pages. 1997.

Vol. 1661: A. Tralle, J. Oprea, Symplectic Manifolds with no Kähler Structure. VIII, 207 pages. 1997.

Vol. 1662: J. W. Rutter, Spaces of Homotopy Self-Equivalences – A Survey. IX, 170 pages. 1997.

Vol. 1663: Y. E. Karpeshina; Perturbation Theory for the Schrödinger Operator with a Periodic Potential. VII, 352 pages. 1997.

Vol. 1664: M. Väth, Ideal Spaces. V, 146 pages. 1997.

Vol. 1665: E. Giné, G. R. Grimmett, L. Saloff-Coste, Lectures on Probability Theory and Statistics 1996. Editor: P. Bernard. X, 424 pages, 1997.

Vol. 1666: M. van der Put, M. F. Singer, Galois Theory of Difference Equations. VII, 179 pages. 1997.

Vol. 1667: J. M. F. Castillo, M. González, Three-space Problems in Banach Space Theory. XII, 267 pages. 1997.

Vol. 1668: D. B. Dix, Large-Time Behavior of Solutions of Linear Dispersive Equations. XIV, 203 pages. 1997.

Vol. 1669: U. Kaiser, Link Theory in Manifolds. XIV, 167 pages. 1997.

Vol. 1670: J. W. Neuberger, Sobolev Gradients and Differential Equations. VIII, 150 pages. 1997.

Vol. 1671: S. Bouc, Green Functors and *G*-sets. VII, 342 pages. 1997.

Vol. 1672: S. Mandal, Projective Modules and Complete Intersections. VIII, 114 pages. 1997.

Vol. 1673: F. D. Grosshans, Algebraic Homogeneous Spaces and Invariant Theory. VI, 148 pages. 1997.

Vol. 1674: G. Klaas, C. R. Leedham-Green, W. Plesken, Linear Pro-*p*-Groups of Finite Width. VIII, 115 pages. 1997.

Vol. 1675: J. E. Yukich, Probability Theory of Classical Euclidean Optimization Problems. X, 152 pages. 1998.

Vol. 1676: P. Cembranos, J. Mendoza, Banach Spaces of Vector-Valued Functions. VIII, 118 pages. 1997.

Vol. 1677: N. Proskurin, Cubic Metaplectic Forms and Theta Functions. VIII, 196 pages. 1998.

Vol. 1678: O. Krupková, The Geometry of Ordinary Variational Equations. X, 251 pages. 1997.

Vol. 1679: K.-G. Grosse-Erdmann, The Blocking Technique. Weighted Mean Operators and Hardy's Inequality. IX, 114 pages. 1998.

Vol. 1680: K.-Z. Li, F. Oort, Moduli of Supersingular Abelian Varieties. V, 116 pages. 1998.

Vol. 1681: G. J. Wirsching, The Dynamical System Generated by the 3n+1 Function. VII, 158 pages. 1998.

Vol. 1682: H.-D. Alber, Materials with Memory. X, 166 pages. 1998.

Vol. 1683: A. Pomp, The Boundary-Domain Integral Method for Elliptic Systems. XVI, 163 pages. 1998.

Vol. 1684: C. A. Berenstein, P. F. Ebenfelt, S. G. Gindikin, S. Helgason, A. E. Tumanov, Integral Geometry, Radon Transforms and Complex Analysis. Firenze, 1996. Editors: E. Casadio Tarabusi, M. A. Picardello, G. Zampieri. VII, 160 pages. 1998.

Vol. 1685: S. König, A. Zimmermann, Derived Equivalences for Group Rings. X, 146 pages. 1998.

Vol. 1686: J. Azéma, M. Émery, M. Ledoux, M. Yor (Eds.), Séminaire de Probabilités XXXII. VI, 440 pages. 1998.

Vol. 1687: F. Bornemann, Homogenization in Time of Singularly Perturbed Mechanical Systems. XII, 156 pages. 1998.

Vol. 1688: S. Assing, W. Schmidt, Continuous Strong Markov Processes in Dimension One. XII, 137 page. 1998.

Vol. 1689: W. Fulton, P. Pragacz, Schubert Varieties and Degeneracy Loci. XI, 148 pages. 1998.

Vol. 1690: M. T. Barlow, D. Nualart, Lectures on Probability Theory and Statistics. Editor: P. Bernard. VIII, 237 pages. 1998.

Vol. 1691: R. Bezrukavnikov, M. Finkelberg, V. Schechtman, Factorizable Sheaves and Quantum Groups. X, 282 pages. 1998.

Vol. 1692: T. M. W. Eyre, Quantum Stochastic Calculus and Representations of Lie Superalgebras. IX, 138 pages. 1998.

Vol. 1694: A. Braides, Approximation of Free-Discontinuity Problems. XI, 149 pages. 1998.

Vol. 1695: D. J. Hartfiel, Markov Set-Chains. VIII, 131 pages. 1998.

Vol. 1696: E. Bouscaren (Ed.): Model Theory and Algebraic Geometry. XV, 211 pages. 1998.

Vol. 1697: B. Cockburn, C. Johnson, C.-W. Shu, E. Tadmor, Advanced Numerical Approximation of Nonlinear Hyperbolic Equations. Cetraro, Italy, 1997. Editor: A. Quarteroni. VII, 390 pages. 1998.

Vol. 1698: M. Bhattacharjee, D. Macpherson, R. G. Möller, P. Neumann, Notes on Infinite Permutation Groups. XI, 202 pages. 1998.

Vol. 1699: A. Inoue,Tomita-Takesaki Theory in Algebras of Unbounded Operators. VIII, 241 pages. 1998.

Vol. 1700: W. A. Woyczyński, Burgers-KPZ Turbulence,XI, 318 pages. 1998.

Vol. 1701: Ti-Jun Xiao, J. Liang, The Cauchy Problem of Higher Order Abstract Differential Equations, XII, 302 pages. 1998.

Vol. 1702: J. Ma, J. Yong, Forward-Backward Stochastic Differential Equations and Their Applications. XIII, 270 pages. 1999.

Vol. 1703: R. M. Dudley, R. Norvaiša, Differentiability of Six Operators on Nonsmooth Functions and p-Variation. VIII, 272 pages. 1999.

Vol. 1704: H. Tamanoi, Elliptic Genera and Vertex Operator Super-Algebras. VI, 390 pages. 1999.

Vol. 1705: I. Nikolaev, E. Zhuzhoma, Flows in 2-dimensional Manifolds. XIX, 294 pages. 1999.

Vol. 1706: S. Yu. Pilyugin, Shadowing in Dynamical Systems. XVII, 271 pages. 1999.

Vol. 1707: R. Pytlak, Numerical Methods for Optical Control Problems with State Constraints. XV, 215 pages. 1999.

Vol. 1708: K. Zuo, Representations of Fundamental Groups of Algebraic Varieties. VII, 139 pages. 1999.

Vol. 1709: J. Azéma, M. Émery, M. Ledoux, M. Yor (Eds), Séminaire de Probabilités XXXIII. VIII, 418 pages. 1999.

Vol. 1710: M. Koecher, The Minnesota Notes on Jordan Algebras and Their Applications. IX, 173 pages. 1999.

Vol. 1711: W. Ricker, Operator Algebras Generated by Commuting Projections: A Vector Measure Approach. XVII, 159 pages. 1999.

Vol. 1712: N. Schwartz, J. J. Madden, Semi-algebraic Function Rings and Reflectors of Partially Ordered Rings. XI, 279 pages. 1999.

Vol. 1713: F. Bethuel, G. Huiksen, S. Müller, K. Steffen, Calculus of Variations and Geometric Evolution Problems. Cetraro, 1996. Editors: S. Hildebrandt, M. Struwe. VII, 293 pages. 1999.

Vol. 1714: O. Diekmann, R. Durrett, K. P. Hadeler, P. K. Maini, H. L. Smith, Mathematics Inspired by Biology. Martina Franca, 1997. Editors: V. Capasso, O. Diekmann. VII, 268 pages. 1999.

Vol. 1715: N. V. Krylov, M. Röckner, J. Zabczyk, Stochastic PDE's and Kolmogorov Equations in Infinite Dimensions. Cetraro, 1998. Editor: G. Da Prato. VIII, 239 pages. 1999.

Vol. 1716: J. Coates, R. Greenberg, K. A. Ribet, K. Rubin, Arithmetic Theory of Elliptic Curves. Cetraro, 1997. Editor: C. Viola. VIII, 260 pages. 1999.

4. Lecture Notes are printed by photo-offset from the master-copy delivered in camera-ready form by the authors. Springer-Verlag provides technical instructions for the preparation of manuscripts. Macro packages in TEX, LATEX2e, LATEX2.09 are available from Springer's web-pages at

http://www.springer.de/math/authors/b-tex.html.

Careful preparation of the manuscripts will help keep production time short and ensure satisfactory appearance of the finished book.

The actual production of a Lecture Notes volume takes approximately 12 weeks.

5. Authors receive a total of 50 free copies of their volume, but no royalties. They are entitled to a discount of 33.3 % on the price of Springer books purchase for their personal use, if ordering directly from Springer-Verlag.

Commitment to publish is made by letter of intent rather than by signing a formal contract. Springer-Verlag secures the copyright for each volume. Authors are free to reuse material contained in their LNM volumes in later publications: A brief written (or e-mail) request for formal permission is sufficient.

Addresses:

Professor F. Takens, Mathematisch Instituut,
Rijksuniversiteit Groningen, Postbus 800,
9700 AV Groningen, The Netherlands
E-mail: F.Takens@math.rug.nl

Professor B. Teissier, DMI, École Normale Supérieure
45, rue d'Ulm,
F-7500 Paris, France
E-mail: Teissier@ens.fr

Springer-Verlag, Mathematics Editorial, Tiergartenstr. 17,
D-69121 Heidelberg, Germany,
Tel.: *49 (6221) 487-701
Fax: *49 (6221) 487-355
E-mail: lnm@Springer.de